TROPICAL SOILS

Classification, Fertility and Management

F. S. C. P. Kalpagé

Visiting Professor of Soil Science, University of Malaya

M

First published in India in 1974 by
THE MACMILLAN COMPANY OF INDIA LIMITED

Revised impression published in the UK in 1976 by
THE MACMILLAN PRESS LTD
London and Basingstoke
Associated companies in New York Dublin
Melbourne Johannesburg and Madras

SBN 333 19276 1

Made and printed in Great Britain by
Butler & Tanner Ltd, Frome and London

TROPICAL SOILS

To

CHITHRA, SANJAY AND PRAVIN

Contents

List of Plates, Figures and Maps

FIGURES

MAPS

Preface

MODERN SOIL studies were initiated in Russia during the latter part of the nineteenth century and soon spread to Western Europe and to the USA. During the last four decades the guiding principles of such studies have been applied to tropical regions. Considerable work has been done in all parts of the tropical belt and a large number of new soils and a vast array of facts about them have been discovered. Some of this information has been incorporated into textbooks dealing with tropical agriculture but much of it is found scattered throughout the extensive literature that has accumulated.

A suitable textbook on tropical soils for use at the undergraduate level has been a long felt need. Books hitherto available have either been written for use in temperate countries or do not cover the entire tropical region adequately. During more than 22 years as a teacher of soil science, the author has felt the absence of an adequate textbook covering all aspects of the study of soils in the tropics. The books recommended for reading, for one reason or another, did not provide that coverage or depth necessary for university students. For this reason, it was considered worthwhile to collect together in one book the results of investigations into tropical soils.

This book is an attempt to present the available information to students of soil science and others interested in tropical agriculture. It assumes an understanding of the fundamentals of soil science and the basic principles are, therefore, not considered here. It does, however, emphasize the main differences between tropical and temperate zone soils.

Tropical Soils—Classification, Fertility and Management originated some time ago and evolved over many years. However, it took its present form and was completed for publication when the author was Guest Professor in Soil Science at the Technological University of West Berlin. Grateful thanks are due to the governing authorities of the Technological University of West Berlin for making this assignment possible, Prof. Knud Caesar for his personal interest and his readiness to assist in many ways, Prof. H. P. Blume for some suggestions regarding the text, Dr. Wilfried Kantor for sparing much time to give the benefit of his experience on tropical soils in East Africa, and Dr. S. Selvanayagam for assistance with maps and criticism of the chapter on the tropical environment.

The author wishes to express his deep gratitude to Dr. O. Talibudeen, Senior Principal Scientific Officer, Rothamsted Experimental Station, Harpenden, Herts., England, for his detailed comments and helpful suggestions during the preparation of the manuscript, Mr. C. J. Ekanayake and Miss W. I. Perera for typing the manuscript with efficiency and speed, and Mr. P. Munasinghe for helping with the illustrations.

Peradeniya Sri Lanka F.S.C.P. KALPAGE
June 1974

Conversion Factors

B→A	A	B	A→B
		LENGTH	
X 0.394	Inches	Centimetres	X 2.54
X 3.281	Feet	Metres	X 0.305
X 1.094	Yards	Metres	X 0.914
X 0.621	Miles	Kilometres	X 1.609
		AREA	
X 0.155	Square inches	Square centimetres	X 6.452
X 10.764	Square feet	Square metres	X 0.093
X 1.196	Square yards	Square metres	X 0.836
X 2.471	Acres (4840 square yards)	Hectares	X 0.405
X 0.386	Square miles (640 acres)	Square kilometres	X 2.590
		VOLUME	
X 0.061	Cubic inches	Cubic centimetres	X 16.387
X 35.315	Cubic feet	Cubic metres	X 0.028
X 1.308	Cubic yards	Cubic metres	X 0.765
		CAPACITY	
X 0.880	Imperial quarts	Litres	X 1.136
X 1.057	US quarts	Litres	X 0.946
X 0.220	Imp. gal. (1.201 US gal.)	Litres	X 4.546
X 0.264	US gal. (0.833 Imp. gal.)	Litres	X 3 785
X 2.750	Imp. bushels (1.032 US bu.)	Hectolitres	X 0.364
X 2.838	US bushels (0.969 Imp. bu.)	Hectolitres	X 0.352
		WEIGHT	
X 15.432	Grains	Grams	X 0.065
X 0.035	Ounces (Avoirdupois)	Grams	X 28.350
X 2.205	Pounds (Avoirdupois)	Kilograms	X 0.454
X 1.968	Hundredweights (112 lb)	Quintals (100 kg)	X 0.508
X 1.102	Short tons (2,000 lb) (English tons)	Metric tons (1000 kg)	X 0.907
X 0.984	Long tons (2,240 lb)	Metric tons	X 1.016
	Maunds (82.28 lb)	Kilograms	X 37.64
	Bushels (rice)	Kilograms	X 20.412
	Bushels (rice)	Pounds	X 45

CONCENTRATION

B→A	A	B	A→B
X 10^4	ppm	percentage	X 10^{-4}
X 0.5	ppm	pounds per acre	X 2
X 0.160	Ounces per Imp. gallon	Grams per litre	X 6.236
X 0.134	Ounces per US gallon	Grams per litre	X 7.490

YIELD

B→A	A	B	A→B
X 0.892	Pounds per acre	Kilograms per hectare	X 1.121
X 0.008	Hundredweights per acre	Kilograms per hectare	X 125.54
X 0.446	Short tons per acre	Metric tons per acre	X 2.240
X 0.040	Short tons per acre	Quintals per hectare	X 22.42
X 0.045	Long tons per acre	Quintals per hectare	X 25.11
X 0.089	Gallons per acre	Litres per hectare	X 11.21
X 1.980	Bushels per acre (rice)	Quintals per hectare	X 0.502
X 1.549	Bushels per acre (maize)	Quintals per hectare	X 0.627

PRESSURE

B→A	A	B	A→B
	Atmospheres	Pounds per square inch	X 14.696
	Atmospheres	**Millimetres of mercury at 0°C**	X 760
	Atmospheres	Centimetres of water	X 1036
	Pounds per square inch	Kilograms per square metre	X 703.07
	Pounds per square foot	Kilograms per square metre	X 4.882

TEMPERATURE

B→A	A	B	A→B
X 9/5 + 32	Degrees Fahrenheit	Degrees Centigrade	−32 X 5/9

FERTILIZER INGREDIENTS

B→A	A	B	A→B
X 0.437	Phosphorus (P)	Phosphoric acid (P_2O_5)	X 2.289
X 0.326	Phosphorus (P)	Phosphate (PO_4)	X 3.066
X 0.830	Potassium (K)	Potash (K_2O)	X 1.205
X 0.715	Calcium (Ca)	Calcium oxide (CaO)	X 1.399
X 0.603	Magnesium (Mg)	Magnesium oxide (MgO)	X 1.667

1 The Tropical Environment

1.1 Geographical Location

In the tropical world, climate, vegetation, geomorphology, lithology, and consequently, the soils, vary more than in the temperate and arctic regions. The Tropics of Cancer and Capricorn encircle the globe at latitudes 23°27' N and S respectively. The tropical belt comprising about one-third of the land surface of the earth should strictly extend into each hemisphere as far as Cancer and Capricorn but the actual limits of the tropical world do not coincide with these boundaries and are difficult to define precisely. They have been taken as coinciding with the 21°C (70°F) or 20°C (68°F) isotherms or as extending into those regions where the coolest month has a temperature of not less than 18°C (64°F). Again, the area between the 20° latitudes has been considered tropical and those between 20° and 40° subtropical to warm temperate. Temperate vegetation is found on tops of high mountains in the tropics such as in South America and in east Africa, while tropical floras are frequently found beyond the confines of Cancer and Capricorn. Countries within the tropics with basic data pertaining to agriculture in these countries are given in Tables 1 and 2 respectively. The principal economies of this region are shown in Map I.

1.2 Climate

A variety of tropical climates (Papadakis, 1961) result from the operation of the planetary pressure belts, the resulting system of winds and their associated air masses. The distribution of atmospheric pressure belts on the globe arises from the decrease in temperature from the equator to the poles and the effect of the rotation of the earth in deflecting the consequent air movements. The general circulation is, however, greatly modified by the arrangement of the continents and oceans, the relief and the nature of the ground cover. Detailed accounts of the dynamics of atmospheric pressure belts and the manner in which the different climatic regimes arise have been given by a number of authors (Blumenstock and Thornthwaite, 1941 ; Beckinsale, 1957; Blumenstock, 1958 ; Strahler, 1960 ; Kendrew, 1961).

The notion 'tropical climate' embraces such greatly differing climates as those of the great dense rain forests, the savannas, the dry deciduous forests, the thorny bush, the desert steppes and the alpine meadows in the high mountains. Five main types of climate govern vegetation: the equatorial, tropical,

TABLE 1 : COUNTRIES WITHIN THE TROPICS

America	*Africa*	*Asia/Australia*
Wholly within the tropics		
*Barbados	Angola	Bangladesh
British Honduras (Bèlize)	*Burundi	Borneo ‡
*Colombia	*Cameroon	*Cambodia (Khmer Republic)
*Costa Rica	*Central African Republic	*Sri Lanka
*Cuba	*Chad	Fiji
*Dominican Republic	*Congo (Brazzaville)	Hawaii
*Ecuador	*Dahomey	Hongkong
*El Salvador	*Equatorial Guinea	*Laos
*Guatemala	*Gabon	*Indonesia
Guisanbourg (Fr. Guiana)	*Ethiopia	*Malaysia
*Guyana (British Guiana)	*Gambia	*Maldives
*Haiti	*Ghana	New Guinea †
*Honduras	*Guinea	Pacific Islands
*Jamaica	*Ivory Coast	*Philippines
*Nicaragua	*Kenya	Portuguese Timor
*Panama	*Liberia	*Singapore
*Peru	*Malawi	*Thailand
Puerto Rico	*Mali	Vietnam
Surinam	*Niger	*Yemen (Arab Republic)
*Trinidad and Tobago	*Nigeria	*Yemen (Peoples Democratic Republic)
*Venezuela	Portuguese Guinea	
Virgin Islands	*Rwanda	
	Rhodesia	
	*Senegal	
	*Sierra Leone	
	*Somali	
	*Sudan	
	*Tanzania (United Republic of)	
	*Togo	
	*Uganda	
	*Upper Volta	
	Zaire (Congo, K.)	
	Zambia	
Partly within the tropics		
*Argentina	*Algeria	*Australia
*Bolivia	*Botswana	*Burma
*Brazil	*Libya	*China
*Chile	*Madagascar	*India
*Mexico	*Mauritania	Oman
*Paraguay	*Mauritius	*Pakistan
	*Mozambique	*Saudi Arabia
	*South Africa	*Taiwan
	South West Africa	
	*United Arab Republic	

*States Members of the United Nations

†New Guinea : the eastern part belongs to Indonesia while the western part is a protectorate of Australia

‡Borneo : Sabah and Sarawak form part of Malaysia, the Southern portion of Indonesia

TABLE 2: BASIC DATA ON AGRICULTURE IN TROPICAL COUNTRIES (FAO, 1970 c)

Country	*Period*	*Population in agriculture*		*Arable land per person in agriculture*
		000	Per cent of total	Hectares per caput
America				
Argentina	1965	4,150	20	6.44
Barbados	1960	48	24	0.54
Bolivia	1950	1,890	63	1.64
Brazil	1960	36,244	52	0.82
Chile	1965	2,349	28	1.92
Colombia	1960	6,554	46	0.77
Costa Rica	1960	675*	51	0.92*
Cuba				
Dominican Republic	1960	1,830	61	0.58
Ecuador	1965	2,955	57	0.98
El Salvador	1960	1,500+	60	0.43+
Guatemala	1950	1,727	62	0.85
Guyana				
Haiti	1965	3,517	80	
Honduras	1960	1,273	67	0.78
Jamaica	1965	788	44	0.39
Mexico	1965	22,200	52	1.23‡
Nicaragua	1965	960	58	0.91§
Panama	1960	507	48	1.11
Paraguay	1965	1,018	50	0.91
Peru	1965	5,775	50	0.47
Surinam	1965	76	26	0.61
Trinidad and Tobago	1960	160	20	1.09
Uruguay	1960	390+	14	5.77
Venezuela	1960	2,337+	31	2.23
USA	1965	11,700	6	15.37
Africa				
Algeria	1965	7,150	60	0.95
Cameroon	1965	4,368	84	
Congo, Dem. Rep. of the	1965	10,945	70	5.39§
Ethiopia	1965	20,120	89	0.62
Gabon	1965	388	84	0.55‡
Ghana	1965	4,642	60	0.55
Ivory Coast	1965	3,105	81	0.66
Kenya	1965	7,821	84	
Liberia	1965	856	80	4.77
Madagascar				
Malawi	1965	3,158	80	0.40
Mauritius				
Morocco	1965	7,295	55	1.08
Nigeria	1965	46,196	79	
Rhodesia	1965	3,195	75	
Senegal	1965	2,605	75	
South Africa	1960	6,995	44	1.72
Tanzania	1965	10,932	95	1.09

Country	*Period*	*Population in agriculture*		*Arable land per person in agriculture*
Togo	1965	1,295	79	1.67
Uganda	1965	6,870	91	0.55
Zambia	1965	3,005	81	
Libya	1965	647	40	
Sudan	1965	10,426	77	
UAR	1965	16,225	55	0.17
Asia/Australia				
Cambodia	1965	4,695	75	0.63
Sri Lanka	1965	5,582	50	0.34
China				
India	1965	340,655	70	0.48
Malaysia (West)				
Pakistan	1965	83,842	74	0.34
Philippines	1965	18,738	58	0.44
Taiwan	1965	5,846	47	0.15
Thailand	1965	24,001	78	0.48
Vietnam, Rep. of	1965	13,705	85	0.21
Saudi Arabia	1965	3,105	72	0.12
Australia	1965	1,117	10	33.26
Other Countries				
Japan	1965	23,685	24	0.25
Netherlands	1965	1,030	8	0.95
U.K.	1965	2,024	4	3.68
World				

*1963
†1961
‡1960
§1950

monsoon, desert, and mountain climates (Map 2). The characteristics of these climatic types are summarized in Table 3.

The tilting of the earth on its axis to form an angle of 66.5° with the plane in which it revolves about the sun causes the rays of the sun to fall more directly on tropical than on polar regions. Tropical latitudes are therefore warmer than temperate latitudes and temperatures are generally higher in the tropics than in temperate regions. They show little seasonal variation, the monthly means being around 27.8°C (82°F).

In the tropics *temperature* variations are important only when altitude causes a drop in temperature. Generally, temperature decreases by 6°C for every 1,000 m rise. In Kenya, the tropical crop production of areas at, or near, sea level contrasts sharply with the extensive semi-temperate farming of the highlands above 2,000 m. Above some altitudes frost is regularly experienced at certain seasons of the year; frost damage to some crops occurs above about

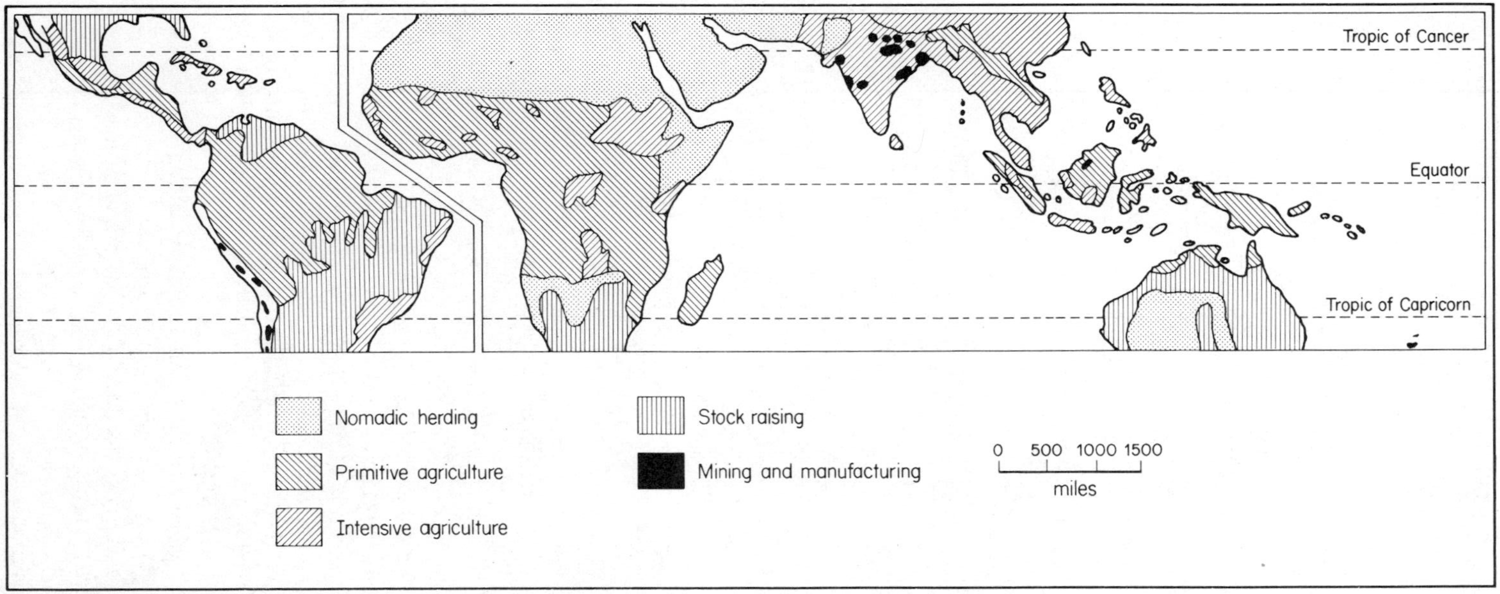

Map 1 : Principal economies of the tropics

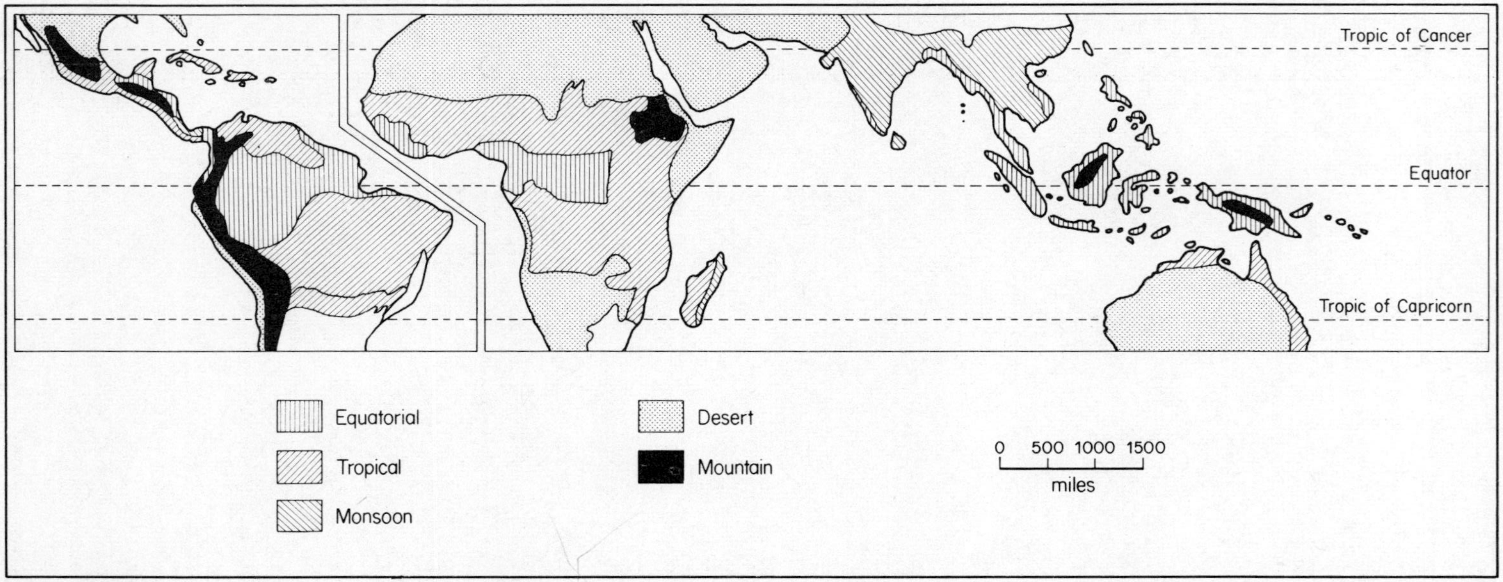

Map 2 : Climatic regions of the tropics

TABLE 3: CHARACTERISTICS OF THE MAIN CLIMATIC TYPES IN THE TROPICS (from Papadakis, 1961)

	Evapotranspiration[1] (cm)																		
	J	*F*	*M*	*A*	*M*	*J*	*J*	*A*	*S*	*O*	*N*	*D*	*An*	*HI*[2]	*Ln*[3]	*Lm*[4]	*Hu.Se*[5]	*Dr.Se*[6]	*Wi. se. & S. he*[7]	*HR*[8]/*TR*[9] *Climatic classification*[10]
											1. *Equatorial*									
Belem *(Para)* ***Brazil***	9	8	8	8	9	10	10	10	11	10	11	11	115	2.11	130	M	12-8	0	Ec, g	Eq, Hu humid semi-hot equatorial
Douala, Cameroons	7	7	8	7	7	6	4	5	4	5	7	6	73	5.66	334	M	2-12	0	Ec, g	Eq, Hu humid semi-hot equatorial
Penang, Malaysia	12	13	14	13	10	11	11	10	9	10	9	10	132	2.06	145	M	3-12	0	Ec, g	Eq, Hu humid semi-hot equatorial
											2. *Tropical*									
Maxcanú, Mexico	10	11	14	17	16	12	13	13	10	9	9	9	143	0.59	5	60	6-9	11-4	Tp, G	TR, Mo dry hot tropical
Wau, Sudan	25	28	28	24	18	15	12	12	14	16	21	24	237	0.46	21	M	6-9	11-4	Tp, G	TR, Mo dry hot tropical
Bangalore, India	13	16	20	20	17	11	10	9	9	10	10	11	156	0.56	17	83	7-10	1-4	Tp, G	TR, Mo dry hot tropical

3. *Monsoon*																			
Cienfuegos, Cuba																			
9	10	11	11	11	11	12	11	11	10	9	9	125	0.85	24	M	5-10	1-4	Tp, g	TR, MO moist /dry monsoon semi-hot tropical
Daru, Sierra Leone																			
14	18	16	13	13	10	8	7	5	11	11	10	136	1.76	147	M	4-11	1-3	Tp, G	TR,MO moist monsoon hot tropical
Semarang, Indonesia																			
8	8	9	10	11	13	15	16	18	17	13	9	147	1.35	94	M	11-15	9	Ec, G	EQ, MO moist monsoon hot equatorial
Cooktown, Australia																			
9	9	8	8	7	7	7	7	8	9	9	9	97	1.85	106	M	12-4	9-10	Tp, g	Tr, MO moist monsoon semi-hot tropical
4. *Desert*																			
Arica, Chile																			
8	8	7	6	5	5	5	5	5	5	6	7	72	m	0	0	0	2-1	tp, M	da, tr very cool tropical desert
Assab, Ethiopia																			
10	10	12	15	18	19	21	19	17	14	11	10	176	0.01	0	0	0	2-1	Ec, G	da, EQ hot equatorial absolute desert

TABLE 3. CHARACTERISTICS OF THE MAIN CLIMATIC TYPES IN THE TROPICS (from Papadakis, 1961)

				Evapotranspiration[1]			(cm)												
J	*F*	*M*	*A*	*M*	*J*	*J*	*A*	*S*	*O*	*N*	*D*	*An*	*HI*[2]	*Ln*[3]	*Lm*[4]	*Hu. Se*[5]	*Dr. Se*[6]	***Wi. se. & S. he***	*HR*[8] *TR*[9] *Climatic classification*[10]
Alice Springs, Australia																			
24	22	20	15	11	8	9	11	15	19	22	24	200	0.13	0	0	0	2-1	Ci, G	do, SU monsoon hot subtropical desert
										5. *Mountain*									
Santa Helena, Venezuela																			
13	13	14	12	10	8	8	8	10	10	10	9	125	1.31	57	M	4-12	0	Tp, c	Tt, Hu humid tierra templada
Banyo, Cameroons																			
18	17	14	11	10	8	8	8	8	9	13	17	141	1.18	92	M	4-10	12-3	Tp, c	Tt, MO moist monsoon tierra templada
Mali, Guinea																			
11	12	13	11	10	5	3	2	3	4	7	10	91	1.93	146	M	5-10	1-4	Tp, c	Tt, MO moist monsoon tierra templada

1. Evapotranspiration, month by month and annual (An)
2. HI—annual humidity index = annual rainfall/annual evapotranspiration
3. Ln—'normal' leaching rainfall = rainfall minus evapotranspiration
4. Lm—'maximum' leaching rainfall = 2x normal rainfall minus evapotranspiration
5. Hu. Se.—humid season, beginning and end, indicated by month
6. Dr. Se.—dry season, beginning and end, indicated by month
7. Wi.se. S.he.—winter severity, summer heat
8. HR—humidity regime as defined by Papadakis (1961)
9. TR—temperature regime as defined by Papadakis (1961)
10. Climatic classification according to Papadakis (1961)

c	coffee type
Ci	sufficiently mild for citrus, but not entirely frostless
da	absolute desert
do	monsoon desert
Ec	mean annual minimum temperature above 15° C (59° F)
Eq	semi-hot equatorial
EQ	hot equatorial
g	cool cotton summer type (Papadakis, 1961 p. 14)
G	warm cotton summer type (Papadakis, 1961 p. 14)
Hu	humid
m	
M	sufficiently warm for maize, but not for cotton
Mo	dry monsoon
MO	moist monsoon
Tp & tp	entirely frostless but mean annual minimum temperature under 15° C (59° F)
tr	cool tropical
Tr	semi-hot tropical
TR	hot tropical
Tt	tierra templada
SU	hot subtropical

2,300 m in Kenya, above 1,700 m in Malawi and above 1,500 m in Rhodesia. Monthly differences are very small at the equator and increase gradually towards the two Tropics. Differences in temperature between the seasons are only very slight.

Tropical climates are influenced mainly by *rainfall*. Precipitation is most abundant and regular in the equatorial zone, approximately between 7° N and S, and decreases towards the two bands of subtropical desert regions. Rainfall is perhaps the most important climatic factor influencing agriculture in the tropics and subtropics. It determines crops grown and farming systems adopted as well as the nature, timing and sequence of farming operations.

Soil moisture supply (S) depends not only on precipitation (P) but also on evapotranspiration (E), surface runoff (R), infiltration and drainage (D) and is given by the equation :

$$P = E + R + D + S$$

Average annual rainfall is of limited value in tropical agriculture. It does not indicate the seasonal variation, the annual reliability, the shower intensity, nor the loss by evapotranspiration, all of which are quite different from those in temperate regions.

Variation in *daylength* is not much in the Tropics, no more than half an hour near the equator and just under three hours at the tropics. A heavy cloud layer, particularly in certain rainy zones can greatly reduce the amount of solar radiation received. Cloud intensity decreases north and south of the equator, and is least near the tropics.

Winds are responsible for the circulation of cloud-masses and the distribution of rainfall, and influence the character of the vegetation. The monsoon winds affect parts of south-east Asia, Indonesia, north-western Australia and west Africa. Typhoons and cyclones sweep across the eastern shores of the continents and the islands nearby during the hot season and are accompanied by torrential rains. The trade winds blow from the desert regions just outside the tropics towards the equator.

1.2.1 Equatorial Climate

This type of climate is found as a belt stretching between 5° N and S of the equator. It is characterized by abundant rains spread throughout the year with two well-marked peak periods, a relatively high and uniform temperature and a high relative humidity. The annual rainfall is usually between 2,000 mm and 3,000 mm. Rainfall in almost any period of two or three weeks considerably exceeds evapotranspiration. Mean monthly temperatures are 28°C and mean monthly maximum and minimum temperatures 31°C and 25°C respectively, with little seasonal variation. Daylength is approximately 12 hours all the year round.

Equatorial climates are found in the Congo (Tables 3-5) and Amazon basins, the Caribbean, Indonesia, the Malay Peninsula, New Guinea and some of the Pacific Islands. Somalia and the eastern Ethiopian coasts are in the rain shadow of the Ethiopian highlands and do not receive rain from the south-westerly wind; these coasts receive small amounts from winds blowing across the Gulf of Aden.

1.2.2 Tropical Climate

In this climate, found on either side of the equatorial belt, the rains are concentrated into one time of the year and there is therefore a dry season of variable duration, increasing in length with the distance from the equator. The annual rainfall is usually less than in the equatorial regions. There is a tendency towards thermal seasons and the difference between the hottest and coolest months may be as much as 6°C. This climatic type is found in the African and South American continents. Sudan is a typical example of this climate, sometimes referred to as the Sudan type (Tables 3-5).

1.2.3 Monsoon Climate

The monsoon climate is similar to the tropical climate, but the rainfall is caused in a different way. It is characterised by three seasons: a cool season with little rain followed by a hot season, and finally, the rainy season. This climate is found in countries bordering the Indian Ocean especially in the countries of

TABLE 4: RAINFALL DISTRIBUTION AT SELECTED STATIONS IN THE TROPICAL ZONE (after Kendrew, 1961)

Region (altitude above mean sea level)		*Rainfall*													*Climatic type*
		J	*F*	*M*	*A*	*M*	*J*	*J*	*A*	*S*	*O*	*N*	*D*	*Mean annual*	
Duala, Cameroons (7.9m)	in	1.8	3.7	8.0	9.1	11.8	21.2	29.2	27.3	20.9	16.9	6.1	2.5	158.5	Equatorial
	mm	46	94	203	231	299	539	715	690	529	428	154	63	4018	
Wau, Sudan (438 m)	in	<0.1	0.2	0.9	2.6	5.3	6.5	7.5	8.2	6.6	4.9	0.6	<0.1	43.3	Tropical
	mm	<2.5	5	23	66	134	165	190	208	168	124	15	<2.5	1095	
Colombo, Sri Lanka (7.3 m)	in	3.5	3.5	5.8	9.1	14.6	8.8	5.3	4.3	6.3	13.7	12.4	5.8	93.1	Monsoon
	mm	89	89	147	231	370	224	134	109	160	346	314	147	2360	
Arica, Chile (29 m)	in	0	0	0	0	0	0	0	0	0	0	0	0	0	Desert
	mm	0	0	0	0	0	0	0	0	0	0	0	0	0	
Quito, Ecuador (2850 m)	in	4.2	4.3	5.2	7.4	5.0	1.5	0.9	1.5	3.0	3.7	3.8	3.8	44.1	Mountain
	mm	106	108	132	188	127	38	23	38	76	94	96	96	1120	

TABLE 5: TEMPERATURE DISTRIBUTION AT SELECTED STATIONS IN THE TROPICAL ZONE (after Kendrew, 1961)

Region (altitude above mean sea level)		Temperature														
		J	*F*	*M*	*A*	*M*	*J*	*J*	*A*	*S*	*O*	*N*	*D*	**Mean** *annual*	*Range*	*Climatic type*
Duala, Cameroons (7.9 m)	°F	80	80	80	80	80	78	75	75	77	76	78	79	78	6	Equatorial
	°C	26.6	26.6	26.6	26.6	26.6	25.5	23.9	23.9	25.0	24.5	25.5	26.1	25.5	3.3	
Wau, Sudan (438 m)	°F	80	82	85	86	83	81	79	79	80	81	82	80	82	7	Tropical
	°C	26.6	27.8	29.5	30.0	28.4	27.2	26.1	20.1	26.6	27.2	27.8	26.6	27.8	3.9	
Colombo, Sri Lanka (7.3m)	°F	79	79	81	82	82	81	81	81	81	80	79	78	80	4	Monsoon
	°C	26.1	26.1	27.2	27.8	27.8	27.2	27.2	27.2	27.2	26.6	26.1	25.5	26.6	2.2	
Arica, Chile (29 m)	°F	71	72	70	67	64	62	60	60	61	63	66	68	65	12	Desert
	°C	21.7	22.3	21.1	19.4	17.8	16.7	15.6	15.6	16.1	17.2	18.9	20.0	18.4	6.7	
Quito, Ecuador (2850m)	°F	55	54	54	54	55	55	55	55	55	55	54	55	55	0.4	Mountain
	°C	12.8	12.2	12.2	12.2	12.8	12.8	12.8	12.8	12.8	12.8	12.1	12.8	12.8	0.22	

south and southeast Asia (Tables 3-5)

1.2.4 Desert Climate

The more arid types of tropical and monsoon climates pass imperceptibly into that of desert regions which are very hot and dry. These regions lie mainly in the high-pressure belts just outside the tropics. The rains have no seasonal rhythm and there are long periods when there is no rain. The temperatures in some places reach over 50°C (122°F). In deserts of higher latitudes, the nights are fairly cool. There are two groups of desert regions. The hot deserts include the Sahara and the Kalahari in Africa, the Arabian Desert, the Great Australian desert and the Atacama Desert in South America (Tables 3-5). The mid-latitude deserts such as the Gobi and Colorado are found on plateaux outside the tropics. These are much cooler in the cold season.

1.2.5 Mountain Climate

Altitude influences temperature (Table 5) and hence has an indirect bearing on vegetation. Rainfall too increases with height up to a certain level. Climbing uphill in the tropics is therefore equivalent to moving away from the equator, and if the slope is steep enough it is possible to pass through different climates in a short time. There is perpetual snow above the 'snow-line' between 4,500 m and 5,000 m. Sharp frosts are however frequent from about 2,000 m. The low temperatures are accompanied by very strong solar radiation from a tropical sun. Climates prevailing at such high altitudes are referred to as alpine climates.

1.3 Vegetation

Climate and vegetation are two important and active factors affecting soil formation. For each climatic type there is a characteristic vegetation. (Map 3). Table 6 shows the distribution of climates, vegetation and soils throughout the earth's land surface.

In the humid tropics, vegetation is more exuberant and luxurious than in temperate regions. In North America and in northern Europe, only about sixteen tree species remained after successive glacial periods, whereas in the tropics the number of tree species is roughly 18,000. It is therefore more difficult to establish a relationship between soil and vegetation in the tropics than in temperate regions.

1.3.1 Rain Forest

Rain forests occur in the hot superhumid tropics characterized by an equatorial climate. Moisture and humidity cause large luxuriant evergreen trees of hard wood to grow to heights of 30 m (100 ft) or more. Woody lianas climb up these trees. The forests are sometimes so thick that no sunlight ever reaches the ground and it is dark and gloomy. Where there is some sunlight, a rich

TABLE 6: DISTRIBUTION OF CLIMATES, VEGFTATION AND SOILS

<table>
<tr><td rowspan="5">DRY COLD

DRY HOT</td><td colspan="6">Climate : Perpetual snow and ice
Vegetation : — — —
Soils : — — —</td><td rowspan="5">WET COLD

WET HOT</td></tr>
<tr><td colspan="6">Climate : Tundra
Vegetation : Tundra (Mosses, lichens and sedges)
Soils : Tundra soils</td></tr>
<tr><td colspan="6">Climate : Taiga
Vegetation : Taiga (Coniferous forests)
Soils : Podsols</td></tr>
<tr><td>C : Arid
V : Desert grasses and shrub</td><td>C : Semiarid
V : Steppe</td><td colspan="2">C : Subhumid
V : Grassland</td><td>C : Humid
V : Forests</td><td>C : Wet
V : Rain Forests</td></tr>
<tr><td>S : Sierozems and Desert soils</td><td>S : Chestnut and Brown soils</td><td>S : Chernozems</td><td>S : Prairie soils and Degraded Chernozems</td><td colspan="2">S : Podsols Gray-Brown Podsolic soils Red and Yellow Podsolic soils Latosolic soils</td></tr>
</table>

undergrowth is found.

The area covered by dense primary rain forest is continually decreasing, mainly as a result of man's interference. Shifting cultivation has destroyed the original virgin forest and what is to be seen in most regions today is largely secondary forest.

Rubber and cocoa trees are natives of these regions. The coconut palm is found on sandy sea coasts while the oilpalm takes its place in Africa.

The most suitable crops are those which flourish under continuous hot, wet conditions and do not require a pronounced dry season for harvesting. Perennials such as rubber, oilpalm, bananas, coconut, *liberica* coffee and cocoa are the main commercial crops. In subsistence farming, fruit trees and root crops, especially yams such as cassava and tania, are included. Rice and maize are grown in regions where a brief dry season prevails.

Weed growth, which is extremely rapid and luxuriant, poses a problem. Because of intense leaching, soil fertility is limited by nutrient deficiencies.

1.3.2 Tropical Grassland

Tall grass with scattered trees is the typical vegetation of the tropical climate, to be found in much of Africa. The grass cover, where not grazed, exceeds 80 cm in height, and the grasses are normally cauline-leaved perennials. The grass grows rapidly during the rains but is scorched brown by the sun during the dry season. Strong winds which prevail do not favour the growth of trees. The more open forms of this type of vegetation are referred to as tree and shrub savanna and the closer forms as savanna woodland, or wooded savanna. Tropical grasslands or savannas are found in a wide belt across Africa between latitude 10° and 18° N from the Atlantic to the Indian Ocean, in South central Africa, in the north and east of Australia and in South America where they form the *llanos* of the Orinoco basin and the *campos cerrados* of central and south-east Brazil. Savanna vegetation is subject to frequent grass fires. Many of the woody components are fire-tolerant.

In several parts of Africa, there are large tracts covered by a mantle of small trees of bushy habit, together with lesser bushes and shrubs. The general level of the mantle is between 4.5 and 9 m. In the semi-arid regions the grass cover is generally thin and less than the savanna level and may also be impoverished by surface erosion. The term 'bushland' has been used for this kind of vegetation. Grass fires in bushland areas are normally much less frequent and less severe than in savanna country.

1.3.3 Monsoon Vegetation

The monsoon forests are different from the tall rain forests, but in places they are often dense and difficult to penetrate. These dry forests are formed of rather short trees from 9 to 18 m in height and are influenced by a long dry season. Thorny species are abundant and these are quite bare for most of the year. Lianas are only feebly developed.

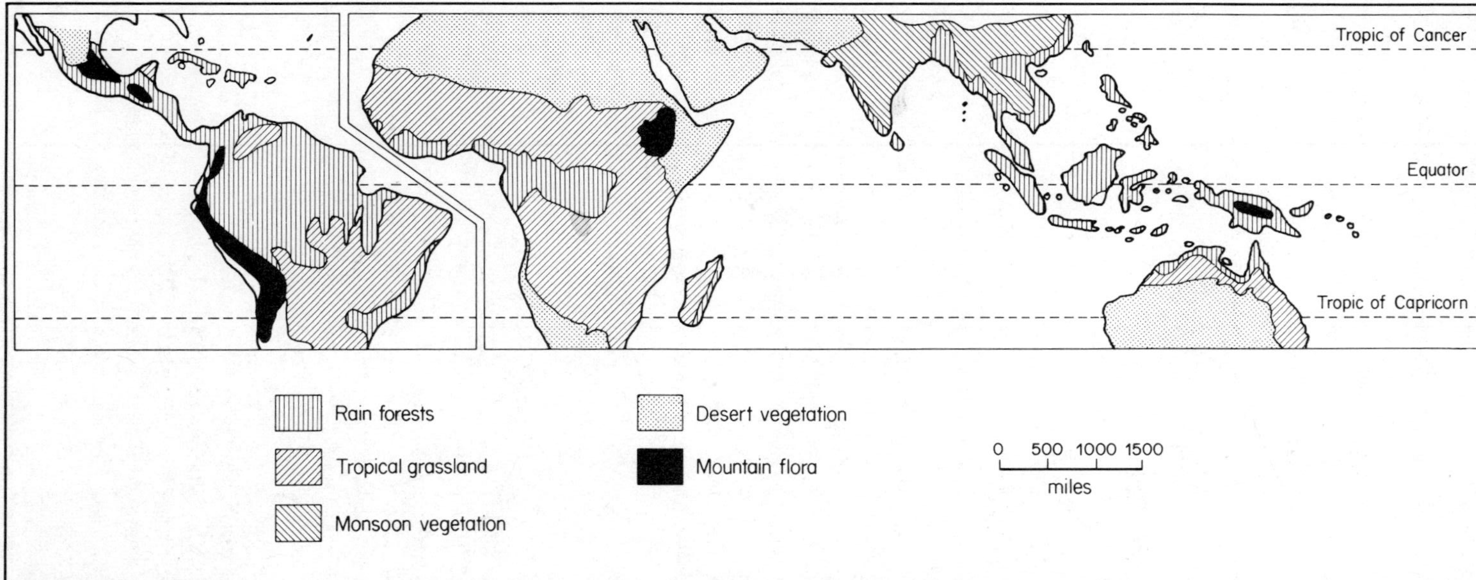

Map 3 : Natural vegetation of the tropics

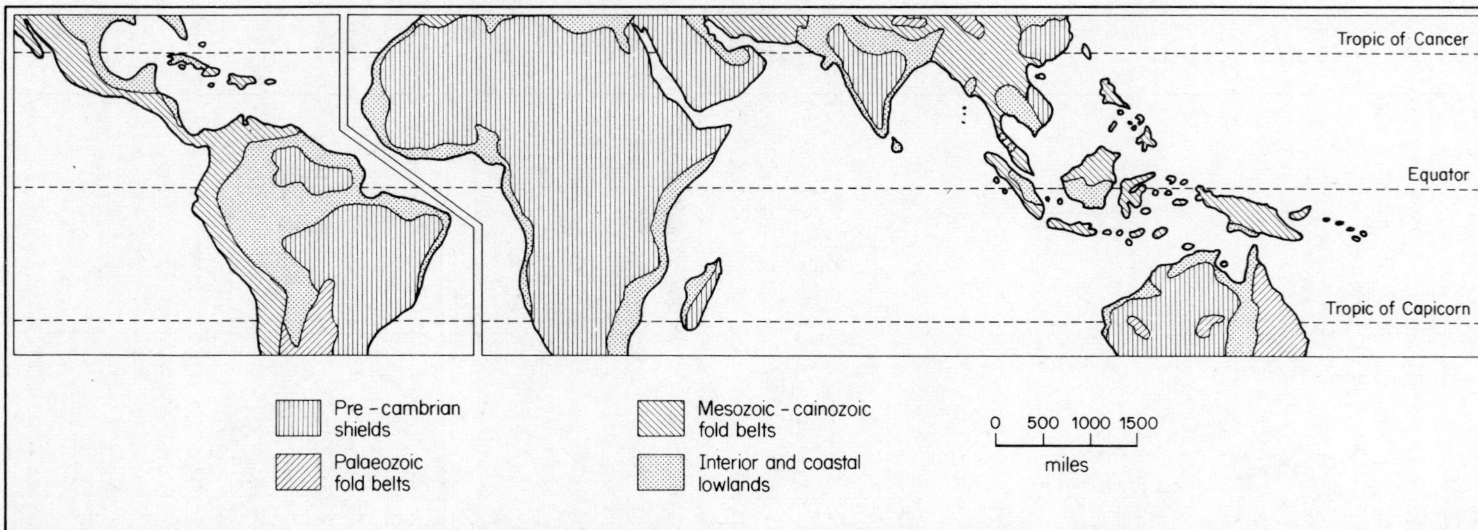

Map 4 : Geological formations of the tropics

1.3.4 Desert Vegetation

Desert vegetation is characteristic of regions where the annual rainfall is less than about 400 mm. On both sides of the equator, tropical grassland and monsoon forests gradually merge into desert vegetation of which the drought-resisting succulents, such as the cactuses, are the most abundant. Spiny umbelliform shrubs and acacias dot these tropical deserts. In the arid plateaux of Mexico are found the graceful yuccas and the agave or maguey, dominating the landscape with their long, broad, curved, blue-green leaves.

1.3.5 Mountain Flora

In the mountains, temperature governs the distribution of the different types of vegetation. The amount of solar heat and the effect of winds are also important. There is a succession of vegetation types but as the tree limit is passed, alpine pastures are encountered, and these in turn give way to scattered alpine vegetation before the snow line is reached.

1.4 Geological History

King (1962) postulated the concept of two primordial continental masses, Laurasia in the north and Gondwana in the south, of similar size and originally in opposed semi-polar positions. The two proto-continents were disrupted in Mesozoic times and their fragments moved radially forming the modern continents. It is a strange coincidence that the continents of the present tropical world were all part of ancient Gondwanaland.

The former unity of Gondwanaland is best demonstrated by the sequences of late Palaeozoic and Mesozoic continental sediments characteristic of South America, Africa, India and Australia. In each of these regions, the several stratigraphic series total some thousands of feet in thickness, and cover several thousands of square miles in the same horizontal position in which they accumulated.

Rocks of late Silurian and of Devonian age are now known in four of the five original Gondwana continents. Beginning with this epoch it is possible to reconstruct, stage by stage, the history of Gondwanaland. All the southern continents, including India, present concrete evidence of glaciation, which was not however contemporaneous. The climate was cooling over Australia towards glaciation in the late Carboniferous while it steadily warmed up in South America because the glaciation was past.

From the early Carboniferous until the mid-Permian, there was a clear progressive movement, first of glacial phenomena and then of cold-temperate coal measure accumulation, from west to east across Gondwanaland. No marked physiographical changes occurred between the early and late stages of the Permian period in Gondwanaland. There was a general increase of temperature indicated by changing lithologies and fresh fossil forms (e.g., reptiles) and cold conditions continued in Australia as contrasted with the semi-tropical conditions of India and the Congo.

Slow denudation and deposition continued throughout the Triassic period but this quiescence was rudely shattered by basaltic eruptions in the Rhaeatic.* Basaltic magma was widely distributed throughout Gondwanaland and the essential similartity of the extrusive and dyke rocks dating frcm this period is quite evident.

Gentle arching and doming with concomitant erosion, seem to have been characteristic of the central portions of Gondwanaland during the Jurassic period with depressions in northern Brazil, Congo, India and eastern Australia. Following the mid-Jurassic, the sea began to penetrate Gondwanaland along modern coast-lines and by the middle of the Cretaceous all the present continental outlines were established, fragmentaticn and drift had occurred, and Gondwanaland had ceased to exist.

King (1962) concludes that the post-Gondwana morphologies of all the fragments of Gondwanaland are essentially identical. Frcm late Mesozoic to Recent times, the tropical continents have experienced similar types of tectonic displacement occurring at closely synchronous epochs. This indicates that the fundamental subcrustal forces operating in these regions are similar. The geological formations of the tropics are shown in Map 4.

1.5 Geomorphology and Lithology

1.5.1 Tropical America

The ancient nucleus of the South American land mass is the Brazil-Guiana massif, composed of Archaean crystalline rocks. This plateau, bisected by the Amazonian lowlands, is a structural unit which extends from the Orinoco River to the Sierra de Tandil, south of the River Plate. Moreover, pre-Cambrian rocks in Peru which underlie the younger structures of the Andean cordillera and the basement rocks in Chile and Western Argentina show that the continental basement has extended from the east to the west coasts of South America. Resting upon the Archaean are several series of less ancient sediments, including the Late Palaeozoic and Mesozoic sequence of the Santa Catarina system, similar to the Karoo system of southern Africa and the Gondwana system of India.

A variety of geological formations and landscapes are found in tropical Central and South America. The mountain ranges are among the highest and longest and include the most important volcanoes in the world. Also, the plateaux, river valleys, forest areas, savannas and grasslands are among the most extensive.

About half of Mexico and all of Central America, the Antilles and the Bahamas constitute Middle America, which forms a distinct geological break between North and South America. The geological structures and surface forms of North America continue southward into Mexico to a little south of latitude 20°N where they are abruptly terminated by a north-west to south-east chain of towering volcanoes. Southern Mexico, Guatemala, and Honduras are

*Topmost stage of the Triassic

similar in structure to south-eastern Cuba, Puerto Rico and the Virgin Islands and this region is connected to South America by two chains of volcanic ridges and peaks; the lesser Antilles and the parallel highlands of El Salvador, south western Nicaragua, Costa Rica, Panama and western Colombia, further south.

There are three main surface divisions in South America : the relatively young Andes range on the west, the much older Brazilian and Guiana highlands on the east, and the central portion which contains the plains of the Orinoco, the Amazon and the Paraguay-Paraná-Platà, filled with debris from the erosion of the highlands on either side.

The Andes are relatively narrow, scarcely three hundred kilometres wide, except in Bolivia where the width is doubled. The peaks reach altitudes from 5,500 to nearly 7,000 metres and include Mt. Aconcagua (6,959m), the highest mountain in the western hemisphere. In general, the Andes are formed by folded and faulted structures with three distinct groups of active volcanoes : in southern Colombia and Ecuador; in middle and southern Peru and along the border of Bolivia and Chile; and in the southern part of middle Chile, south of Capricorn.

The greater part of the surface features east of the Andes is made up of highlands extending from southern Colombia with few interruptions to the northern bank of the Plata River. In the highlands there are ancient crystalline rocks, stratified formations, and sheets of dark-coloured diabase of volcanic origin.

The plains of South America occupy a relatively small proportion of the continent. The Orinoco Plain in the north is separated from the Amazon Plain by a belt of highlands. The Amazon Plain is wide along the eastern base of the Andes but narrows considerably where the river Amazon flows into the Atlantic. It is joined in the south with the plain of the Paraguay-Paraná-Plata system, where much of the area is covered by alluvium brought down from both the Andes and the Brazilian highlands.

1.5.2 Tropical Africa

Much of tropical Africa is a crystalline plateau, largely composed of igneous and metamorphic rocks, two-thirds of which are covered by sediment. This plateau is also modified by volcanic eruptions, ranging from Pre-Cambrian to Recent, and by faults such as the Rift and the Cameroons.

Marine sediments, more recent than Cambrian times, are present in narrow bands on the edges of the continent, on the site of the former Sahara Sea and south-east of the volcanic highlands of Ethiopia. Continental sediments, ranging from Cambrian to Pleistocene, are extensive.

Tropical Africa may be divided into three broad physiographic regions: the west, the centre and the east. The west includes the area west of and including the Cameroon mountains ; the Congo Basin and the Sudan constitute the centre ; in the east, the land is higher and undulating and volcanic deposits cover a relatively small area.

Dominating the plains and plateaux of east Africa, are some of the highest mountains of the African continent, Kilimanjaro (5,895 m), Kenya (5,200 m), Ruwenzori (5,119 m) and Elgon (4,321 m). In contrast, a trough-like depression is formed by the Rift Valley system; Lake Tanganyika—724 km long, 48-72 km wide and 1,434 m deep—occupies one of these depressions.

The rock groups in east Africa range from the crystalline metamorphosed rocks of the various Pre-Cambrian systems with their innumerable intrusions, the sedimentary Karoo rocks of Palaeozoic and Mesozoic times, Jurassic limestones, shales and gypsum-bearing beds, the thin limestones, sandstones and marine deposits of Cretaceous age, and deposits and intrusions of more recent origin. Volcanic material is found in the Rift Valley and the extensive lava plains associated with the major central volcanoes such as Mounts Elgon, Kenya and Kilimanjaro and the Aberdare range.

1.5.3 Tropical Asia/Australia

The landscape in tropical Asia is younger than in Africa. The Indian subcontinent consists of three segments: the Deccan massif, the fold mountain belt, and the geosynclinal basin occupied by the Indus and Ganges river systems.

The Deccan plateau is a thick sequence of crystalline metamorphic Archaean rocks. In some river valleys of the mid-Deccan region, Permian-Carboniferous glacial conglomerates of the Gondwana system occur. The north-western section of the Deccan, the Deccan Trap, consists of basalts, with almost horizontal layers, up to about 3,000 metres in thickness and covering an area of some 518,000 sq km along the coast of Bombay. On weathering, these rocks give rise to the black cotton soils (regur or vertisols).

The encircling belt of fold mountains was thrown up late in the Tertiary, the mountain-building movements culminating in the Miocene. In the Himalayan chain there is a central core of crystalline rocks, flanked by sedimentary rocks of all ages from Cambrian to late Tertiary. Younger rocks occur more in the foothills and in the more open folds of Baluchistan and the India-Burma divide.

The alluvium of the Indus and Ganges river systems is the youngest geological formation laid down by material brought from the Himalayas and the regions through which the rivers pass. This alluvium is more than 900 m deep in parts and consists of the older formations characterized by the presence of lime concretions (*kankar*), in contrast to the newer deposits called *khadar*. The Aravalli range divides the Indo-Gangetic plain into the fertile eastern area and the western expanse of windblown desert sands.

South-east Asia is a great semi-submerged platform bounded by the Philippines, Indonesia and Burma. The geological link between the chief parts of the south-east Asian region and its core is the Sunda Platform, the southernmost continental block of Asia, which underlies Borneo, eastern Sumatra, northern Java and Malaya (Dobby, 1960).

Palaeozoic rocks occur to a small extent in Borneo; Triassic marine rocks are relatively more widespread. In the Philippines, assemblages of volcanic and sedimentary rocks of unknown age form the foundation. In Indonesia, the sediments are of the post-Eocene period.

The islands of the Pacific may be grouped into two: the continental islands extending eastwards in a broad zone from Asia and Australia, and the oceanic islands distributed over the central and eastern Pacific. Among the continental islands are New Guinea (761,000 sq. km), the New Hebrides, Solomon Islands, Tonga, and Fiji. Volcanic activity and earthquakes are characteristic of these islands, which contain two-thirds of the world's active volcanoes. Many rock types and formations occur in the continental islands ranging from the simple volcanic and coral islands of Tonga to the mountain and valley systems of the larger islands with their complex metamorphic schists, volcanic andesites, ash deposits, and alluvial and marine sediments. The oceanic islands are small, a few exceeding 160 sq. km and only one (Hawaii Island) with an area greater than 1,600 sq. km. These islands represent the summits of huge ocean ridges mostly volcanic in origin.

New Guinea is an area of great geomorphological contrasts. Rugged mountain ranges, mostly 2,000-4,000 m high, either rise abruptly from the sea or from young aggrading and often swampy flood-plains. The mountains in the centre of the islands are however bounded in the south-east by large undulating, stable and deeply weathered plains which form part of the Australian continental shelf. The hills were thrown up during Pliocene-Pleistocene times. There are some rocks of Silurian age but most of the rock formations are of Tertiary to Pleistocene age, with considerable areas of Palaeozoic rocks in the central ranges. Large masses of intrusive rocks commonly stand out in the landscape. Limestone is found at all altitudes. Volcanism is integral to the landscape, the highest of the volcanoes rising to over 4,000 m.

Intertropical Africa and parts of Australia are analogous in some respects. Much of the southern part of western Australia consists of a peneplain about 300 to 365 m above sea level and is underlain principally by the Pre-Cambrian shield. But high mountains are not found in Australia; less than half the continent has an altitude of more than 300 m and about 5 per cent exceeds 600 m while the highest point is only 2,240 m above sea level.

The Archaean foundations of Australia are exposed in the western part of the continent. Towards the middle of the continent, the basement sinks down and is concealed beneath a mantle of Cretaceous and Cainozoic deposits, mostly of non-marine origin. Permian glacial rocks and other terrestrial formations of the Gondwana series are found in western Australia.

2 Factors Affecting Soil Formation

DIFFERENCES in pedogenesis between temperate and tropical regions are due primarily to differences in climate and biosphere. Within the tropics, in any given climatic belt, differences among soils are caused mainly by topography and parent material. On an undulating topography the patterns of soil on each type of rock are related to the topography. Some of the soils are very old and have developed under past climatic regimes.

2.1 Climate

Rainfall, temperature, radiation and illumination, humidity and winds are among the climatic factors which affect soil formation.

2.1.1 Rainfall

In the tropics, in regions with an equatorial or a monsoonal climate the annual rainfall is quite high. But unlike in temperate regions, where evapotranspiration is only moderate and rainfall is usually of low intensity, in the tropics evapotranspiration is much higher and the rains fall in storms of high intensity causing considerable surface runoff. The intensity of rainfall generally varies with the altitude, the lower altitudes having a higher intensity. Where a monsoon type of climate prevails there is much seasonal variation in rainfall distribution which in turn affects soil formation. Outside the equatorial regions, rainfall can also vary greatly from year to year much more so than in temperate zones. In east Africa, for example, many meteorological stations have recorded maximum annual rainfalls of more than four times the minimum. There is also a wide variation in the time of year when the rains occur. Rainfall Probability Maps have been prepared showing the minimum annual rainfall to be expected, say in four years out of five, to assist forecasting in these regions.

In tropical areas where the leaching efficiency of precipitation is great, soil materials are poor in plant nutrients. Such soils tend to be acidic. At the other extreme, an excess of evapotranspiration over precipitation produces an upward movement of water through capillary action resulting in deposition of salts upon evaporation of water from the surface. There is a variety of conditions in between.

The battering of the soil surface by raindrops during tropical storms disrupts the soil aggregates ; this results in the blocking of soil pores by fine particles. The soil surface can thus get sealed and infiltration rate could be greatly reduced. This in turn will increase surface runoff and erosion and reduce the water.

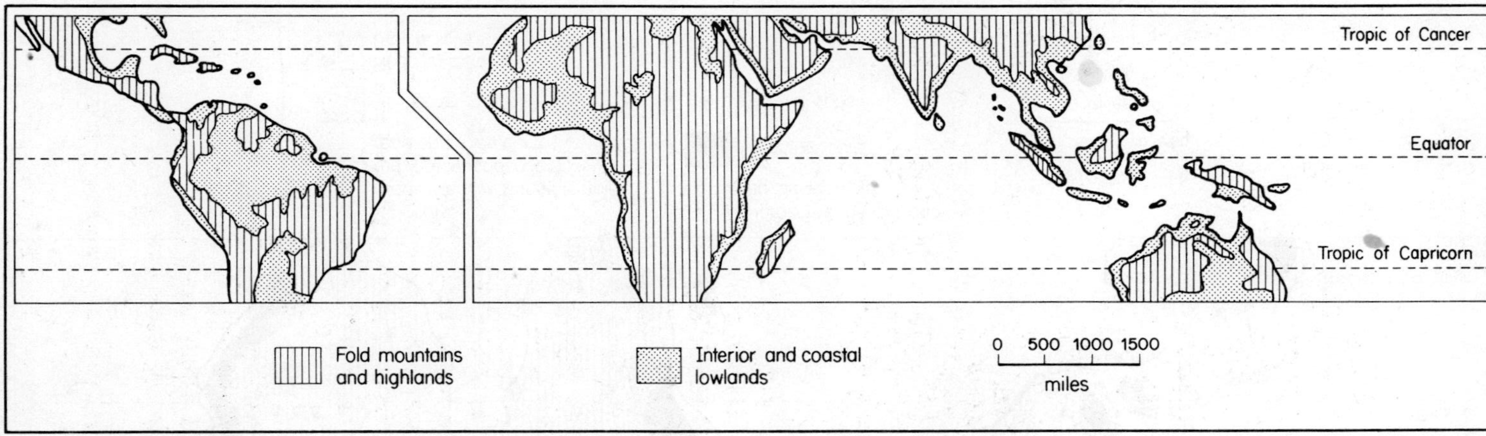

Map 5 : Geomorphology of the tropics

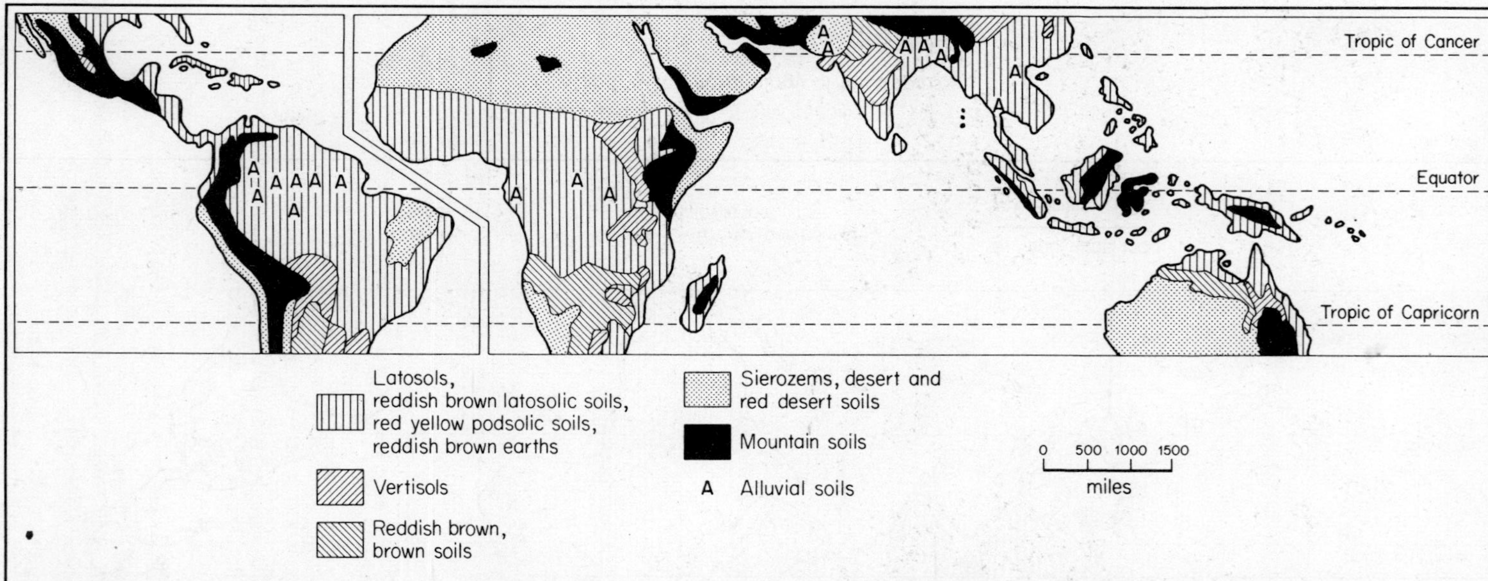

Map 6 : Great soil groups of the tropics

available for percolation through the soil. Ellison (1952) has calculated that the energy in 7.6 cm (3 in) of rain falling in one hour with a drop velocity of 9.1 m/sec (30 ft/sec) is equivalent to that required to plough the land twenty-nine times. It is quite common for such intensities to be reached, and exceeded, over even shorter periods ; a rainfall intensity of 12.3 cm per hour has been reported from the Congo (Vandenplas, 1943). For east Africa, however, Russell (1962) reports that rarely does more than 20.3 cm fall *per day* on the coast, 15.2 cm over high land (above 1,800 metres) and 10.2 cm over plateaux.

2.1.2 TEMPERATURE

The weathering potential of water increases five-fold between 10°C and 30°C. High temperatures together with much rain in the wet tropics therefore result in very intensive chemical weathering and the minerals are in an advanced stage of weathering. In the drier areas and especially in deserts, physical weathering is more important than chemical weathering.

High temperatures are also conducive to a rapid organic matter turnover and a low soilhumus content. Organic matter build-up occurs in the colder altitudes and also under anaerobic waterlogged conditions.

2.1.3 HUMIDITY

Atmospheric humidity varies with total rainfall as well as with its seasonal distribution. Humidity is usually high in equatorial and monsoon climates and seasonally falls to low levels during the dry season. There is also marked diurnal fluctuation. In the rain forests it may remain at around 100 per cent during the night but falls to as low as 70 per cent during daylight on dry days. Humidity influences evapotranspiration and thereby the leaching efficiency of precipitation.

2.1.4 SUNLIGHT

Sunlight is more intense in tropical and subtropical regions, since the rays are nearly vertical and their shorter path through the atmosphere permits a greater proportion of ultraviolet and blue-violet wavelengths to reach the earth. The intensity of solar radiation increases rapidly with altitude as the amount of dust in the atmosphere decreases. The intensity and, hence, the brief duration of tropical rainfall, favours a fair amount of sunshine in some areas. Beckinsale (1957) has reported a 6 hr/mean day for Singapore and a 6.8 hr/mean day for Georgetown, compared to a 4 hr/mean day for London.

In the low altitude tropics, under the rain forest, a rich and varied shade flora develops. Sunlight falling on bare soils can lead to water evaporation and organic matter breakdown.

2.1.5 WIND

Wind erosion in the tropics occurs mostly in arid regions and in extensively cleared areas with a long dry season. Excessive or ill-timed cultivation can increase wind erosion. Measures for its control aim at reducing wind velocity at ground level and also at increasing the moisture content of the soil.

Wind-borne material is sometimes added to the surface soil horizons. Clayey and ferruginous dust blown from the Sahara has added large amounts of iron oxide to the soils in parts of west and east Africa, contributing to laterite formation (Vine, 1949).

2.2 Biosphere

2.2.1 Vegetation

The nature of the vegetation is largely determined by soil, climate and topography, which together form one integrated ecosystem. But vegetation itself influences the formation and nature of soil. Vegetation supplies the soil with organic matter and protects the soil from rainfall, sunlight and wind. Vegetation can also influence climate, though to a limited extent.

A lush and luxuriant vegetation in tropical rain forests has been thought to indicate soil of high fertility. But this is often an illusion. When the jungle cover is removed the nutrient balance formerly maintained between soil and vegetation is disturbed and, after leaching, the soils become rapidly impoverished.

2.2.2 Organisms

Among the soil organisms in temperate climates, bacteria and earthworms may be considered important. Under conditions of extreme drought in the tropics bacteria generally become immobilized but get reactivated again with the onset of the rains. Insects like ants and termites are active in breaking down organic material and in aerating tropical soils.

2.3 Parent Material and Topography

Climate is often considered to be a major factor determining the formation of the Great Soil Groups in the temperate zone. In humid tropical and subtropical regions, on the other hand, high temperature and heavy precipitation promote deeply weathered, well-leached soils which differ from one another particularly because of differences in parent material and topography. The importance of parent material and topography in soil development in the humid tropics has been stressed by several workers.

For instance in the Caribbean, where the climate is fairly uniform with moderate rainfall and well-marked dry seasons, the soils are largely determined by differences in parent material and topography (Hardy, 1949). In Cuba, the soil types are largely determined by the nature of weathering of the parent rock, which consists mainly of transported calcareous and siliceous alluvium. In Puerto Rico, however, marked correlations are seen between soil reaction and rainfall. In the granite regions of Malaysia, the soil types conform to the different types of granite (Coulter, 1964).

Ellis (1958), while reviewing the genetical approach to soil classification in which climate is considered as the major soil forming factor, emphasized the need for re-examining the importance of parent material. Basic igneous rocks tend to resist the acidifying and unsaturating effects of tropical climate. Within any particular climatic regime in the tropics, Ellis regarded parent material to be the best and most practical method of classification, particularly when basic and ultra-basic igneous rocks give rise to the parent materials concerned. Thus in many parts of central Africa, it is possible, within a radius of a mile and under similar conditions of vegetation, topography and age to find podsolic soils derived from acidic igneous granite, red latosolic kaolinitic clays from basic igneous epidiorite, black cracking montmorilloritic clays from basic norite, greyish-brown heavy clays from serpentine and yellow, grey or brown silty or sandy loams, frequently latosolic, as well as fine sands from sedimentary rocks.

Milne (1935*a*) introduced, and Bushnell (1942) elaborated, the catena concept as a 'unit of mapping convenience' to emphasize the relationship between the different soils in the topographical sites of certain parts of East Africa. Morison (1949) referred to the topographical catena concept as 'the fundamental concept around which to build future attempts to classify tropical soils.' He found that in the province of Equatoria in the Sudan in Africa, three soil complexes could be distinguished within a catena, each of which was associated with a broad topographical site.

The *eluvial* complex occupies the high level site and is the parent complex which by the loss of water as well as soluble and suspended matter will provide the material from which the soils of the depressions are built up. One of the soils of the eluvial complex is a true zonal soil and is the most complete reflection of all the factors of the climate. The *colluvial* complex occupies slope sites and receives material from the eluvial complex and loses some of it to the illuvial. The *illuvial* complex occupies low level sites and consists of a mosaic of soil types depending mainly on the amount and nature of the drainage. When the rainfall varies from 890-1140 mm(35-45 in) with a monsoon climate, the development of vegetation and soil is largely controlled by the water factor. Morison (1949) regarded this as the critical climate under which the number of variants within each complex is greatest.

Webster (1960) has argued that climatic effects in Africa are masked by, and subsidiary to, those of geomorphology and age. Climate itself is often dependent on geomorphology. The dependence of climate on topography is shown in the mountains of east Africa where the effect is so great that between the base of a mountain and its summit the soils range from tropical to alpine within a distance of twenty miles (Milne, 1935 *b*).

In east Africa, parent materials vary from igneous, metamorphic, sedimentary rocks to transported materials of aeolian, pyroclastic, lacustrine, ialluval and colluvial origin. Many different kinds of soils are found on these parent materials. Those high in quartz such as granites and sandstones give rise to sandy soils while those low in quartz, such as lavas, produce clayey soils. In the neighbourhood of the Rift Valley extensive volcanic activity during Tertiary to Pleistocene times has deposited an ash mantle over the older soils and newer soils are found developing on these ash deposits.

Berry and Ruxton (1959) stressed the importance of drawing a clear distinction between soil profiles and weathering profiles. In the semi-arid and humid tropics, under suitable conditions, a thick cover (9-10 m) of weathering debris may be formed *in situ* over fresh solid granitic bedrock. They divided the weathering profiles on granite at the foot of concave hillslopes into four zones:

zone 1 : a structureless mass of reddish brown sandy clay;

zone 2 : a pallid silty sand with the structure of the parent granite preserved and containing a few corestones;

zone 3 : where corestones were dominant;

zone 4 : weathered bedrock.

In such a case, soil formation will proceed differently on terraces, convex hills and concave hillslopes. The soil catena which develops will not only be a function of drainage and slope but also depend on the state of the weathered debris (Fig. 1).

2.3.1 Mineral Weathering in the Tropics

In the humid tropics the chemical decomposition of minerals in soils is rapid and generally near completion. The secondary minerals formed are responsible for the physical and chemical properties of the soil.

Several workers have developed the concept of a sequence of soil development which is related to the formation of secondary minerals. Sherman and Ikawa (1968) emphasized that soil genesis results from the combined action of weathering and leaching. Sherman (1952) pointed out that smectite clay minerals would develop where primary minerals decompose under conditions of poor leaching. Under such conditions, the concentration of bases, soluble silicate, and available aluminium hydroxides remain high in the weathering system. As the leaching intensity increases, bases and SiO_4^- are depleted, leading to the formation of minerals of the kandite group, which remain stable under acid conditions so long as the SiO_4^- concentration is sufficient. When the concentration of SiO_4^- falls below a critical value, the kandite minerals decompose to form free hydroxides and hydrous oxides of aluminium, which do not appear in appreciable concentration until the soluble SiO_4^- concentration becomes very limited. On the other hand, iron, manganese and titanium have a low affinity for combining with SiO_4^- and therefore appear as amorphous hydroxides and hydrous oxides in the very early stages of weathering. Soil genesis in the tropics is thus basically the process of chemical decomposition of minerals. Physical weathering paves the way for chemical reaction by increasing the specific surface area.

Climate and time influence the sequence of soil development. As rainfall increases, a succession of soils with peak concentrations of salts, gypsum, carbonate, hydrous mica, montmorillonite, kaolinite, iron oxides, ferruginous bauxite, and gibbsite are formed, each attaining its peak concentration according to the prevailing weathering and leaching conditions. The accumulation of free iron oxide increases through all these stages until it rea-

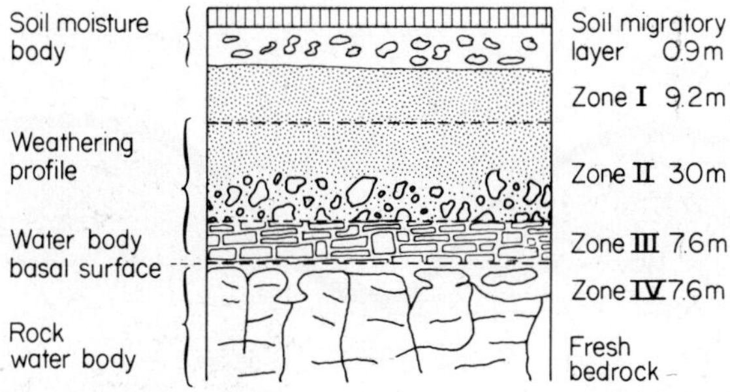

a

Fig. 1 a : Weathering profile

TERRACE

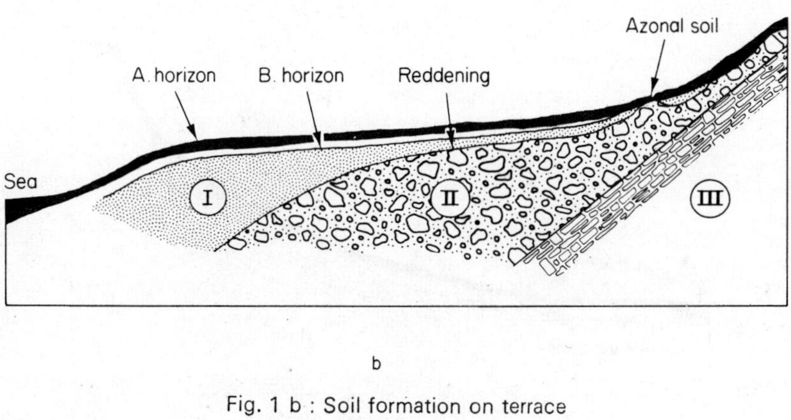

b

Fig. 1 b : Soil formation on terrace

Fig. 1 : Influence of topography on soil formation (after Berry and Ruxton, 1959)

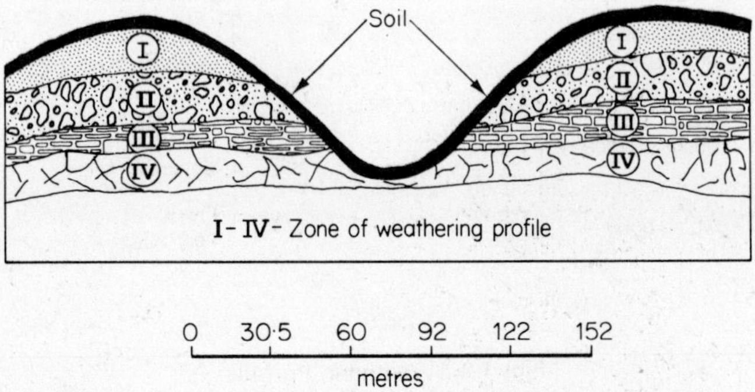

c

Fig. 1 c : Soil formation on convex hill

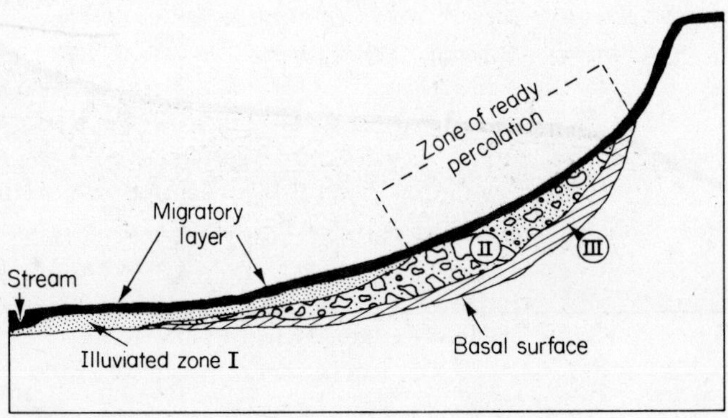

d

Fig. 1 d : Soil formation on concave hillslope

ches its peak in the iron oxide stage, declines thereafter in the ferruginous bauxite, and disappears in the gibbsite soils of high rainfall regions. Manganese follows a similar course to iron reaching its peak accumulation in the kaolinite soils and disappearing with the appearance of free aluminium hydroxide.

The amorphous minerals in tropical soils are important, because they determine the specific surface area of the soil system. In the early stages of weathering, specific surface area increases. Later, because of mineral aggregation, the effective specific surface area decreases, as the amorphous hydroxides and hydrous oxides crystallize to form inert larger mineral aggregates. Soils with amorphous minerals have a low bulk density (around 1.0), a low particle density (around 2.2), and a large water-holding capacity.

Soils with montmorillonite clays are physically unstable, because they disperse easily, especially when the sodium or magnesium content is high. Kaolinite clays have an inherent stability due to their non-swelling properties. Their stability increases as the content of free aluminium hydroxides and hydrous oxides increases. However, they become friable and erodible when the hydrous oxides of iron and quartz increase. Bauxitic soils are very stable even on extremely steep slopes.

2.4 Age of Parent Materials and Soils

Most European soils developed since Pleistocene, and some of them with well developed profiles may be considered to have reached maturity. A number of tropical soils, on the other hand, have had a history which began much earlier. The history of certain African soils, for example, began in Jurassic times. They are past maturity; they have reached old age. They have a polycyclic history and the unicyclic concepts of Dokuchaev do not apply. Spurr (1954) studied soil catenas in the Southern Highlands of Tanzania where the topography is composite and comprises the remnants of several surfaces which are representative of successive erosion cycles which have occurred since Jurassic times. Senile soils cover large areas of Tanzania.

Soils formed during past erosion cycles also form catenas, irrespective of the nature, age or origin of parent material (J. P. Watson, 1964). Many of what appear today to be catenas formed under existing conditions include among their members truncated profiles of senile soils developed during past erosion cycles and in an entirely different environment from that in which they now appear. The only soil catenas which have developed under the topographical and drainage conditions seen today are those developed on land forms or parent material younger in age than the Tertiary period.

In general, the fertility of the soils comprising each catena would vary according to the age of the catena; the youngest catena would possess the most fertile soils. However, when the fertility of soils composing catenas developed from different parent materials is considered, the soils of the youngest catena are not necessarily the most fertile. The nature of the parent material is the controlling factor. This determines the physical properties of the soil and ultimately its degree of fertility. Thus, not only the age of the soil, but the nature and origin of the parent material on which the soils are forming, will determine their fertility status.

2.5 Other Factors

2.5.1 Human Interference

Human interference with soil is dealt with in other chapters. Generally both the traditional practices such as shifting cultivation as well as the more modern methods of farming involving the use of fertilizers affect soil formation in a very marked manner.

2.5.2 Fire

Both forest and grassland are usually cleared by cutting and burning. Nearly all the carbon, nitrogen and sulphur in the fallows and litter are lost to the atmosphere in gaseous form but not the amounts in the soil humus. The ash spread on the surface of the soil contains carbonates, phosphates and silicates of cations. The alkaline ash raises the *p*H and availability of cations in the surface soil. The heating of the soil surface has some direct effect on the microbial population, on the physical and chemical properties of soil colloids and on nutrient availability. Changes in *p*H and nutrient availability may change the microflora. Thus, Meiklejohn (1955) found that on a Kenya upland soil after burning the numbers of micro-organisms, especially fungi, were reduced. Aerobic nitrogen fixers and nitrifying bacteria were killed but anaerobic nitrogen fixers appeared to survive.

The direct effects of heat on the soil are confined to local spots where the wootd is piled. Temperatures at soil level rise very steeply to between 100° C and 850° C, varying with wind and the height, density and type of vegetation but except in localized spots these return to normal values within a few minutes. Soil temperature at a depth of about 2 cm varies at most by 14° C but more often by as little as 3°-4° C or less (Cook, 1939).

Intense fires occur in some areas as, for example, in the Guinea Savanna Zone in Ghana (Ramsay and Innes, 1963). The extreme drought prevailing from mid-October to about mid-April is further aggravated by the effects of the dry north-easterly *harmattan* blowing out of the Sahara from December to April. The aerial portions of grasses wither and die during this period and the entire zone is subjected to fierce sweeping fires during the dry season. Fires may occasionally be started by lightning but most of them are due to human activities. Late dry-season fires are particularly destructive to trees and shrubs of all except the most fire-resistant species (Willimoth and Anthony, 1958). In the Ethiopian highlands farmers burn the topsoil after it is shovelled together in heaps, an operation referred to as 'Guie'. Wehrmann and Johannes (1965) estimated that by burning, 24,000 kg organic matter and 1,000 kg nitrogen are lost per hectare. At the same time, part of the burnt organic nitrogen was fixed in the upper heap layers as 100 kg NH_4-N. Plant available P and K increased.

2.5.3 Overgrazing

Resistance of soil to erosion is lowered through overgrazing. This occurs particularly in semi-arid and arid regions where the bare soil tends to get heavily eroded with the onset of the rains (Pereira *et al.*, 1961).

3 Processes of Soil Formation

THE PROCESSES by which soils develop arise from the operation of different combinations of the factors of soil formation. Since an infinite number of such combinations is possible the processes are also infinitely varied and give rise to a large variety of soils. It is not always easy to sort out the genetic factors responsible for any particular soil throughout its history. Even if the dominant factor or factors now operating can be identified these may not have been of much significance during earlier stages of development of that profile.

During the past two decades, soil studies in tropical regions have led to the discovery of new kinds of soil and to a better understanding of the processes involved in the formation of these as well as of other previously known soils. Before classifications of tropical soils are considered it would be both appropriate and useful to identify some of the more important soil forming processes. This does not mean that such processes are mutually exclusive and can be rigidly compartmentalized. Often these processes merge into one another or are superimposed on each other.

This discussion can be approached in one of several ways. If categorizing soils at the highest level into *zonal*, *intrazonal* and *azonal* is accepted, the main processes responsible for the formation of the soils in each order can be described. But, as Thorp and Smith (1949) have pointed out, the characteristics used to separate intrazonal from zonal soils are not mutually exclusive. Moreover, the concept that any given group of zonal soils is restricted to a single bio-climatic zone has many exceptions. And, some intrazonal and azonal soils show many zonal characteristics. The defects of this categorization are thus evident.

In spite of the limitations mentioned earlier, for convenience of arrangement, soil-forming processes will be considered as being dominated by one or other of the main soil forming factors and discussed under the following divisions:

I. Processes in which *climate* is dominant (climatogenic)
   1. Humid regions: (*a*) Ferrallization (laterization)
      (*b*) Fersiallitization
      (*c*) Podsolization
   2. Subhumid semi-arid and desert regions

II. Processes in which *parent material* is dominant (lithogenic)
   1. Montmorillonitic parent material (*vertisols*)
   2. Volcanic ash parent material (*andosols*)
   3. Calcareous parent material (*rendzinas*)
   4. Soluble alkali salts (*halomorphic soils*)

III. Processes in which *hydromorphism* is dominant (hydromorphogenic)
IV. Processes in which *topography* is dominant (topogenic)

3.1 CLIMATOGENIC PROCESSES

3.1.1 HUMID REGIONS

3.1.1.1 FERRALLIZATION: Ferrallization is the main soil forming process in hot humid tropical regions. It has also been referred to as laterization, lateriza-tion, kaolinization or latosolization. Plinthization is the formation of plinthite (*plinthos*—Greek for brick) or mottled clay, formerly called laterite. Buchanan (1807) suggested the term 'laterite' for a highly ferruginous material he observed at Angadipuram (Stephens, 1961) in Malabar. The word is derived from the Latin (*later*—a brick), and relates to the use of the soil as a building material.

Ferrallization consists of an intensive continuous weathering with hydrolysis of silicates. The bases and silica are leached resulting in a relative accumulation of sesquioxides. Chemical changes are active and vigorous throughout the year. Organic matter is often rapidly mineralized. Kandite clays, such as kaolinite, halloysite and metahalloysite, with 1:1 lattice structures are formed. Many ferrallitic soils have at least 15 per cent clay of which more than 90 per cent is 1:1 lattice clay. The clay particles are usually cemented by sesquioxides. The silt/clay ratio is less than 0.25.

The resulting soils—ferrallitic soils or latosols—have a low SiO_2/R_2O_3 ratio (<2) and a low SiO_2/Al_2O_3 ratio (<2), a low cation exchange capacity (cec), a low base saturation (<35 per cent), a low *p*H (4.5 to 5.5), a low content of most primary minerals, low content of soluble constituents, a high degree of aggregate stability and usually a red colour.

3.1.1.2 DISTRIBUTION OF LATERITE (PLINTHITE): Laterite and lateritic materials are widely distributed in the tropics and subtropics (Map 7) (Prescott and Pendleton, 1966). In India, the Malabar coast from Ratnagiri southwards is composed chiefly of laterite, which is found capping the summits of hills and plateaux on the highlands of central and western India. On the east coast of India, from Cape Comorin to the Rajmahal hills, laterite is found at a lower level in the topography.

Other parts of Asia where laterite is found are Sri Lanka, Malaysia, Indonesia, the shores of Burma, Thailand, and Cambodia. In all these countries laterite is quarried for building. Laterite is associated with iron ore deposits in the north-eastern Surigao region of Mindanao in the Philippines. In Indonesia, true lateritic soils occur on old geological formations.

In Australia extensive laterite deposits occur in the west (Carroll and Jones, 1947), in the south near Adelaide, in the Blue Mountains of New South Wales, near Brisbane, and in the vicinity of Darwin. The laterites of Queensland are likely to be of Pliocene age (Bryan, 1939).

Laterite is extensively distributed in Africa. In the west, various types of laterites are widely distributed in Nigeria. Laterite is also found in Sierra Leone, in the Ivory Coast, in eastern Guinea and in Ghana. In east Africa, laterite occurs in Malawi, in Mozambique and in Rhodesia. In the Sudan, indurated ironstones are associated with flat-topped hills. Over much of Uganda, concretionary ironstone layers, called *murram*, of varying thickness occur at different depths. Massive ironstone or laterite is displayed in the cliffs on the shores of Lake Victoria. Laterite formations are also found in the Belgian Congo and in South Africa.

True laterite is virtually unknown in North America. It is uncommon in Central America probably due to the rough terrain and to volcanic activity. In South America, there are important occurrences in Brazil and in the Guianas. Incipient forms of laterite have been noted in eastern Peru and laterite is widely scattered in the Amazon valley (Sombroeck, 1966). Marbut (1932) termed the indurated ferruginous illuvial horizons in some of the soils of the Amazon basin (Marbut and Manifold, 1926) "ground-water laterite".

3.1.1.3 Laterite Formation (Plinthization): A number of theories have been proposed to explain the formation of laterite (plinthite). But it is becoming increasingly clear that although certain fundamental conditions are prerequisite to the reactions that occur, these can be brought about by a variety of local combinations of weathering reactions, water relationships, and other factors (Sivarajasingham *et al.*, 1962).

Minimum amounts of water are necessary for weathering, the removal of bases and combined silica, and the relative accumulation of sesquioxides. Periods of drying would favour the crystallization of goethite or similar minerals, which is probably associated with hardening. Sivarajasingham *et al.* (1962) suggest that alternating wet and dry seasons would favour crust (*cuirass de fer* or *ferricrete*) formation.

Laterite is found where temperatures are warm or are believed to have been warm at the time of formation and it can be reasonably concluded that warm temperatures favour the formation of laterite.

Although Glinka (1927) considered a forest vegetation necessary for laterite formation, later observations have shown that laterite is found in regions with rain-forest vegetation, under low forest and in open savanna.

Laterite can develop on a variety of parent materials. These may be residual over basic igneous, acid or sedimentary rocks including limestone. The laterites on the banks of the Amazon have developed presumably on alluvium. Laterite is also found in colluvial deposits. Whatever the source of material in which laterite forms, an adequate supply of iron appears to be essential.

Laterite is generally associated with a level or gently sloping topography. Sivarajasingham *et al.* (1962) state that laterite has been identified with four physiographically distinct landscapes: (1) high level peneplain remnants, (2) colluvial footslopes subject to water seepage, (3) low level plains having high water tables or receiving water from higher land and (4) residual uplands other than peneplain remnants.

Many existing laterites, such as those of Queensland (Australia) and the

Congo, are clearly relics of earlier geological periods. However, many examples of laterites believed to be forming currently have also been reported.

All the processes associated with laterite formation involve almost complete removal of bases and substantial losses of the combined silica of primary minerals. Under some circumstances gibbsite is the first identifiable crystalline mineral to appear. But kaolin can form first and give rise later to gibbsite by desilication. Resilication of gibbsite to kaolin is also possible.

The genesis of laterite requires enrichment of a material in iron or, sometimes, aluminium. This can take place in a number of ways : (1) downward movement in solution, (2) enrichment by capillary rise, (3) enrichment by a fluctuating water table, and (4) enrichment by laterally moving water. Iron may be mobilized by any of several mechanisms while its immobilization can also occur in any one, or a combination, of a number of different ways.

In permanent moist conditions, laterite is soft but when it dries, by exposure to the atmosphere or by a fall of the water-table, the material hardens irreversibly by dehydration and hard laterite or indurated ironstone forms. Hard laterite may occur as concretions, cemented concretions and hardpans. When formed *in situ* hard laterite overlies soft laterite and there is a gradual transition. Hard laterite may occur at the surface as a result of the overlying material being eroded or if the landscape is uplifted by tectonic movement.

3.1.1.4 FERSIALLITIZATION: Ferrallization together with other processes such as argillation and rubefaction result in fersiallitic soils. *Argillation* is a type of podsolization and consists in the mobilization, transportation and accumulation of fine clay particles. The sudden wetting of a dry soil causes an appreciable rise in *p*H and a temporary mobilization of clay. The mobilized clay is translocated over a short distance and accumulates as clay skins (cutans) on ped surfaces lower down in the profile forming an argillic horizon. *Rubefaction* or ferrugination is the process whereby hydrated ferric oxide, by dehydration and crystallization in the dry season, is transformed into bright red haematite.

Da Costa (1959) defined tropical fersiallitic soils as characterized essentially by the fact that the clay fraction consists predominantly of kaolin minerals associated with some illite and variable amounts of iron oxide. They have a higher SiO_2/Al_2O_3 ratio than ferrallitic soils, the term fersiallitic suggesting their intermediate character between the ferrallitic soils (with low SiO_2/R_2O_3 ratios) and siallitic soils (with high SiO_2/R_2O_3 ratios).

In the French classification system, fersiallitic soils are considered as those soils which are rich in iron oxides with a SiO_2/Al_2O_3 ratio>2 and a base saturation >40 per cent (Sys, 1967). A SiO_2/R_2O_3 ratio >2 is also required in the Portuguese system of Botelho da Costa *et al*; (1964).

Fersiallitic soils can be sub-divided into (*i*) those with low base-saturation in the subsoil and (*ii*) those with medium to high base-saturation in the subsoil.

Subgroup (*i*) constitutes the red-yellow podsolic soils of South America and south-east Asia, including Sri Lanka. They show deposition of clay in a B-horizon, which is heavier in texture than the layers above. Possible mechanisms of movement of secondary alteration products in the profile were discussed by McCaleb (1959). They have markedly less stable aggregation than latosols of similar texture, lower permeability, a well-developed blocky structure and a higher cation exchange capacity of the clay fraction than latosols. The clay fractions of red-yellow podsolic soils in Puerto Rico (Roberts, 1942), formed on shales of low base content have a SiO_2/Al_2O_3 ratio of 2.9 to 3.4 and a SiO_2/R_2O_3 ratio of 2.2 to 2.8. In Trinidad, Red-Yellow Podsolic soils have clays with a cation exchange capacity of 50 me per 100 g clay.

Fersiallitic soils with medium to high base-saturation in the subsoil include the 'reddish-brown earths' which are the most extensive soils of Sri Lanka occupying much of the lowland dry zone with a rainfall of 1300 to 1800 mm.

The reddish-brown earths of Sri Lanka are formed from Archaean rocks containing considerable amounts of ferromagnesian minerals, and to a smaller extent from limestone. The SiO_2/Al_2O_3 ratio of the clay fraction is greater than 2.2, *p*H values range between 6 and 7, the B-horizon has a strong subangular blocky structure, and the cation exchange capacity is between 45 and 55 me per 100 g clay.

Similar soils have been reported from India (Nagelschmidt, Desai and Muir, 1940), Brazil, Hawaii, and Ethiopia.

3.1.1.5 PODSOLIZATION: There is evidence that typical podsols occur in the hot, humid tropics (Russell, 1961; Klinge, 1965). Hardon (1937) refers to a podsol in the lowlands of Borneo, Jenny (1948) to podsols in Colombia, Tan *et al.* (1970) to podsols in Indonesia. Richards (1941) has given other examples of podsols in tropical regions. These soils usually occur in sandy parent material carrying a heath vegetation. There is usually an A-horizon (10-20cm), bleached A_2 (20-100 cm) and a rather thin B-horizon enriched with iron and aluminium and sometimes indurated, often occurring at some depth. These podsols have developed in regions where the mean annual temperatures are around 25°C and the temperature of the percolating water is higher (22°C to 26°C) than that in temperate regions (15°C or less). Podsols also occur in the uplands of tropical regions at elevations of 1800 m or over, where the mean annual temperature is about 15°C. Table 7 gives the analytical data for a tropical Podsol profile.

3.1.2 SUBHUMID, SEMI-ARID AND DESERT REGIONS

About one-third of the earth's land surface is arid or semi-arid, but much of this land is in the subtropical and warm temperate areas which are strictly outside the tropics. In these areas of the arid or subhumid tropics, where evaporation exceeds precipitation, soluble salts are deposited in the soil profile. Salt formation can also occur where irrigation waters contain dissolved salts, and in depressions that collect run-off from surrounding land or that have a

TABLE 7: ANALYTICAL DATA FOR A TROPICAL PODSOL (from Jenny, 1948)

LOCATION: near Alban (Cundinamarca), Colombia, 2460 m. CLIMATE: rainfall 2010 mm; temperature 17 °C. VEGETATION: now pasture, originally forest with dense cover of ferns. TOPOGRAPHY: gently sloping fan, slope 5-10 per cent. PARENT MATERIAL: rock debris consisting of sandstones and shales of Tertiary age.

Depth *cm.*	*Horizon*	*Texture*	*Colour*		*p*H	C %	N %	C/N
0-46	A_1	very fine sandy loam	greyish brown	10YR 4/2	5.3	9.75	0.88	11.1
46-77		sandy loam	pale brown	10YR 6/3	5.4	4.97	0.32	15.5
77-114	A_2	sandy loam	light grey	2.5Y 7/2	5.6	3.68	0.19	18.8
114-142		fine sandy loam	white	2.5Y 8/3	5.7	2.40	0.15	16.0
1.3 cm thick	chocolate	coloured band						
142-152	B_1 iron-horizon	fine sandy loam	yellowish red	5YR 5/8	5.6	4.21	0.10	42.1
152-203	B_2	very fine sandy clay loam	yellow	10YR 8/6	5.3	1.45	0.09	16.1
203+	C	sandstone	white inside, coated reddish yellow					

shallow water-table. Where much sodium carbonate is present, salt accumulation (salinization) is followed by the entry of sodium into the soil exchange complex (alkalization). Such soils have been studied outside the tropical belt in the USSR, in Hungary, in Holland and in the USA. Kelley (1951) has discussed the processes offormation of these soils and classified them. They are considered later (p.38) as they are intrazonal soils whose development is dominated by the salts present and the processes of soil formation are therefore more lithogenic than climatogenic.

Soils formed in the drier regions of the tropics and sub-tropics, where the leaching action of the rain is restricted to the upper part of the profile have layers of accumulation of $CaCO_3$ (calcification), and possibly also $CaSO_4$(gypsification). These have been called pedocals by Vine (in Webster and Wilson, 1966). Stephens (1953) has described the extensive upland and slope soils of regions in Australia with less than about 500 mm rainfall as pedocals having '*deflated* neutral to alkaline eluvial horizons and calcareous and/or gypseous illuvial horizons'. The soils shown in the CCTA soil map of Africa (d'Hoore, 1960) as 'brown tropical soils of arid and sub-arid regions with up to about 500 mm of rain', are probably soils of this group. They stretch in an east-west strip south of the Sahara from Senegal to beyond the Nile and are also found in extensive areas in east Africa, the Kalahari and the Lowveld of southern Africa, as complexes with other soils.

Milne *et al.* (1936) had provisionally recorded the occurrence of calcareous or non-calcareous 'plains soils', described as light-coloured pedocals, in regions in east Africa with rainfall of between 380 to 760 mm extending from Somalia across Kenya into Tanzania. In most arid regions in east Africa with less than about 380 mm of rainfall, layers of $CaCO_3$ and $CaSO_4$ are common and *p*H values are usually above 7.

Desert soils are mainly the product of physical weathering, as rainfall is very low ($<$ 100 mm av. annual) and temperature is the dominant factor in soil formation. The soils are exposed to strong solar radiation. They are generally coarse-grained sandy soils in which a grain-size sorting has been effected by wind. Strong winds frequently cause sand and dust storms. The surface soil is often blown away, or wind blown (aeolian) material from elsewhere has accumulated on top of former soils.

Effective rainfall is even less than the meagre total rainfall. Much water is lost as run-off because of a surface seal, formed by wetting with rain, action of raindrops and rapid drying of the surface, in which soil particles are cemented by lime.

The sparse vegetation does not influence soil formation very much. Organic matter contents are low, 0.2 to 0.5 per cent in arid soils and 1 to 2 per cent in semi-arid soils.

Vast areas of desert and semi-desert soils are found in Africa, Asia, and Australia and smaller areas in South America and in the south-western United States.

Desert soils in the tropics are exemplified by soils derived from the Kalahari and Sahara sands, extending from the Union of South Africa across the equator to Nigeria and Sudan on the outskirts of the Sahara desert. These soils are derived under fairly constant temperature but under varying rainfall.

The Kalahari and Sahara desert soils are similar in that they are derived from uniform aeolian parent material. They are all very sandy, with poor water-retaining capacity. Iron compounds do not accumulate to any depth and generally there is no horizon differentiation.

The soils and their development are, however, different. As the rainfall ranges from 150 to 1500 mm the colour changes from brick-red under arid conditions to white with increasing rainfall, the reaction also varying simultaneously from slightly alkaline to acid. The colloidal material content seems to decrease with increasing rainfall (Van der Merwe, 1954). The red colour of the soils in the drier regions is imparted by a thin film of iron oxide on the sand grains.

The Goz soils of the Sudan are also sandy soils derived from sandstone covering a large part of the Libyan desert and extending southwards into the Sudan. These soils have a hardened top surface which is usually yellowish red in colour, the red colour fading with depth.

The Damaturu soil of Nigeria is a loose, grey-brown sand merging into orange-brown sand overlying bright yellowish sand. Rainfall in these regions is between 500-600 mm. The *p*H is slightly acidic, between 5.5 and 6.5.

3.2 Lithogenic Processes

3.2.1 Montmorillonitic Parent Material (Dark Clay Soils)

Dark clay soils are widely distributed not only in tropical and subtropical regions but also in warm-temperate areas. There are about 98 million hectares in Africa ; 40 million in the Sudan, 17 million in the Chad and 10 million in Ethiopia. About 71 million hectares are found in Australia, 62 million in Asia of which about 60 million are in the Deccan plateau of India (Tamhane, 1950; Simonson, 1954 a and b). Out of the 17 million hectares in South America, 6 million are in Argentina and 5 million in Brazil. These soils occur in about 6 million hectares in North America and about 1 million hectares in Europe.

These dark clay soils have been known by at least forty different names in the past—*regur*, *black cotton*, *black earth*, *margalite*, *rendzina* and *tir* being a few of the better known terms (Dudal, 1963). Oakes and Thorp (1951) distinguished Houston Black Clay from the rendzina soils of North America and western Europe and suggested the name *grumusol* (from Latin *grumus*—a little heap or crumb, and *sol*—soil) to include all the dark heavy cracking soils known at that time, often displaying a 'gilgai' relief. Finck (1961) compared the characteristics of Gezira clay with other grumusols. Dudal and Bramao (1965) have summarized the available knowledge of 'dark clay soils of tropical and subtropical regions'.

Dark clay soils occur under a wide range of climatic conditions ; in hot and in cool climates, in arid as well as under humid conditions with average annual temperature ranges of 15.5°-26.5°C. The summer temperatures in these regions are invariably higher than 20°C and may be as high as 35°C. Mean annual rainfall varies between 150-2000 mm, the rainfall distribution being markedly seasonal. In fact, dark clay soils occur only in climates with a pronounced dry season, the dry period commonly ranging from four to eight months.

The available information indicates that the vegetation of uncultivated dark clay soils is at present mostly bush or tall-grass savanna, although the original climax vegetation may have been some form of forest in many areas.

Dark clay soils are derived from a variety of parent materials ranging from basalts, shales, limestones to unconsolidated lacustrine clays and marls. But the parent materials are usually fine-textured (at least 30 per cent clay) and are high or moderately high in alkaline earths like calcium and magnesium. The parent materials are either basic in character containing plagioclase felspars, ferromagnesium minerals, calcium and magnesium carbonates; or alkaline earth elements are added by seepage or by flood waters.

Vertisols occur mainly below 300 metres above sea level (Buringh, 1970). Some vertisols are developed *in situ* on lower slopes, but most form in depressions.

The heavy texture and the dominance of swelling clays are the main factors governing soil formation. The presence of calcium and magnesium ensure the formation of montmorillonitic clays. Climatic conditions permit intermittent weathering of the rocks but do not cause rapid leaching. The downward movement of soluble compounds is also impeded by the slow permeability and poor internal drainage of the heavy-textured parent materials.

Swelling and shrinking of the clays with changes in moisture content cause wide deep cracks of a polygonal pattern, each polygon having a horizontal diameter ranging from 1 to 4 metres. These cracks extend from the surface into the soil mass. Soil material can fall or be washed down into these fissures. The sliding of soil material causes polished, grooved surfaces called 'slickensides.' On re-wetting, the soil mass swells up above the surface giving rise to the low hillocks and shallow depressions known as 'gilgais'. The whole soil mass is churned to some depth and the churning causes these dark clay soils to be 'self-ploughing' or 'self-mulching'.

Below the influence of the churning process, there may be a horizon of lime accumulation or even a calcic horizon. Drying, and the combining with iron and manganese may harden the lime into small lime concretions (*kankar*), which are found throughout the A-horizon.

The humus content of dark clay soils is generally low (not more than 0.5-1.5 per cent) and the dark colour is due not to the total amount of organic matter but rather to its nature (Brammer and de Endredy, 1954). It has been suggested (Singh, 1956) that the dark colour is due to the formation of a clay/organic matter complex and that soils with poor drainage are usually darker in colour than the better drained ones or those which develop in drier climates (Johnson *et al.*, 1962).

In the older U.S. classification (USDA 1938), dark clay soils were grouped as Rendzinas. Oakes and Thorp (1951) introduced the term 'grumusol' for dark clay soils which have a crumb structure on the surface layer. But this would not include the 'margalitic' soils of Indonesia (Mohr and Van Baren, 1954) with a crusty, low self-mulching surface horizon or even the 'black turf' soils of South Africa (Theron and Van Niekerk, 1934).

All dark clay soils can perhaps be included in the order of Vertisols according to the 7th approximation (USDA, 1960). Subdivisions are then possible into suborder as Aquerts and Usterts ; those saturated with water at some season and showing gley phenomena, and those restricted to drier climatic regions respectively. Subdivision at the great group level may be into the granulated and self-mulching Grumaquerts and Grumusterts and the crusty surface horizoned, low self-mulching Mazaquerts and Mazusterts.

3.2.2 Volcanic Ash Parent Material

Dark soils, characterized by a very high humus content (8 to 30 per cent humus to a depth of 30–60 cm (1-2 ft), looseness and very high porosity, are found in the Japanese islands, in Indonesia (Tan, 1965), in the Philippines, in Hawaii, in Africa and in Central America. They have been called *ando* soils (Thorp and Smith, 1949) and are also known as andosols. They occur under moderate to very high rainfall in the humid tropics, are strongly leached and of low base status. Their clay fraction always consists of allophane ($Al_2O_3 . SiO_2 . 5H_2O$) or allophane-like clay minerals, which are usually the first product of weathering of sub-basic volcanic ashes. The type and speed of weathering is slightly lower in acid material (e.g. dacite) than in intermediate (e.g. andesite) or basic (e.g. basalt) material. On maturing, allophane is transformed into metahalloysite and finally into kaolinite (Buringh, 1970).

The brown coloured soils in the volcanic islands of the West Indies, having dark upper layers, 30 cm or more thick, also contain allophane-like clay minerals and belong to this group of andosols.

3.2.3 Calcareous Parent Material

These are comparatively young soils where the effect of climate has not been fully felt. They are weekly acidic to neutral in reaction, loamy or clayey in texture ; usually fertile and very productive. The Montserrat series in Trinidad, derived on calcareous Upper Miocene glauconitic sandstone, is an example of this type of soil. It is reputed to be one of the most fertile of tropical soils.

No satisfactory explanation has been given for the relatively great thickness of soil developing on rock which is almost 100 per cent $CaCO_3$. The Matanzas soils of Cuba, for example, exhibit a uniform profile even up to 15 metres depth overlying the limestone. Soil layers of 1 metre or more are, however, quite common. Harrison (in van Baren, 1928) calculated that it would require 150 metres of rock to produce 1 metre of soil if the soil is produced only on the residuum after the limestone has been dissolved away. Mohr and van Baren (1954) incline to the view that air-borne materials play a decisive part in the formation of limestone soils, especially if the latter are developing in the vicinity of volcanoes (van Baren, 1927; Grange, 1949).

3.2.4 Soluble Salts (Saline and Alkali Soils)

Salts usually present in soils originate during weathering and, in humid regions, these salts are leached out into the drainage water and finally into the sea. Sometimes, volcanic gases are responsible for the salts found in soils; chloride from hydrogen chloride gas, sulphate from sulphur and the oxides of sulphur. Chloride and sulphate can also arise from wind-borne mists and sprays from the ocean surface. Nitrates are chiefly of biological origin and result from the decomposition of nitrogenous organic matter while small amounts of nitrate are found in rainfall. Calcium carbonate is the most abundant relatively insoluble carbonate in soils and it is widely distributed in soils of arid regions. The most common soluble carbonate is sodium carbonate, whose formation arises from cation exchange reactions in calcareous soils, by separation from soil solutions with a large HCO_3: Ca ratio, by the reduction of sodium sulphate under anaerobic conditions, or by the decay of plants rich in sodium. Wherever irrigation is practised, irrigation waters can add considerable quantities of salts to soils.

In whatever manner they may originate, soluble salts accumulate in soils when the drainage water evaporates, and especially so wherever evaporation exceeds precipitation. A high ground water level is conducive to the rise of water by capillarity and the ground water of dry regions is usually more saline than that of humid regions.

As a result of salt accumulation (*salinization*) a number of changes take place in a soil. If much soluble sodium is present, sodium ions can displace calcium ions and enter the exchange complex (*alkalization*). The most pronounced physical effect of such entry is a reduction in the permeability of the soil. Clay particles become dispersed and move downwards to settle as dense subhorizons with a columnar structure. Kelley (1951) recognized three types of soils formed under conditions where salts accumulate. Their more important characteristics are summarized in Table 50. In older alkali soils, intensive leaching and degradation gives rise to *solodization*. Exchangeable sodium is replaced by hydrogen ions and the surface horizon becomes acidic. Mobilized clay particles are leached from the A to the B horizons (argillation).

As indicated earlier (p. 32), saline and alkali soils are more widespread in regions outside the tropics. But they do occur in the drier tropical regions and under water-logged conditions.

In the north and north-central parts of Argentina, saline and alkali soils occur in the Chaco area and in the pampa plains. The salt problem in Peru and Ecuador is restricted primarily to the semi-arid coastal regions where the average annual rainfall is less than 250 mm. In Brazil, only a small portion of the interior north-east, in Sertao, has salt problems. But, further north, in Mexico, saline soils occur in all of the agricultural areas of northern Mexico, along the west coast and in north-central Mexico.

The problem of salinity is not widespread in equatorial and tropical Africa.

Most of the saline soils in the African continent are found north of the Tropic of Cancer (Hayward, 1956). Williams (1968) mapped the distribution of saline soils in the west-central Gezira of the Sudan.

In India and Pakistan, saline and alkali soils occur chiefly throughout the Indus Valley, in the valleys and basins of Pakistan and western India, in the Gangetic valley west of about 80° E longitude, on the uplands of the Deccan plateau and in the saline marshes of the sea coast and of river deltas.

In Australia, the problems associated with saline soils are confined largely to the western and southern parts of the continent, but saline soils also occur to a limited extent in subtropical northern areas of Queensland.

3.3 Hydromorphogenic Processes

These are soils formed under conditions of poor drainage and displaying prominent grey or mottled colouration of the subsurface layers indicating lack of aeration at certain times. They include such soil groups as Humic Gley, Low-Humic Gley and Planosols.

Humic Gley soils (Thorp and Smith, 1949) refer to a 'group of poorly to very poorly drained hydromorphic soils with dark-coloured organo-mineral horizons of moderate thickness underlain by mineral glei horizons. . .' Many of them are medium acidic to mildly alkaline in reaction. Few are strongly acidic.

Strongly acidic soils, called acid sulphate soils or 'cat-clays', are found in the marine lowlands of the tropics, in south-east Asia, in tropical Africa and in South America.

Particularly large areas are found in south-east Asia—in Vietnam (more than 1 million ha according to Moorman, 1963), Bangladesh (formerly East Pakistan) and Indonesia. In tropical Africa they have been located both on the west and east coasts. In South America they have been described in Surinam, in British Guiana, and they also occur in tidal areas such as the delta of the Amazon river.

Acid sulphate soils occur in mangrove areas naturally flooded with brackish water during the dry season and with fresh water when the rivers flood with the rains. On drainage and aeration, they become strongly acidic because of the oxidation of sulphides to sulphates. The acid soil material in its oxidized form shows straw-yellow mottlings and streaks of basic ferric sulphate.

Low-Humic Gley soils are 'imperfectly to poorly drained soils with very thin surface horizons, moderately high in organic matter, over mottled gray and brown glei-like mineral horizons with a low degree of textural differentiation' (Thorp and Smith, 1949).

Planosols include Ground-Water Podsols and Ground-Water Laterites. Thorp and Smith (1949) defined them as 'intrazonal soils having one or more horizons abruptly separated from and sharply contrasting to an adjacent horizon because of cementation, compaction, or high clay content'. Ground-Water Podsols developed from highly leached materials on old surfaces of planation have been reported from Trinidad, the Guianas and Indonesia (Richards, 1964). Ground-Water Laterites were first observed in the Amazon Basin by Marbut (1932) and have been subsequently reported from Thailand (Pendleton, 1943), Sierra Leone (Kellogg, 1949) and Malaysia (Arnott, 1957).

3.4 Topogenic Processes

Variations in topography cause erosion, colluviation and sedimentation, which affect soil formation. The effect of altitude to air temperature has already been referred to.

Nye (1955 a) studied in detail the genesis of a catenary sequence of soils in the west African forest at Ibadan, south-west Nigeria. The main differences among soil catenas were due to variations in the rocks from which they were derived. The rainfall, of from 1270–3750 mm (50–150 in) over most of the forest, was sufficient to maintain thorough leaching of the soil and did not give rise to any marked change in the direction of soil-forming processes. Soils formed under higher rainfall were in general more leached and more deeply weathered than those in lower rainfall areas. The patterns of soil on each type of rock were related to the topography.

A profile section of one catena studied is shown in Fig. 2(*a*) and the essential features of the succession of profiles are shown diagrammatically in Fig. 2(*b*). The catena was developed over banded biotite gneiss There are two primary genetic horizons in each profile:—*Cr*, the horizon of soil creep and *S*, the sedentary horizon.

The *Cr* horizon is further subdivided into three quite distinct genetic horizons:

CrW–the horizon formed from worm-cast material
CrT–the horizon formed largely by termites
CrG–the horizon of gravel accumulation.

The *S* horizon can be subdivided into :

S_1–where little decomposed rock is visible
S_2–with many decomposed rock fragments
S_3–which is altered rock retaining the structure of the fresh rock.

In the mature upper slope profile, the lowest horizon S_2 (137-183+cm) contains much evidence of the structure of the parent gneiss but in the horizon S_1 (68-137 cm), the traces of altered rock have almost disappeared leaving a red sandy clay. At the top of this horizon there is a well-marked change to the *CrG*-horizon containing abundant quartz and concretionary iron gravel. The quartz gravel derived from quartz veins is concentrated in this horizon by eluviation of clay, further decomposition of the silt and sand fractions, and removal by termites of fine earth fractions into the horizon above

Above the *CrG* horizon there is a fairly sharp transition to the *CrT* horizon of gravel-free loamy coarse sand about 15 cm thick. This layer of gravel-free soil is an invariable feature of west African soils. Doyne and Watson (1933) suggested that this horizon may be detritus from another source, or more probably may be due to a secondary weathering of the concretion layer itself. Vine (1941) thought it was due to earthworms. Nye (1954) agreed with Charter (1949) that this horizon is formed by the soil fauna which bring up fine earth, leaving behind those particles too large for them to carry or digest. He showed that termites are mainly responsible for the *CrT* horizon, there being a sharp limit to the maximum size of particle in this horizon at 4 mm, which corresponds to the maixmum size of particle in the mound of the principal and largest termite, *Macrotermes nigriensis*.

The uppermost *Cr*W horizon (2.5cm) of fine sandy loam consists mainly of humus and breakdown material from the worm casts of *Hippopera nigeriae* which abound on the surface of the soil. These worms sort out the fine material from the *CrT* horizon and void it on the surface.

In the Immature Profile the nearly fresh rock is exposed at a depth of 1.5m. Below the *CrG* horizon there is only the S_2 horizon. The *CrG* horizon is of about the same thickness as in the upper slope profile, but it contains in addition to quartz and concretionary gravel, much decomposing rock.

In the upper mid-slope profile the *CrT* horizon has increased in thickness from about 15 to 50 cm. This is because this horizon will include the material passed down from the portion above and also any material brought up by termites. The thickness of the *CrG* horizon has also increased somewhat, probably for the same reason.

In the lower mid-slope, the *CrT* horizon remains at about the same thickness but the *CrG* horizon has increased considerably to a thickness of 1.8 metres. At the same time the quartz gravel in this horizon tends to accumulate at the base. This tendency increases in the lower slope and valley profiles and, in fact, in the latter it appears as a marked stone line.

There are differences in the drainage pattern as well as in colour, texture, structure and consistence down the slope. The upper slope profile is well drained and the reddish soil is very permeable. In the upper mid-slope profile, manganese dioxide concretions appear below 1.2 m suggesting moister conditions. The prominent colours of the lower mid-slope profile have brown instead of red hues and there are many manganese dioxide concretions. At 2.38 m, the colour changes abruptly to a reduced grey. This reduced zone is met at 2.13 m in the lower-slope profile and the soil below 0.76 m hardens to form an iron pan. In the valley profile, much affected by seepage, the colour below 1.07 m is a bluish green.

Variations in texture indicate a marked loss of clay in the surface horizons down the slope. This is probably due to both lateral and vertical eluviation.

4 Classification of Tropical Soils

SOILS CAN BE classified in more than one way, according to the differentiating characteristics and the level of abstraction at which the latter are used. These characteristics usually depend on the purpose for which the classification is made. At the higher categoric levels, the causes of the peculiar development of a soil are used, either the soil-forming factors or the soil-forming processes.

> A classification of soils should be based on the known properties of the soils themselves, on a recognition of the processes and regimes which created them, and on a consideration of factors under the influence of which these processes originated and developed. (Tiurin, 1961).

4.1 Early Classifications

In different systems of soil classification different emphases have been laid on these three groups of criteria. In the Russian system, present-day climate defines the two highest categoric levels, intrinsic soil properties being brought in only at the third level. Western systems, used in western Europe and in North America lay stress on soil properties, with priority being given to those properties which most clearly reflect the factors and processes of soil formation.

Kellogg (1949) suggested that the guiding principles of the United States Soil Survey would be useful in surveying tropical soils although the detailed methods employed would be somewhat different from those used in temperate regions. Up to that time, the U.S. Soil Survey had conducted the field research necessary for detailed soil surveys in only two tropical countries, Puerto Rico and Hawaii. In 1938 the American Soil Survey (Soils and Men, USDA, 1938) had recognised the following sub-orders and great soil groups for zonal soils in the tropics.

Sub-order: *Lateritic soils of forested warm-temperate and tropical regions*:
- Laterite soils
- Reddish-brown Lateritic
- Yellowish-brown Lateritic
- Red Podsolic } These were later combined under
- Yellow Podsolic } Red-Yellow Podsolic

Sub-order: *Light-coloured soils of arid regions*:
- Red Desert soils
- Reddish-brown soils

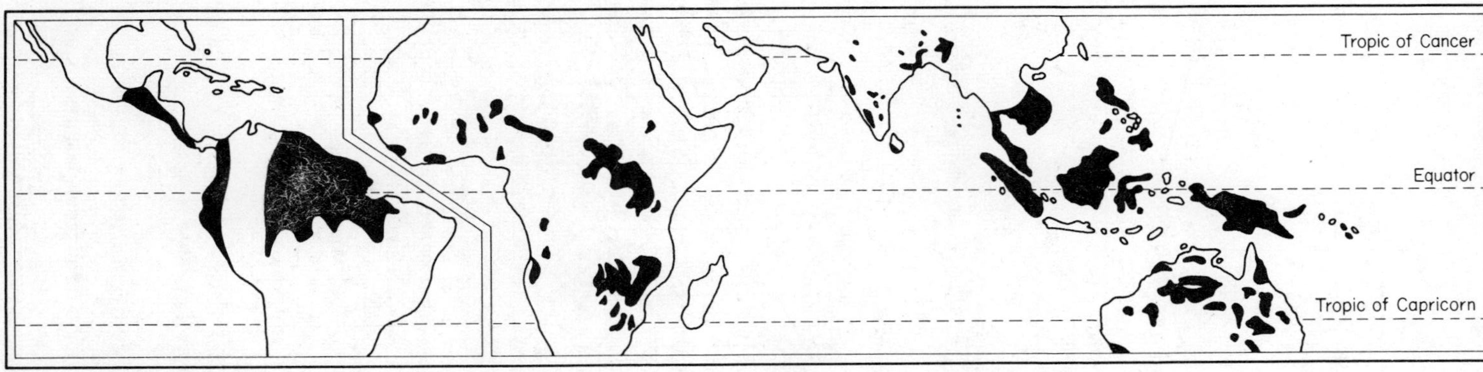

Map 7 : Distribution of laterite

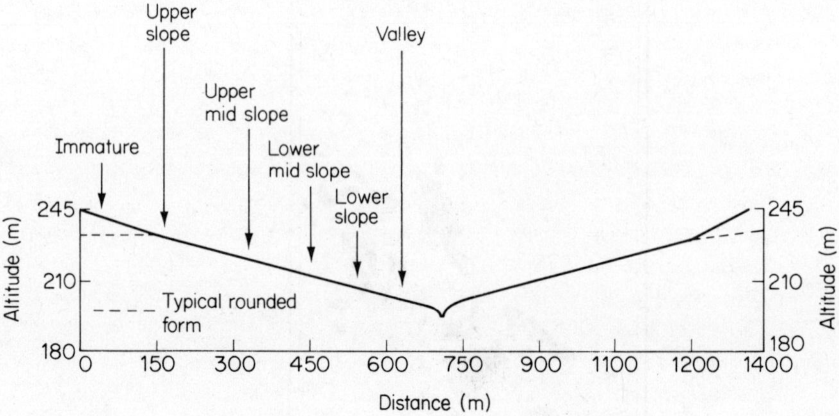

a

Fig. 2 a : Catenary sequence of soils

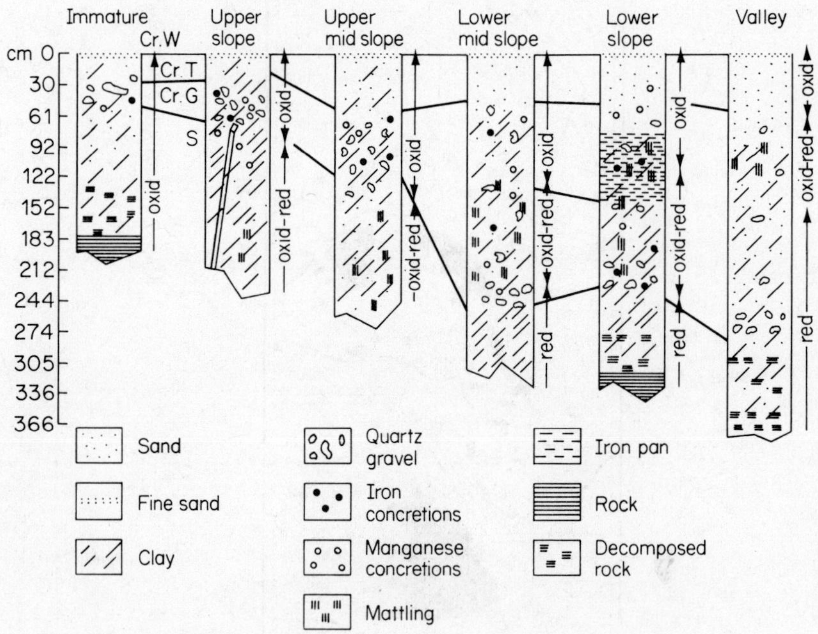

Cr.W, Worm-formed horizon

Cr.T, Termice-formed horizon

Cr.G, Horizon of gravel accumulation

S, Sedentary horizon

b

Fig. 2 b : Soil profile features

Fig. 2 : Genesis of a soil catena in the west African forest (after Nye, 1955a)

Sub-order : *Dark-coloured soils of the semi-arid, sub-humid and humid grasslands* :
Reddish Chestnut soils
Reddish Prairie soils

In addition, it was assumed that azonal and intrazonal tropical soils might be included with several of the great soil groups recognised in temperate regions. These included the azonal groups, Lithosol, Alluvial soils and Dry Sands, and the intrazonal groups, Solonchak, Solonetz, Bog, Half-Bog, Wiesenboden, Planosol, Rendzina, Brown Forest and Ground-water Podsol. Ground-water Laterite was thought to be peculiar to the tropics. Kellogg (1941) and Thorp and Baldwin (1938, 1940) expanded these concepts and definitions.

Kellogg (1949, 1950) proposed that the word 'laterite' should be confined to materials in certain tropical soils which harden on exposure and to the fossil relicts of such materials. Laterite should refer principally to (1) soft mottled clays that change irreversibly to hardpans or crusts when exposed, (2) cellular and mottled hardpans and crusts, (3) concretions, and (4) consolidated concretions.

Instead of the term laterite, 'latosol' was proposed for all zonal soils in tropical regions at the categorical level of sub-order which have the following characteristics :

(1) low silica-sesquioxide ratios in the clay fraction ;
(2) medium to low cation exchange capacity of the mineral fraction ;
(3) low activities of the clay, e.g. swelling, exchange capacity ;
(4) low content of primary minerals, except highly resistant types ;
(5) low content of soluble materials ;
(6) high degree of aggregate stability ;
(7) red colour or reddish shades of other colours.

At the invitation of the National Institute for the Study of the Agronomy of the Belgian Congo (INEAC), Kellogg (1949) made a comprehensive survey of the soils of the Belgian Congo and defined a number of Great Soil Groups. But he emphasized that not all the Great Soil Groups had been discovered and that provision will have to be made for others as they were recognised. The warning against attempts to fit every soil found in the field into one or the other of the existing Great Soil Groups was repeated by Pendleton (1949) and others. Pendleton stressed the need to map observable soil differences on available base maps and/or aerial photographs. The mapping units had to be described carefully and similarities and differences between them noted. It was necessary, moreover, to correlate the soil types of different countries, arranging these types into broader categories and these into the ultimate Great Soil Groups. Bramao (1963) indicated how soil correlation, at the Great Soil Group level, was carried out in Latin America, in Asia and the Far East, and in Africa.

Later systems of classifying soils were based on the morphology of the soil profile and on genetical factors or processes of soil formation. In classifications of temperate zone soils, climate has been shown to be a major factor which indirectly influences the nature of the vegetation and, together with the biosphere, is jointly responsible for the formation of great soil groups like podsols and chernozems.

In humid tropical and sub-tropical regions on the other hand, high temperature and heavy precipitation lead to deeply weathered, well-leached soils which differ from one another particularly because of differences in parent material and topography. The importance of parent material and topography in categorising tropical soils has been stressed by several workers (see p. 22).

At the Fourth International Congress of Soil Science held in Amsterdam in 1950, H.A. Middleburg (1950) prepared a tentative scheme for the classification of tropical and sub-tropical soils (Table 8). Middleburg listed eigh-

TABLE 8: CLASSIFICATION OF TROPICAL AND SUB-TROPICAL SOILS BY MIDDLEBURG (1950)

Order	*Sub-order*	*Great Soil Group*	
Zonal	humid regions	1(a) podsolic yellow soils with black humic topsoil	
		(b) red-yellow podsolic soils	
		2. non-laterized red soils, $SiO_2/R_2O_3 > 2$	
		3. laterized red soils, $SiO_2/R_2O_3 < 2$	
		4. degraded grey and black clay soils	catenary association
		5. illuvial grey and black clay soils	
	arid regions	6. red desert soils	catenary association
		7. dark desert soils	
Intrazonal	calcimorphic	8. red (calcareous) soils over limestone	catenary association
		9. black (calcareous) soils over limestone	
		10. red (calcareous) soils over marls	catenary association
		11. black (calcareous) soils over marls	
	halomorphic	12. saline soils	
	hydromorphic	13. groundwater lateritic soils	
		14. planosols	
		15. organic soils	
Azonal		16. lithosols	
		17. regosols (ash, loess, dry sands)	
		18. alluvial soils	

teen different Great Soil Groups and indicated the different names under which soils belonging to each of these Great Soil Groups have been described. Bramao and Dudal (1958) summarized the information on the main soil groups found in the humid tropics. They did not list tropical soils according to a definite classification scheme but merely pointed out the identifying characteristics, general distribution and use potentialities. In the stage of knowledge of tropical soils at that time this was most necessary and probably was the best that could have been done.

The classification of soils of the United States proposed by Thorp and Smith (1949) has been the basis of soil classification in a number of tropical countries (Table 9). Thorp and Smith excluded the terms pedalfer and pedocal and added a few new or modified Great Soil Groups, such as Red-Yellow Podsolic soils and Low-Humic Gley soils, to those referred to earlier by Marbut (1927). The classification system of Thorp and Smith is both genetic, in the recognition of orders, and morphological, in the criteria used for the recognition of the different Great Soil Groups.

The azonal order was defined in terms of soil characteristics. The zonal and intrazonal orders were defined primarily in genetic terms. In placing new Great Soil Groups in the system, the proper order had to be selected genetically. In the case of ando soils, if the dark surface was due to the fibrous bamboo roots, then these soils can be considered zonal. If, however, the dark colours were due to the volcanic ash parent material, ando soils are intrazonal soils. Since the genetic cause of the dark colour was not known at that time (1949), the ando soils were excluded from the classification.

Stephens (1950) published a classification of soils which followed the main outline of the American system but indicated more specifically the basis on which Orders, Sub-orders and Great Soil Groups were to be sub-divided. Differences in Orders are associated with the presence or absence of lime and/or gypsum in the A and B horizons. Sub-orders were differentiated according to the position of horizons containing organic matter, clay, sesquioxides, lime and gypsum. The Great Soil Groups were determined by the differences in profile colour and texture and the presence or absence of halomorphic, calcimorphic or hydromorphic features in the profile.

Soils of Ghana were classified by Brammer (1962) according to a scheme first worked out by C. F. Charter. In this, the division into Orders, Suborders, Great Soil Groups and Great Soil Subgroups is retained in a form modified to suit Ghanaian conditions.

4.2 The Seventh Approximation

Up to 1950, soil classifications gave more emphasis to virgin soils. Yet the greatest need is for a classification of arable soils of which at least the upper horizons have been mixed. There are other soils which do not have any clear-cut horizons, e.g. alluvial soils or eroded soils or soils formed on transported materials.

It thus became necessary to classify soils not merely on genesis but also on behaviour, and on the known existing or demonstrable properties of the soils themselves. A new nomenclature has been developed based primarily on

TABLE 9: SOIL CLASSIFICATION BY THORP AND SMITH (1949)

Order	*Sub-order*	*Great Soil Group*
Zonal soils	(1) Soils of the cold zone	Tundra soils
	(2) Light-coloured soils of arid regions	Desert soils Red Desert soils Sierozem Brown soils Reddish Brown soils
	(3) Dark-coloured soils of semi-arid, subhumid & humid grassland	Chestnut soils Reddish Chestnut soils Chernozem soils Prairie soils Reddish Prairie soils
	(4) Soils of the forest-grassland transition	Degraded chernozem Noncalcic Brown or Shantung Brown soils
	(5) Light-coloured podsolized soils of the timbered regions	Podsol soils Gray Wooded, or Gray Podsolic soils Brown Podsolic soils Gray-Brown Podsolic soils Red-Yellow Podsolic soils
	(6) Lateritic soils of forested warm-temperate and tropical regions	Reddish-Brown Lateritic soils Yellowish-Brown Lateritic soils Laterite soils
Intrazonal soils	(1) Halomorphic (saline & alkali) soils of imperfectly drained arid regions and littoral deposits	Solonchak, or Saline soils Solonetz soils Soloth soils
	(2) Hydromorphic soils of marshes, swamps, seep areas and flats	Humic Gley soils (includes Wiesenboden) Alpine Meadow soils Bog soils Half-Bog soils Low-Humic Gley soils Planosols Ground water Podsol soils Ground water Laterite soils
	(3) Calcimorphic soils	Brown Forest soils (Braunerde) Rendzina soils
Azonal soils		Lithosols Regosols (includes Dry Sands) Alluvial soils

Greek and Latin roots to connote the most significant soil characteristics.

The new system of soil classification was called the 'Seventh Approximation', or the seventh draft of a system produced by workers in the U.S. Soil Survey. It was presented to the Seventh International Soil Science Congress held at the University of Wisconsin, Madison, USA, in August 1960. Later (1970) it has been published as a new system of soil classification. Some of the fundamental objections to the Seventh Approximation were discussed by Webster (1968).

In the Seventh Approximation six categories of classification are defined. Starting with the broadest group, these are I. Order ; II. Sub-order ; III. Great Soil Group ; IV. Subgroup ; V. Family ; and VI. Series.

Ten Orders are recognised. These are broader but more specific than the old Great Soil Groups (Table 10). The Orders are : (1) Entisols ; (2) Vertisols ; (3) Inceptisols ; (4) Aridisols ; (5) Mollisols ; (6) Spodosols ; (7) Alfisols ; (8) Ultisols ; (9) Oxisols ; and (10) Histosols.

TABLE 10 : THE SEVENTH APPROXIMATION AND GREAT SOIL GROUPS

Order	*Approximate equivalents in classification of Thorp and Smith* (1949)
Entisols	Azonal soils, and some Low Humic Gley soils
Vertisols	Grumusols
Inceptisols	Ando, Sol Brun Acide, some Brown Forest, Low-Humic Gley and Humic Gley soils
Aridisols	Desert, Reddish Desert, Sierozem, Solonchak, some Brown and Reddish Brown soils, and associated Solonetz
Mollisols	Chestnut, Chernozem, Brunizem (Prairie), Rendzinas, some Brown, Brown Forest, and associated Solonetz and Humic Gley soils
Spodosols	Podsols, Brown Podsolic soils, and Ground water Podsols
Alfisols	Gray-Brown Podsolic, Gray Wooded soils, Noncalcic Brown soils, Degraded Chernozem, and associated Planosols and some Half-Bog soils
Ultisols	Red-Yellow Podsolic soils, Reddish-Brown Lateritic soils of the U.S., and associated Planosols and Half-Bog soils
Oxisols	Lateritic soils, Latosols
Histosols	Bog soils

1. *Entisols* : Soils with little or no profile development in deep regolith, except a ploughed surface or other evidence of man's activities. Soils on recent alluvium are commonly Entisols.

2. *Vertisols*: Heavy clay soils, with swelling clays of the montmorillonite group. Peds show slickensides, gilgais are common, and the soils crack widely when dry. These include the Grumusols and Margalitic soils.

3. *Inceptisols*: Young soils with limited profile development, the horizons present being those most readily formed, such as the A_1. Structure has developed and iron oxide may form a coloured horizon, but there has been little clay illuviation to form a textural B horizon.

4. *Aridisols*: Soils of arid areas, usually dry, but may be moist because of seepage, when they are also saline.

5. *Mollisols*: Certain of those soils with a surface horizon which is dark in colour, has a low carbon-nitrogen ratio (less than 14.5) and base saturation above 50 per cent with Ca dominant. These are the famous Prairie and Chernozem soils of the American and Russian grasslands.

6. *Spodosols*: Soils with an illuvial accumulation of sesquioxides with organic matter, or organic matter, or iron oxide. These soils are usually formed on sandy material under humid conditions and coniferous forest in the northern hemisphere.

7. *Alfi sols*: Soils with a clay B horizon derived at least in part from illuvial clay and saturation with metallic cations exceeding 35 per cent.

8. *Ultisols*: Soils with a clay B horizon derived at least in part from illuvial clay and saturation with metallic cations generally less than 35 per cent. They are acidic, leached soils of humid areas, usually on ancient land surfaces.

9. *Oxisols*: Soils with sesquioxides of clay-size dimensions, mixed with silicate clays usually of 1:1 lattice structure. Free sesquioxides should exceed 12 per cent of the total clay fraction.

10. *Histosols*: Soils rich in organic matter such as Peats.

4.3 FAO/UNESCO Soil Map of the World Project

The classification of tropical soils has been the subject of numerous conferences, organised by FAO/UNESCO in connection with the Soil Map of the World, the first sheets and explanatory text of which, for South America, have already been published.

The key to soil units for the Soil Map of the World (Dudal, 1970) lists 103 soil units, classified in 26 larger groupings. The soil mapping units correspond most closely to the Great Soil Group level, while the larger main groupings (Fluvisols, Regosols etc.) may be considered similar to soil sub-orders. The names from a number of earlier classifications have been used derived from languages such as Russian, Latin, Greek, Austrian and Japanese. The FAO/UNESCO system has been devised more as a tool for the preparation of small scale soil maps than as a comprehensive classification system. Table (11) gives the correlation between the FAO/UNESCO and the U.S. Seventh Approximation Soil Classification Systems.

TABLE 11: APPROXIMATE CORRELATION OF THE FAO/UNESCO AND THE U.S. SEVENTH APPROXIMATION SOIL CLASSIFICATION SYSTEMS
(Adapted from Drosdoff *et al.*, 1972)

FAO/UNESCO	*Seventh Approximation*
FLUVISOLS	Fluvents
REGOSOLS	Psamments Orthents
ARENOSOLS	
Ferralic A.	Oxic Quartzipsamments
GLEYSOLS	
Eutric G. Dystric G.	Tropaquepts
Humic G.	Humaquepts
Plinthic G.	Plinthaquepts
ANDOSOLS	Andepts
PLANOSOLS	
Eutric P.	Paleudalfs
Dystric P.	Paleustalfs
CAMBISOLS	
Dystric C.	Dystropepts
Eutric C.	Eutropepts
Humic C.	Humitropepts
LUVISOLS	Tropudalfs Paleudalfs Paleustalfs
ACRISOLS	
Rhodic A.	Rhodudults Rhodustults
FERRALSOLS	Oxisols
LITHOSOLS	Lithic subgroups

TABLE 12 : COMPARISON OF CLASSIFICATION SYSTEMS OF TROPICAL SOILS

System	*USSR*	*USDA*	*ORSTOM*	*INEAC*	
Level 1	*Belt/Formation*	*Order*	*Classe*	*Order*	
			Sols minéraux bruts	Sols minéraux bruts	
				Matériaux récents	Matériaux kaolinitiques
	Polar	Entisols	Sols peu évolués	Sols tropiceaux récents	
	Boreal	Vertisols	Vertisols	Terres noires tropicales	
	Sub-boreal	Inceptisols		Sols brun tropicaux	
	Tropical	Aridisols			
	Equatorial	Mollisols	Sols isohumiques		
		Spodosols	Sols à humus évolué	Sols Podzoliques	Sols Podzoliques
		Alfisols	Sols à humus brut	Sols récent lessivés	
		Ultisols			Kaolisols lessivés
		Oxisols	Sols à sesquioxydes		Kaolisols
		Histosols		Sols organique	
			Sols calco-magnesimorphes		
			Sols halomorphes		
			Sols hydromorphes		
Level 2	*Facies/Class*	*Sub-order*	*Sous-classe*	*Sous-ordre*	
	for intertropical regions		*Sous-classe* of Sols à sesquioxydes		
	Continental monsoon facies	Aquerts	Sols rouges méditerranéens		Hygro-Kaolisols
		Aquepts	Sols ferrugineux tropicaux		Humic-Hygro-Kaolisols

	Oceanic monsoon facies				
	Tropical facies with prolonged dry seasons	Tropepts	Sols á humus grossier		Hygro-Xero-Kaolisols
	Equatorial facies with medium dry season	Ustalfs			Humic-Hydro-Xero-Kaolisols
		Ustults			Xero-Kaolisols
	Equatorial facies without dry season				Hydro-Kaolisols
Level 2′	*Sub-class*	—	—	—	
	Biogenic				
	Bio-lithogenic				
	Bio-hydrogenic				
Level 3	*Type*	*Great Soil Group*	*Groupe*	*Grand Groupe*	
Level 4	*Subtype*	*Subgroup*	*Sous-groupe*	*Petit groupe*	
Level 5	*Genus*	*Family*	*Famille*	*Famille*	
Level 6	*Species*	*Series*	*Série*	*Série*	
Level 7	*Variety*	—	*Type, phase*	*Type, phase*	

4.4 Comparison of USSR, USDA, ORSTOM and INEAC Systems of Soil Classification as Applied to Tropical Soils

Table 12 summarizes the main features in four systems which have been used to classify tropical soils (d'Hoore, 1968) : the USSR system elaborated by the Dokuchayev Soil Institute, USSR Academy of Sciences (Ivanova and Rozov, 1960), the USDA system, prepared by the Soil Conservation Service (USDA, 1960, 1964; Smith, 1965), the ORSTOM system of the French Office de la Recherche Scientifique et Technique Outremer (Aubert, 1964), and the INEAC system of the former Belgian Institut National L'Etude Agronomique au Congo (Sys, 1960). Duchaufour (1963) discussed the USDA and the French systems, Sys (1967) compared the more important classifications used in Africa. Sys (1967) also attempted a tentative correlation of tropical soils in the American Seventh Approximation.

At Level 1, the USSR system classifies soils according to climatic belts, each climatic belt corresponding to a characteristic *Soil Formation*. Soils of tropical and sub-tropical regions are all classified only in the last two formations, the Tropical and Equatorial.

In the USDA system, ten Orders are distinguished according to the kinds and relative strengths of the processes tending to develop horizons. These have been described earlier (p. 47). Soils representative of each Order are found in the tropical area ; the Oxisol Order is almost exclusively tropical.

Separation into Classes in the ORSTOM system is made according to mode and intensity of evolution of the soils, as reflected by the degree of mineral weathering, profile development, the presence of carbonates, salts or hydromorphism (Duchaufour, 1963). The *Sols minéraux bruts* are raw mineral soils with traces of organic material only in the surface horizon. *Sols calcomagnesimorphes, Sols halomorphes* and *Sols hydromorphes* are separated at the highest level and are soils dominated by special parent material or drainage conditions. *Sols peu évolués* correspond to Entisols and are slightly developed soils with an A-C horizon sequence. The term Vertisol is similar to that used in the USDA system. *Sols isohumiques* (Bocquier and Maignien, 1963), *Sols à humus évolué* and *Sols à humus brut* are distinguished on the basis of the quality and distribution of organic matter. *Sols isohumique* are comparable to the Mollisols while *Sols à humus évolué* and *Sols a humus brut* approximate to the Spodosols. *Sols à sesquioxydes* correspond to the Oxisols but also include the Red Mediterranean Soils.

In the INEAC system, classification at the highest level is based on profile development and horizon sequence but the stage of weathering of soil parent material is also emphasized. The category of *Sols minéraux bruts* is retained for 'non-soil' while two kinds of parent materials are distinguished : the more recent *matériaux récents* and the older deeply weathered loose materials, the *matériaux kaolinitiques*. Soils formed on each of these parent materials are separated (Table 12). The *Kaolisols* and *Kaolisols lessivés*, corresponding to the Oxisols and Ultisols respectively, are found only on the *matériaux kaolinitiques* parent materials. The *kaolisols* include both the ferrallitic and fersiallitic tropical soils (Sys, 1967). *Sols organique* are organic soils corresponding to the Histosols.

At Level 2, the USSR system subdivides climatic Belts into Facies defined by summation values of day temperatures exceeding $+10°$ C, and rainfall-over-evaporation ratios. Soils of the same Facies belong to one soil Class. Here too, the criteria adopted are climatic rather than those pertaining to the soils. Five Facies are provided for tropical regions (Table 12).

In the USDA system, soils are separated into Sub-orders based either on the water regime or on genetic differences caused by climate and vegetation. The *Aquerts* and *Aquepts* are hydromorphic sub-orders while the *Tropepts, Ustalfs* and *Ustults* are tropical soils which are not Oxisols.

Within each Class of the ORSTOM system, *Sous-classe* are defined mainly according to pedoclimatic conditions such as soil temperature and soil humidity. The *Sols á sesquioxydes* Classe, for example, is subdivided into *Sols rouges méditerranéens* in which the iron is present in the form of siliceous complexes and *Sols ferrugineux tropicaux* which contain neither siliceous iron complexes nor free alumina but free iron oxides.

In the INEAC system, criteria used to separate the *Sous-ordres* are pedoclimatic such as soil temperature and, indirectly, the organic matter content. For each Ordre developed on *matériaux récents* a *hydromorphique* and *non-hydromorphique* Sous-ordre is separated while six Sous-ordres are recognised for Ordres developed on *matériaux kaolintiques*.

Classification at *Level* 2′, is appropriate only to the USSR system and marks the transition from purely geographic and climatic criteria to soil properties. In each Class, three Subclasses are recognised : the *biogenic subclass* containing zonal soils, the *bio-lithogenic subclass* in which parent material dominates profile development, and the *bio-hydrogenic subclass*, comprising hydromorphic soils.

Below *Level 3*, the classification systems become increasingly similar. In neither the USSR nor in the USDA systems are tropical soils adequately categorized at level 3, the Russian literature being mainly bibliographical and the USDA system until recently tentative (d'Hoore, 1968). Some correlation has been established between INEAC and ORSTOM soil groups in the legend of the Soil Map of Africa by d'Hoore (1964).

4.5 A Classification of Tropical Soils

A relatively simple classification of tropical soils at the higher levels of categorization is given in Table 13. At the first level, classification into Zonal, Intrazonal and Azonal Orders is adopted as in the scheme proposed by Thorp and Smiths (1949). Separation of Zonal soils into Sub-orders is according to Marbut'd proposal in 1928; pedalfers are those soils which show a differentiation in the clay complex resulting in the accumulation of the sesquioxides of iron and

TABLE 13: A SUGGESTED CLASSIFICATION OF TROPICAL SOILS (after Vine, in Webster and Wilson, 1966)

Order	*Sub-order*	*Great Soil Group*
Zonal		
		Latosols, Reddish-Brown Latosolic, Yellowish - Brown Latosolic, Red - Yellow Latosolic
	Pedalfers	Podsols Fersiallitic soils: Red-Yellow Podsolic, Reddish-Brown Earths, Noncalcic Brown soils
	Pedocals	Reddish-Brown and Brown soils, Plain soils, Sierozems, Desert soils
Intrazonal		
	Calcimorphic	Rendzina
	Halomorphic	Solonchak, Solonetz, Solod
	Hydromorphic	Humic Gley, Low-Humic Gley, Groundwater Podsols, Groundwater Laterites, Peat/Muck, Acid sulphate soils
	Vertisols	Grumusols Margalite
	Allophane soils	Andosols
Azonal		
	Lithosols	
	Regosols	
	Alluvial soils	

aluminium while pedocals are characterized by a zone of calcium carbonate accumulation. It is unlikely that some of the sub-orders such as 'dark-coloured soils of semi-arid, subhumid and humid grasslands' are found in the tropics (Vine in Webster and Wilson, 1966); however Noncalcic Brown soils of 'the forest-grassland transition' in Thorp and Smith's classification have been reported from tropical countries (e.g. Moorman and Panabokke, 1961). The Intrazonal soils include not only the calcimorphic, halomorphic and hydromorphic soils of Thorp and Smith's classification of 1949 but also the vertisols and allophane soils recognised subsequently. Lithosols, Regosols and Alluvial soils show little profile development. The use of these terms is in accordance with generally accepted defiinitions (See Glossary).

The Great Soil Groups in Table 13 are the main types found in tropical regions. Latosols (or ferallitic soils) are of four distinct types (Bennema, 1963). Podsols and Fersiallitic soils are also pedalfers. Soils referred to as 'red-yellow podsolic' in the tropics do not have the pale-coloured ashy grey surface layers characteristic of podsols, but they have heavy-textured B horizons. They are non-latosolic, have less aggregate stability than latosols and a higher cation exchange capacity. These 'red-yellow podsolic soils' as well as the Reddish-Brown Earths and Noncalcic Brown soils are best grouped together as fersiallitic soils, with SiO_2/Al_2O_3 ratio greater than 2.0 in the clay fraction. All pedal fers which are neither podsols nor latosols can be termed 'fersiallitic soils' (Vine in Webster and Wilson, 1966). They approximate to the *sols ferrugineux tropicaux* of the ORSTOM classification. The Reddish-Brown Earths and the Noncalcic Brown soils of Ceylon (Moorman and Panabokke, 1961) and the *terra roxa estruturada* soils of Brazil can also be considered to be 'fersiallitic' rather than 'latosolic' but they have a higher base saturation in the subsoils than Red-Yellow Podsolic soils and are found in regions of lower rainfall. Red-Yellow Podsolic soils are found in the wet zone (greater than 1,900 mm rainfall) while Reddish Brown Earth and Noncalcic Brown soils occur in the dry zone (1,260-1,780 mm rainfall) of Sri Lanka.

Groundwater Podsols and Groundwater Laterites have profile characteristics resembling podsols and latosols respectively but the conditions of their formation under conditions of impeded drainage suggest that they are more appropriately classified under the hydromorphic subgroup of Intrazonal soils. Grumusols and Margalite soils are both Vertisols, the former being self-mulching and the latter containing crusty low self-mulching surface horizons.

After soils have been sorted out at the Great Soil Group level further classification into the lower categories can be carried out according to generally accepted criteria (USDA, 1951).

5 Profile Features and Fertility Characteristics of the Great Soil Groups

THE GREAT SOIL GROUPS of the tropics and sub-tropics vary widely in their fertility characteristics and in their agricultural potential. Knowledge of these soils is increasing but there is much more to be learnt for the effective development of agriculture on them.

Each of the Great Soil Groups contains many thousands of local soils having important differences in the way they respond to fertilizer application and management practices. But a broad view of the profile features and the fertility status of modal soils in these Great Soil Groups would be useful in understanding soil fertility problems in tropical regions. The classification given on p. 54 is used here. Nutrient status is discussed in chapters 7 and 8.

5.1 Zonal Soils

5.1.1 Latosols

Kellogg and Orvedal (1969) estimated that Latosols and Red-Yellow Pod solic soils together with associated hydromorphic soils, Lithosols and Regosols cover nearly 3,214 million hectares, or between 24 and 25 per cent of the land area of the world.

Latosols have good physical properties. They are friable, have good aggregate stability, and a stabilized micro-structure which makes for good aeration and free drainage. The subsoil is deep, porous, friable, and favourable for root development. The latosols of SriLanka, however, are generally structureless and may differ considerably from latosols that have been described in other Asian countries (Moorman and Panabokke, 1961). Latosols are less susceptible to erosion than many other soils with the same slope. The shallow latosols and the more sandy latosols are exceptions to this rule. Owing to their friable condition, latosols are easy to work with. Their moisture equivalent is generally high to medium if the texture is not too sandy. But they dry out easily if located in a climate with a dry period.

Their chemical fertility is normally medium to very low. They have only small amounts of primary minerals (except quartz) and plant nutrients (Table 14). They fix phosphate ions strongly and are acid to very acid. Correction of acidity may be necessary to grow crops, but liming is difficult because they have a low effective base exchange capacity resulting in very limited residual values from added liming materials. Cultivated latosols have a low organic matter content but this may be high at higher altitudes. Latosols developed

TABLE 14: ANALYTICAL DATA FOR A RED-YELLOW LATOSOL (Moorman and Panabokke, 1901)

LOCATION : Mannar district, Sri Lanka. CLIMATE: rainfall 1300 mm, temperature 28° C. VEGETATION: low tropical evergeen forest and shrubs. TOPOGRAPHY: flat to slightly undulating terrace or old coastal shelf; elevation 30 m.

Depth cm	*Horizon*	*Stone & gravel* %	*Sand* $>50\mu$	*Silt* $50\text{-}2\mu$	*Fine silt* $20\text{-}2\mu$	*Clay* $<2\mu$	pH H_2O 1:1	pH 1N KCl 1:1	*Cond.* m mhos/cm	C %	C/N	P %	CEC	Ca	Mg	K	Na	BS %
													me/100g			soil		
0-38	A_1	Nil	65.1	21.3	1.2	14.1	6.8	5.7	0.04	0.4	7	—	4.3	0.8	0.4	0.19	tr	35
38-81	B	,,	46.2	24.8	0.8	29.9	5.5	5.0	0.03	0.1	2	—	3.5	0.6	0.3	0.26	tr	30
81-183	B	,,	49.4	19.1	1.6	31.0	5.6	5.2	0.03	—	—	—	7.6	0.6	0.8	0.31	tr	20
183-242	B	,,	45.4	25.3	1.0	29.0	5.7	5.1	0.03	—	—	—	4.7	0.6	0.8	0.32	tr	36
242-355	B	,,	46.2	24.6	2.8	28.3	5.9	5.4	0.03	—	—	—	4.4	1.1	0.6	0.27	0.01	39
355-381	B	,,	44.6	21.7	1.8	33.3	6.0	5.4	0.03	—	—	—	4.3	1.3	0.6	0.24	0.01	44

on basic crystalline rock materials in a semi-arid to sub-humid climate with a pronounced dry period have small amounts of organic matter (Sherman and Alexander, 1959).

Bennema (1963) classified latosols into five distinct groups four of which are of limited occurrence while the fifth is a broader group. Among these is the *terra roxa legitima* of Brazil and the Red-Yellow latosols.

5.1.1.1 TERRA ROXA : There are two main Terra Roxa soils of Brazil; the Terra Roxa of Ribeirao Preto (*terra roxa legitima*) and the Terra Roxa of Parana (*terra roxa estruturada*). These soils are developed from basalt, diabase or similar rocks. Both have a peculiar surface colour of reddish-purple. The structure of the top horizon is characterized by granular aggregates of extra coarse size. The Terra Roxa soils of Ribeirao Preto appear to be an end product of laterisation and contain little or no primary or secondary minerals except oxides. The clay fraction may comprise 10-15 per cent of kandite minerals. Under natural luxuriant forest vegetation, the A horizon may contain as much as 5-6 per cent of organic matter.

The deep soil profile is a uniform dusky reddish purple regardless of the high organic matter content of the surface horizon. While the surface horizon has a coarse granular structure, the soil mass below is fluffy and has a fine granular structure. The *terra roxa legitima* has a high iron oxide content and a relatively high content of titanium and manganese oxides.

The Terra Roxa of Parana differs from the Terra Roxa of Ribeirao Preto in having a well developed coarse prismatic structure instead of a fine granular structure in the B horizon. The soil mass of the Terra Roxa of Parana is friable and not fluffy. Clay in the B horizon of the Terra Roxa of Parana accumulates with clay films and subangular blocky structures which are absent in the Ribeirao Preto soils. The Terra Roxa of Parana are not considered as latosols (Bennema, 1963), but are Red-Yellow Podsolic soils (Bennema in Turk and Crowder, 1967).

The *terra roxa legitima* rank among the best latosols for agricultural production and are often intensively used for coffee and sugarcane plantations. However, application of fertilizers presents many problems because of their high iron content, relatively high manganese content, and low effective base exchange capacity.

5.1.1.2 RED-YELLOW LATOSOLS : The most widespread of latosols are the dark-*red* and *red-yellow latosols* which are coloured red to yellow and are found on acid parent materials in large areas throughout Central and South America, Africa and south-east Asia. The dark red latosols are found on gneisses with more ferromagnesian minerals, and also on some sedimentary rocks. These soils occur under a wide range of climatic conditions and under a great variety of natural vegetation. Organic matter content varies with rainfall and soil texture.

These are very deep, extremely well-drained soils uniform throughout their depth. They are moderately coarse-textured, friable when dry or moist, non-sticky when wet.

The B horizon is frequently 9-12 metres in depth and is very uniform. Texture is almost the same as in the A horizon. Structure is weak subangular blocky to structureless. Consistence is very friable when dry or moist and non-sticky when wet.

In these latosols the clay fraction consists mainly of a mineral of the kaolinitic type with a medium amount of iron oxide (10-15 per cent). There may be little or no free aluminium oxide or only moderate amounts. Table 14 gives the analytical data for a Red-Yellow Latosol profile.

Although in many parts of the tropics they are widely used for intensive agriculture, their fertility is low and the use of fertilizers would increase yields considerably. Not only the primary nutrients (N, P, K) but also secondary nutrients (Ca, Mg and S) as well as some micro-nutrients (Zn, Mo, B) can be deficient in these latosols (Bennema in Turk and Crowder, 1967).

5.1.1.3 REDDISH BROWN LATOSOLIC SOILS : Reddish Brown Latosolic soils are deep to very deep, well drained soils which occasionally have a stone line at some depth. They are latosolic soils with a reddish brown to dark reddish brown granular A horizon, 30 cm or more in thickness. There is red friable clay in the B horizon which may extend to more than three metres in uneroded profiles. Structure is strong to moderate subangular blocky. Consistence is friable to firm when dry and slightly sticky when wet.

The C horizon consists of somewhat weathered material and decomposing rock with a high content of unweathered minerals. The reddish brown latosolic soils of Sri Lanka are formed from some of the less siliceous rocks in the wet zone on parent material consisting mainly of old slope colluvium in rolling to hilly terrain. In cultivated soils, especially on slopes, there is much erosion. They occur on hilly and sharply rolling terrain. In the intervening valleys they are replaced by the wet gley soils. They have good drainage, a high degree of aggregate stability and no mottled layer.

5.1.2 FERSIALLITIC SOILS

5.1.2.1 REDDISH BROWN EARTHS : Reddish Brown Earth soils occupy the largest extent of the land surface of Sri Lanka, being confined mainly to the dry zone. They occur on rolling landscape being confined to the well-drained higher topographical aspects with yellowish brown earths in the mid aspects and Low Humic Gley soils in the valleys.

The A horizon, usually less than 25 cm in thickness is a dark brown to dark reddish brown sandy clay loam. It is friable when moist and soft to slightly hard when dry. The B horizon is between 60 to 90 cm thick and usually contains a quartz gravel or ironstone gravel layer. It is distinctly redder than the A horizon and its texture is heavier, mostly clay loam or clay. The soil is slightly acid, neutral, or slightly alkaline, ranging between 6 and 7. The dominant clay mineral is kaolinite. Analyses of the clay fraction indicate the 'non-lateritic' character of these soils.

The organic matter and nitrogen status of Reddish Brown Earths are low. Phosphorus content is small while the potassium status varies from medium to high. Their cation exchange capacity is good (50 me per 100 g) and they are well supplied with calcium and magnesium (Table 15).

With efficient weed control and correct tillage practices, Reddish Brown Earths could be productively cropped both under rainfed systems and irrigation. Their agricultural potential is good and a wide range of food crops, pastures and orchard crops could be grown on them.

5.1.2.2 NONCALCIC BROWN SOILS : These soils are found in the comparatively drier tropical regions on parent materials rich in quartz. They are shallow to moderately deep, coarse-textured and are fairly well drained. They occur on the higher aspects of the undulating relief in a complex pattern with Reddish Brown Earths.

The A horizons (15-20 cm) are mostly sandy loam, grey brown to yellowish brown in colour. The B horizon is yellowish brown to brown in colour, has a base saturation over 40 per cent and a higher clay content than the A horizon. The C horizon consists of decomposed acid gneisses with a low content of ferro-magnesian minerals.

The physical fertility of these soils is poor. Their poor structural stability makes them more liable to erosion. Their field capacity is low, they dry out more quickly and they have to be irrigated more frequently than the Reddish Brown Earths.

These soils have a slightly acid reaction (Table 16). They are low in organic matter, nitrogen, phosphorus and potassium but are fairly well supplied with calcium and magnesium and have a moderately good cation exchange capacity.

5.1.2.3. RED YELLOW PODSOLIC SOILS: The teım *podsol* has long been used to describe acid soils with bleached coarse-textured A_1 and dark heavy-textured B_2 horizons, which are found in cool humid temperate climates with a coniferous vegetation. It is now thought that low temperatures are not necessary for the formation of podsols. Groundwater podsols have been found in several parts of the tropics and there may also be some podsols formed in conditions of free drainage (Vine, in Webster and Wilson, 1966). There are extensive areas, however, where the processes which produce podsols are superimposed on processes which operate in the more drastic climatic regimes characteristic of the tropics. As a result, soils are formed with certain morphological features similar to those found in podsols. Red-Yellow Podsolic soils are one such group.

Red-Yellow Podsolic soils are 'a group of well-developed, well-drained acid soils having thin (A_0) and organic mineral (A_1) horizons over a light-coloured bleached (A_2) horizon, over a red, yellowish-red, or yellow and more clayey (B) horizon. Parent materials are all more or less siliceous. Coarse reticulate streaks or mottles of red, yellow, brown, and light gray are characteristic of deep horizons of Red-Yellow Podsolic soils where parent materials are thick'

TABLE 15: Analytical Data for a Reddish Brown Earth (Moorman and Panabokke, 1961)

Location: Timbolketiya, Ceylon. **Climate:** rainfall 1550 mm, temperature 27° C, dry season from June to September. **Vegetation:** tropical mixed evergreen forest and shrubs. **Topography:** 2-3 % slope on the side of a low ridge, undulating; 76-92 m elevation. **Patent Material:** semi-recent colluvium over residuum from mica schist.

Depth cm	*Horizon*	*Stones* & *gravel* %	*Sand* >50μ	*Silt* 50-2μ	*Fine Silt* 20-2μ	*Clay* <2μ	*p*H H_2O 1:1	*p*H IN KCl 1:1	*Cond.* *m mhos*/ cm	C %	C/N	P %	CEC	Ca	Mg	K	Na	BS %
													me/100g soil					
0-10	A_1	11.4	58.0	25.2	7.1	18.0	7.1	6.1	0.06	1.2	9		12.7	6.2	2.3	0.82	0.07	77
10-38	B_1t	8.9	52.6	23.0	6.2	23.1	6.8	5.5	0.04	0.5	5		14.4	4.4	2.3	0.41	0.06	51
38-104	II B_2t	28.8	53.0	12.6	4.7	35.6	6.5	5.2	0.04	—	—		17.4	5.2	3.1	0.35	0.09	52
104-129	II B_3t	8.4	51.1	21.0	7.4	28.2	6.2	5.0	0.03	—	—		18.4	7.7	1.1	0.45	0.13	53
129+	II C	14.5	76.7	18.9	3.3	4.7	6.6	4.9	0.07	—	—		8.1	3.7	1.5	0.22	0.13	74

TABLE 16: ANALYTICAL DATA FOR A NONCALCIC BROWN SOIL (da Costa *et al.*, 1959)

LOCATION: Haila district, southern Angola. CLIMATE: rainfall 900 mm; temperature 25°C. VEGETATION: Low and sparse mesophytic woodland. TOPOGRAPHY: Moderately steep. PARENT MATERIAL: Quartziferous eruptive rock.

Depth cm	*Colour* dry	*Particle size analysis* > 2mm	CS	*FS*	S	Cl	CO_3 *as* $CaCO_3$	*p*H	OM %C	N %	C/N	*Total P* $\%P_2O_5$	CEC	BS %	*Free* Fe%	*Moisture Equivalent* %
0-12	10YR 7/4	7.7	62.1	27.1	8.4	3.1	0.0	6.2	0.46	0.028	16.4	0.07	3.6	67	0.27	7.0
12-36	7YR 6/4	5.1	61.3	24.7	8.4	6.4	0.0	5.5	0.33	0.026	12.7	0.07	3.8	59	0.36	7.9
36-57	5YR 6/6	11.8	58.0	20.8	8.3	12.9	0.0	5.4	0.25	0.024	10.4	0.06	3.4	50	0.48	9.0
60-90	5 YR 5/6	23.7	32.5	25.3	2.4	38.3	0.0	5.5	0.25	0.015	15.6	0.05	3.8	63	0.45	10.0
100-130	—	36.1	53.9	18.4	7.2	20.8	0.0	5.4	0.13	0.012	10.8	0.04	4.2	60	0.57	11.4
140-170	—	62.1	46.4	19.2	8.0	26.4	0.0	5.5	0.10	0.013	7.7	0.04	5.0	62	0.66	13.7

(Thorp and Smith, 1949).

Red-Yellow Podsolic soils (Table 17) are highly leached soils in which clay accumulates in the sub-surface horizons. These clays are more active and have higher base exchange capacities than those of latosols. They have a low permeability and the soil aggregates have less stability than those of latosols. Blocky structures are the most common in the sub-surface horizon. They are more susceptible to erosion than most latosols. Some water stagnation in or on the B horizon may also, in extreme cases, hinder root development. They are acid soils with a low *p*H and are usually formed on acid quartz-rich parent materials. The plant nutrient content is low but they are less weathered than latosols and weatherable minerals may be present. Large applications of N, P and K are commonly required. Boron and magnesium may be necessary as well as moderate amounts of lime.

5.1.3 Desert Soils

These soils are mainly the product of physical weathering as rainfall is very scarce. Water, rather than soil fertility, is the limiting factor for agricultural production. Most desert soils, except the very sandy ones, are well supplied with plant nutrients. They are, however, very low in organic matter and where water is available, it is necessary to build up the level of organic matter. Phosphorus is also likely to be deficient, while potassium is usually abundant. Micronutrient deficiencies like those of iron, molybdenum and zinc are common. Their fertility problems are often further complicated by the presence of excessive soluble salts.

5.2 Intrazonal Soils

5.2.1 Rendzina Soils

Rendzinas occur under a variety of climatic regimes from cool to hot, from humid to semi-arid. They are formed on chalk, gypsum, or soft limestone parent material. The native vegetation is chiefly grass.

Rendzina soils contain from 3 to 12 per cent organic matter and varying amounts of calcium carbonate. The *p*H is neutral to alkaline. The profile generally consists of only A and C horizons. The A horizon is black to dark grey, with a high humus content, usually shallow (15-25 cm), although the depth may extend down to 50 cm. The C horizon is a light-coloured, greyish or yellowish, calcareous material.

5.2.2 Halomorphic Soils (see also pp. 38-39)

5.2.2.1 Solonchak Soils : These are also called *white alkali* soils and contain large amounts of the chlorides, sulphates, nitrates, carbonates and bicarbonates of sodium, potassium, calcium and magnesium. Sodium chloride is most abundant. A thin, grey, salty crust on the surface overlies a fine granular soil, below which is a greyish, friable, salty material. The texture is uniform throughout the profile. The influence of salts on crop production has

TABLE 17: ANALYTICAL DATA FOR A RED-YELLOW PODSOLIC SOIL (Moorman and Panabokke, 1961)

LOCATION: Pelmadulla, Sri Lanka CLIMATE: rainfall 3300 mm, temperature 27°C. VEGETATION: rubber estate, local thorny bush and grass. TOPOGRAPHY: 10% slope; lower slope of a low ridge in a sharply rolling to hilly ridge and valley landscape; 152 m. elevation. PARENT MATERIAL: slope colluvium with lateritic gravels over residuum from garnet-sillimanite schists.

Depth cm	Horizon	*Stone & gravel* %	*Sand* >50μ	*Silt* 50-2μ	*Fine silt* 20-2μ	*Clay* <2μ	*pH* H_2O 1:1	*pH* 1N KCl 1:1	*Cond.* m mhos/cm	C %	C/N	P %	CEC	Ca	Mg	K	Na	BS %
													me/100 g soil					
0-8	Ap	33.2	55.1	10.9	4.4	33.5	5.3	4.0	0.04	1.6	10	—	10.1	0.56	1.2	0.12	0.03	20
8-38	A_2	68.4	47.7	11.6	4.0	40.9	5.5	4.1	0.02	0.8	10	—	9.0	0.39	0.3	0.09	0.04	9
38-58	B_1t	63.2	45.4	12.2	5.9	43.3	5.7	4.1	0.02	0.6	9	—	9.0	0.17	0.4	0.08	0.04	7
58-99	II $B_{21}t$	31.5	37.4	14.4	7.8	47.9	5.7	4.2	0.02	—	—	—	10.2	0.09	0.4	0.07	0.03	6
99-158	II $B_{22}t$	17.0	40.2	15.3	8.2	46.1	6.0	4.3	0.02	—	—	—	9.1	0.17	0.4	0.09	0.02	6
158+	II B_3t	9.3	41.9	16.6	9.4	41.2	6.0	4.4	0.02	—	—	—	8.9	0.09	0.4	0.08	0.04	6

been explained by Buringh (1970) who gives critical values. Saline and alkali soils are poor agricultural soils. Measures adopted to make them non-saline and non-alkali are discussed elsewhere.

5.2.2.2 SOLONETZ SOILS: These contain sodium in the colloidal exchange complex (Tables 18 and 19) and are black in colour. Sodium hydroxide and sodium carbonate present dissolve the organic matter and carry it downwards to a lower horizon which is black in colour. On drying, the humic layer develops a columnar structure and looks like domed columns.

Soils high in exchangeable sodium have extremely poor physical conditions. Application of gypsum together with irrigation and drainage helps to make them more productive.

5.2.2.3 SOLOD SOILS: Solod soils are derived from solonetz soils by hydrogen ions replacing sodium ions on the exchange complex. These soils are therefore very acid and the acid A horizon contains only a little colloidal material while the B horizon is heavy-textured.

5.2.3 HYDROMORPHIC SOILS

5.2.3.1 HUMIC GLEY SOILS: These soils have poor to very poor drainage. They have a dark coloured mineral surface horizon of moderate thickness, underlain by a grey to bluish grey horizon. The surface horizon is slightly acid to neutral in reaction. The organic matter content is higher than in the Low Humic Gley soils.

5.2.3.2 LOW HUMIC GLEY SOILS: These are soils found throughout the tropics and sub-tropics in the lower topographical sites. They are usually the lower members of the drainage catena found in association with better drained soils on the higher aspects.

They are characterized by a pronounced mottling of the B horizon which is heavier in texture than the dark A horizon (Table 20).

The chemical fertility of the Low Humic Gley soils is similar to that of the Great Soil Group with which they are associated. Rice is the main crop grown on these soils. Sugarcane, bananas and, on better drained sites, coffee, cocoa and citrus are also grown.

5.2.3.3 GROUNDWATER LATERITE SOILS: These soils are found in hot humid tropical forest regions where drainage is poor due to a high groundwater table. A yellowish grey leached surface horizon overlies a thick reticulately mottled sub-surface which is soft when formed under some 40-100

TABLE 18: ANALYTICAL DATA FOR A NON-SALINE ALKALI SOIL (Raychaudhuri and Murthy, 1960)

LOCATION: Kanganj district Etah, Uttar Pradesh, India. CLIMATE: rainfall 815 mm, temperature 26°C*.
TOPOGRAPHY AND DRAINAGE: Low-lying, with water standing during the monsoon. Moderate drainage. Ground-water not deep.

Depth cm	*pH*	*Elec. cond.* m mhos/ cm sat. extr.	*Anions* CO_3	HCO_3	Cl	SO_4 me/100 g *soil*	Na	*Cations* K	Ca + Mg	Exch. Na %	*Clay* %
0–8	7.3	0.6	tr	2.0	1.7	1.4	5.0	0.1	tr	3.5	9.3
8–53	7.7	1.8	tr	9.5	5.9	2.7	18.0	0.1	tr	10.4	23.3
53–76	8.3	2.0	tr	10.0	4.0	4.2	19.0	1.0	0.2	69.7	25.1
76–92	9.3	1.0	tr	4.0	4.0	3.4	10.0	0.5	0.9	98.6	29.0
92–130	9.0	1.5	tr	5.0	4.2	4.3	11.5	0.1	1.9	83.4	30.0
130–173	8.8	1.0	tr	6.5	2.8	1.7	10.9	0.1	tr	28.7	30.9
173–183	8.1	1.3	tr	6.0	4.8	2.0	10.9	0.5	1.5	12.3	37.5

*from Kendrew (1961)

TABLE 19: ANALYTICAL DATA FOR A SALINE ALKALI SOIL (Raychaudhuri and Murthy, 1960)

LOCATION: Etawah district, Uttar Pradesh, India. CLIMATE: rainfall 815 mm, temperature 26°C*
TOPOGRAPHY AND DRAINAGE: Low-lying depression with very poor drainage. Groundwater not very deep, much fluctuation between dry and wet periods.

Depth cm	pH	*Elec. cond.* m mhos/cm sat. extr.	*Anions* CO_3	HCO_3	Cl	SO_4 me/100 g soil	*Na*	*Cations* K	Ca+Mg	*Exch* Na %	*Clay* %
0-5	9.2	4.0	6.0	7.0	20.8	5.2	38.3	0.5	0.2	43.3	24.0
5-25	9.4	4.5	5.0	7.2	18.0	6.2	40.0	0.5	0.5	73.0	25.7
25-51	9.8	2.5	6.0	4.0	15.0	2.0	26.1	0.4	0.5	68.1	33.6
51-81	9.9	3.9	10.0	5.0	25.9	2.1	41.9	0.7	0.4	39.3	43.6
81-107	10.1	8.5	39.0	14.0	30.7	2.1	81.1	1.2	3.5	52.4	37.0
107-158	10.1	8.0	10.0	28.0	36.0	4.3	74.2	1.2	2.9	47.8	37.0
158-178	9.3	1.3	tr	5.0	5.5	1.5	6.5	0.4	5.1	14.1	26.8
178-188	9.2	3.0	2.5	11.3	12.0	6.9	30.2	0.4	2.0	3.7	18.6

*from Kendrew (1961)

TABLE 20: ANALYTICAL DATA FOR A LOW HUMIC GLEY SOIL (da Silva Teixeira, 1959)

LOCATION: Cacheu, Portuguese Guinea, Africa. CLIMATE: rainfall 2,000 mm, temperature 27^{0}C. Hot rainy season from May to November; dry season, five to six months. VEGETATION: rice field. TOPOGRAPHY: flat. PARENT MATERIAL: alluvium.

Depth cm	Colour	*Mechanical analysis* CS	FS	S	C	*p*H	OM	N	C/N	P	CEC	BS	*Av. water cap.*
			%				%	%				%	%
0-25	YR 6/3 light brown	24.7	62.4	6.5	6.4	4.2	0.8				3.41	56	8.2
25-60	10YR 7/1 light grey	23.9	61.4	3.1	11.6	6.0	0.2				1.85	73	7.5
60-100	10YR 7/2 light grey	37.9	54.2	2.2	5.7	5.9	0.1				1.77	72	4.0
100-120	10YR 6/2 light brownish grey	29.1	61.1	2.3	7.5	4.0	0.1				2.19	54	4.0
120-165	10YR 5/1 grey	19.8	70.1	4.8	5.3	2.5	0.8				11.63	10	—

TABLE 21: ANALYTICAL DATA FOR A GROUNDWATER LATERITE (Obeng and Quagraine, 1960)

LOCATION: Seilo-Tuni Land Planning Area, Ghana. CLIMATE: Rainfall 1120 mm, temperature 36°c (max.) to 19°C (min). VEGETATION: Tall grassland with fire resistant trees. PARENT ROCK: Terrace colluvium associated with phyllites and schists. TOPOGRAPHY: Gently undulating with broad valleys.

Depth cm	*Colour*	*Mechanical analysis* (mm) >20 *stones*	20-6.25 *coarse gravel*	6.25-2.00 *fine gravel*	<2 *fine earth*	<.002 *clay*	*pH* 1:1 H_2O	OM %C	N %	C/N	P *total ppm*	CEC	me/100g *fine earth* Ca	Mg	K	Na	Mn	BS %
0-10	2.5YR 6/2 light brownish grey	<0.1	<0.1	<0.1	100	4	6.3	0.38	0.031	12.26	45	2.44	1.18	0.57	0.09	0.05	0.04	79
10-20	2.5Y 6/4 light yellowish brown	<0.1	<0.1	<0.1	100	5	5.8	0.23	0.024	9.58	30	1.99	0.61	0.37	0.05	0.04	0.03	55
20-46	2.5Y 6/6 olive yellow stained, yellow-red	<0.1	<0.1	<0.1	100	10	5.8	0.15	0.024	6.27	38	2.30	0.71	0.46	0.07	0.05	0.04	58
46-76	10YR 6/6 brownish yellow mottled, reddish yellow	<0.1	21.8	12.8	65	17	6.1	0.13	0.017	7.65	66	3.50	0.06	0.82	0.10	0.07	0.04	60
76-109	Incipient pan	<0.1	21.9	10.5	68	29	6.0	0.12	0.015	9.25	101	4.69	0.50	1.39	0.14	0.07	0.03	67

TABLE 22: ANALYTICAL DATA FOR A GROUNDWATER PODSOL (Wills 1962)

LOCATION: Atuabo, Ghana. CLIMATE: rainfall 2160 mm, temperature 27°C. VEGETATION: short grass savanna. TOPOGRAPHY: flat bottom, 0.2-3 m elevation.

Depth cm	*p*H H_2O 1:2	C %	N %	C/N	*total* P ppm	CEC	Ca	Mg	K	Na
						me/100g				
0-5	5.2	2.0	0.08	24	52	3.5	0.50	0.31	0.05	0.11
5-20	5.6	0.3	0.01	22	43	0.9	0.18	<0.01	0.05	0.14
20-58	5.8	0.1	0.01	13	38	0.8	0.15	<0.01	0.02	0.04
58-79	4.8	3.1	0.11	27	73	25.6	0.35	0.24	0.04	0.09
79-119	4.8	3.6	0.09	41	84	41.3	0.31	0.16	0.02	0.07
119-155	5.9	0.1	0.01	22	14	1.2	0.15	<0.01	0.02	0.03

TABLE 23: ANALYTICAL DATA FOR AN ANDOSOL (USDA, 1960)

LOCATION: Hawaii Island, Hawaii, altitude 980 m. CLIMATE: rainfall 1040 mm, temperature 18.3°C. VEGETATION: Bermuda grass. TOPOGRAPHY: 12 per cent slope, convex. PARENT MATERIAL: Basic, volcanic ash.

Depth cm	*Horizon*	*p*H 1:1 H_2O	OM %C	N %	C/N	P	CEC	Ca	Mg me/100g	K	BS %
0–5	A_{11}	6.2	13.0		11		80.8	34.4	6.6	5.5	58
5–13	A_{12}	7.5	5.1		8		71.8	32.8	7.2	5.8	64
13–20	B_{21}	7.6	4.8		9		74.1	37.7	5.7	6.3	67
20–58	B_{22}	7.6	4.3		11		80.8	43.4	7.3	6.4	71
58–89	B_{23}	7.4	3.2		12		90.8	32.9	22.6	2.5	65
89–119+	B_{24}	7.4	2.9		11		79.2	33.5	13.2	3.7	65

cm of soil but hardens on exposure or is cemented *in situ*. These are acid soils and contain concretions with iron, aluminium and manganese (Table 21).

Groundwater Laterites over shales and granites in the interior savanna zone of Ghana have been described by Brammer (1962).

5.2.3.4 GROUNDWATER PODSOLS: Groundwater Podsols are found in cool temperate to tropical humid forest regions on sandy parent material in imperfectly drained situations. They have been located in Indonesia and in central Africa. A thin layer of raw, acid organic matter overlies a light grey, strongly leached, sandy horizon 60-90 cm thick. Below this is a dark brown illuvial B horizon of organic matter and/or sesquioxides (Table 22).

5.2.4 VERTISOLS

Vertisols include both the old soil group of the Grumusols and the 'margalitic' soils of Indonesia (see p. 36). Grumusols have a thick, dark grey-brown to black clayey, surface horizon, relatively poor in organic matter (1-3 per cent) with a greyish, yellowish or bluish sub-surface. Some grumusols in the southern prairie region of the USA contain 5 per cent of humus to a depth of 15 cm or more. The clay fractions are montmorillonitic, the clay swells when moist and the soil becomes sticky, plastic and impervious. Wide cracks develop on drying and the surface soil falls into these cracks, thus preventing the formation of separate horizons. The physical features of these soils are not as good as those of latosols and they are more susceptible to sheet and gully erosion as well as to creep. There can be self-mulching of the surface, by its splitting into a loose layer of fragments when moistened after drying. This is true of grumusols like the 'regur' soils of India but not of the 'margalite' soils of Indonesia which have a crusty, low self-mulching surface horizon.

Grumusols vary in depth from about 30 cm to 1 metre or more. They usually have a medium to high content of exchangeable calcium and a fair content of magnesium. Calcium carbonate concretions are often found at a depth of 60-90 cm or more. They are low both in nitrogen and phosphorus. They have a neutral to alkaline reaction.

Grumusols occur under natural vegetation consisting of semi-deciduous forests, shrubs and savanna, under a wide range of climatic conditions and on a variety of parent materials. With proper drainage and careful management, grumusols can be made to give good yields of cotton, grain, sugarcane and other crops.

5.2.5 ALLOPHANE SOILS:

The allophane-containing soils rich in organic matter are called *andosols* Andosols are found in climates with moderate to very high rainfall, are strongly leached and of low base status. The parent material is volcanic ash or easily weatherable volcanic glass. Commonly they support coniferous, broad leaf-evergreen and deciduous forest in which bamboo predominates, although the growth of bamboo is not regarded as an essential factor in their development.

They have a relatively thick, crumbly, black or dark brown, medium to coarse-textured, acid surface layer which has a crumb structure and is porous and very friable. The organic matter content is considerable (8 to 30 per cent to a depth of 30 or 60 cm) as is shown in Table 23.

A striking feature of andosols is their exceptionally large water-holding capacity which may be as much as 200 per cent on dry weight, especially in the B horizon. This may be due to the abundant humus and the very hydrous mineral colloids but mostly to the vesicular structure of the original volcanic fragments. The exchangeable calcium content is small or very small but exchangeable aluminium is considerable. Andosols have a high C/N ratio (sometimes as high as 20-30) and a considerable phosphate fixing capacity. The *p*H varies from 4.5 to 5.5.

The sub-soil is brown to dark yellowish brown. Allophane ($Al_2O_3.SiO_2.5H_2O$) or allophane-like clay minerals are a characteristic feature of the colloidal fraction of these soils. A hardpan about 20 cm thick may be present at 40 to 100 cm depth.

Andosols respond to potassium and often also to nitrogen. Their productivity is generally poor unless they are well fertilized.

5.2.6 Organic Soils

5.2.6.1 Peat and Muck Soils: Coulter (1950) termed these soils as a group 'bog soils' in Malaysia and divided them into peat soils with less than 35 per cent mineral matter, 'muck soils' with 35-65 per cent mineral matter and 'organic clay soils' with 65-80 per cent mineral matter. They are found in low-lying ill-drained areas in the humid tropics and are fairly extensive, for example, in Indonesia, in Malaysia, in Mozambique (Ripado, 1954), and in British Guinea (Paul and Shariff, 1954). They consist of relatively thick layers of peat, of acid reaction, overlying a grey mineral horizon. When drained they can be used to grow rice, rubber, pineapples and vegetables.

The fertility of organic soils varies widely according to the plant residues they contain and the extent to which these have been decomposed; also, the additions of mineral nutrients from surface or ground water (Bailey, 1950). They are generally deficient in potassium and many in manganese or copper.

5.2.6.2 Acid Sulphate Soils (cat-clays): On drainage and aeration, acid sulphate soils show a definite and severe acidification because of the oxidation of sulphides, mainly pyrites, producing sulphuric acid. The term 'cat-clay' has been used for the acid soil material in its oxidized form, showing straw-yellow mottlings and streaks of basic ferric sulphate.

These soils are usually clayey in texture. They contain much organic matter. Colours vary from neutral gray or blue-black to brownish with increasing organic matter content. In the oxidized form the characteristic yellow mottling and streaks of basic ferric sulphate may occur anywhere in the profile.

The pH values vary from 1 or 2 to near neutrality under very poor drainage conditions. Seasonal variation of pH values is very pronounced especially in periodically inundated rice fields.

Factors limiting agricultural production on these soils are their acidity, toxicity of iron, aluminium and, possibly, manganese, poor fertility and physical soil conditions. The C/N ratio may be 30 or more; phosphorus is strongly fixed in unavailable forms and micronutrient deficiencies such as that of copper are common. Soil tilth is poor especially if the organic matter content is small.

Acid sulphate soils can be used for sustained rice cultivation if they are not drained and allowed to oxidize. Pineapple is grown on ridged land in acid sulphate soil areas in Vietnam and in Malaysia. If drained, the soils should be limed liberally. A combination of liming and careful, progressive drainage seems to be the best way of reclaiming these soils. Lime requirements for the very acid soils may be so much as to prove uneconomical (Hart *et al.*, 1965; Bloomfield *et al.*, 1968).

5.3 Azonal Soils

5.3.1 Alluvial Soils

Alluvial soils are found in river flood and coastal plains. They have a thin surface organic layer and not much profile development. Although river formed landscapes are similar all over the world, there are some differences between alluvial soils in the tropics and in temperate regions (Edelman and van der Voorde, 1963). In temperate regions alluvial soils are rich in mineral nutrients; the physically weathered glaciated materials laid down by rivers contain a substantial quantity of powdered fresh minerals. In the tropics, rivers generally originate from deeply weathered areas and river alluvium consists mainly of quartz and other resistant minerals. Therefore, except where rivers originate in recent volcanic areas the natural fertility of alluvial soils in the tropics is often poor.

Tropical alluvial soils are usually acid and their cation exchange capacities are less than those of soils found in temperate regions (Table 24). The organic matter is rapidly oxidized on draining the soils.

Alluvial soils vary widely in chemical fertility and in texture. Clay soils are apt to be difficult to drain and to till. But heavy clay soils dry out intensively in the tropics and cracks develop to greater depths. Permeability is thus improved where alternate wet and dry seasons prevail.

With suitable irrigation, alluvial soils are the most desirable in the tropics. Hitherto they have been used mainly for rice but with proper management a number of other crops can be grown.

5.3.2 Lithosols and Regosols

These are soils which have either little or no profile development. In a regosol (Table 26) the parent material is loose and unconsolidated while lithosols

TABLE 24: ANALYTICAL DATA FOR AN ALLUVIAL SOIL (Anderson, 1961)

LOCATION: Mkula, Tanzania. CLIMATE: Rainfall 888 mm, temperature 27°C. VEGETATION: tall grass with shrub. TOPOGRAPHY: slightly undulating, slope 2%. PARENT MATERIAL: well-drained alluvium.

Depth cm	*p*H	OM % C	N %	C/N	P mg %	CEC	Ca	Mg me/100g	K	Na	BS %
0-38	6.2	3.57	0.25	14	0.3	24.0	13.0	5.4	0.74	0.14	81
38-89	6.0	1.64	0.11	15	0.1	17.6	6.0	3.7	0.45	0.41	60
89-178	5.9	0.88	—	—	0.1	—	—	—	—	—	—
178-208	6.0	—	—	—	0.1	—	—	—	—	—	—
208-266	6.0	—	—	—	0.1	9.0	3.0	2.7	0.23	0.21	69
266-310	6.2	—	—	—	0.0	—	—	—	—	—	—
310-482	6.4	—	—	—	0.0	7.1	3.8	3.2	0.06	0.33	100

TABLE 25 : ANALYTICAL DATA FOR A LITHOSOL (Garcia and Cardoso, 1960)

LOCATION: S. Tome Gulf of Guinea. CLIMATE: rainfall 1775 mm, temperature 21-27°C. VEGETATION: Tropical rain forest. TOPOGRAPHY: steep slope. PARENT MATERIAL weathered basalt

epth Dcm	*Horizon*	CS	FS	Si	Cl	*p*H	C %	N %	C/N	CEC	*Exch. bases*	*Exch.* H me/100g	BS %
0-30	(A)	10.3	15.2	33.5	41.0	5.2	1.30	0.15		32.2	25.1	7.1	78
30-170	C	weathered	basalt										

TABLE 26: ANALYTICAL DATA FOR A REGOSOL (Garcia and Cardoso, 1960)

LOCATION: S. Tome Gulf of Guinea. CLIMATE: rainfall 1775 mm, temperature 21-27° C. VEGETATION: Tropical rain forest. TOPOGRAPHY: Gently undulating. PARENT MATERIAL: unconsolidated sand.

Depth cm	*Horizon*	CS	FS	Si	Cl	*p*H	C %	N %	C/N	CEC	*Exch. bases*	*Exch. H* me/100g soil	BS %
0-30	(A_1)	55.6	35.2	6.3	2.9	5.5	1.05	0.12		4.7	1.0	3.7	22
30-69	(A_2)	40.0	55.5	1.0	3.5	6.1	0.31	0.01		3.0	0.6	2.4	21
69-109	C	40.3	53.8	2.0	3.9	6.0							

(Table 25) are found on consolidated or stony parent materials. Lithosols include soils on the slopes of rocky hills of gneisses and igneous rocks as well as recent volcanic ash soils. Regosols occur as coastal dunes, some other sandy accumulations and loess but there are extensive soils that can be called regosolic as their properties are determined mainly by their very considerable content of quartz sand even though they show some similarities to well-developed zonal or intrazonal soils.

6 Physical Properties

SOIL FERTILITY is a complex of physical and chemical properties which enable a soil to provide plants with a favourable environment for growth resulting in better crops and higher yields. Physical properties of major importance are *texture* and *structure* influencing aggregate stability and the ability of the surface soil to withstand the impact of beating raindrops. Infiltration rate will be affected by the physical condition of the surface soil and this in turn will determine the extent of surface run-off and hence erosion. Texture and structure are therefore of primary importance in connection with the ability of a soil to resist wind and water erosion.

Crops also require good soil *aeration* and a soil that is pervious to the root system. Extremes of *temperature* can be harmful to both plant roots and soil organisms. To some extent soil *colour* influences soil temperature. The *water regime* in a soil is of utmost importance as the ability of a soil to supply plants with adequate water throughout the growing season is an important aspect of soil fertility. The presence of impervious layers in a soil can hinder water percolation and lead to adverse effects associated with waterlogging.

Soil depth is an important feature of tropical soils, especially in regions where crop growth is limited by rainfall that is seasonal. The deeper the soil, the greater the proportion of seasonal rainfall the soil can retain for use by crops during the dry season.

6.1 Soil Texture and Structure

A fertile soil must be in a physical condition that enables it to resist accelerated erosion and is favourable for root growth. The physical properties of a soil that determine, to a large extent, the predisposition of that soil to erosion are *texture* and *structure*.

Fine-textured clayey soils containing much montmorillonitic clay will become sticky when wet and form hard clods when dry. If the clods are broken up by tillage implements the resulting fine powdery material can get washed away by water or be blown about by wind. Clay soils with minerals of the kaolinitic group are less sticky when wet and are more friable when dry ; their physical properties make them less liable to erosion. Latosols have good physical properties, while Red-Yellow Podsolic soils have B horizons which are more plastic and more difficult to work with. Grumusols with montmorillonitic clays are poor in their physical condition.

Structure refers to the arrangement, size and stability of soil aggregates or peds. It determines the magnitude and distribution of the soil pore-space and together with texture and organic matter content is responsible for water holding properties and aeration of soils. The structure of the B horizon is affected by the dominant cation. Sodium causes a high degree of dispersion and consequently a hard and compact horizon, while much calcium renders a soil loose and friable. Solonetz soils have compact B horizons while the B horizons of Reddish Brown Earths are loose and friable. A number of tropical soils have good aggregate stability. The Montserrat Series of Brown Earths, derived from calcareous glauconitic sandstone are well aggregated and have a high degree of water stability and permeability (Ahmad *et al.*, 1968 *a*).

Pereira (1955) compared methods of measuring the structural condition of cultivated surface soils in east Africa. Measurements of water stability of crumbs did not provide a satisfactory indication of pore-space, of percolation rates or of rainfall acceptance. Dry-and Wet-sieving tests were inadequate to distinguish between the physical conditions of soils which showed contrasting field behaviour. Of the measurements on soil cores, total pore-space and field capacity (1/3 atmosphere) showed little change, while percolation rate, free-draining pore-space and a specially devised rainfall acceptance test, reflected the observed field behaviour of the soils.

The physical condition of a soil that is favourable for root growth depends on the size and distribution of the pores, which determine the permeability of the soil to rainfall and to roots, as well as the ability of the soil to supply water and air to roots. Experience has shown that a soil will have a very favourable distribution of large pores if it is in the form of crumbs between about 0.5 and 3 mm in size (Russell, 1958). If water from heavy rainstorms is to percolate through the soil without running off the surface a fair proportion of interconnected larger pores, of at least 0.5 mm in size, is necessary. Except in very coarse sandy soils a system of pores larger than 0.03 mm can only exist if the soil particles are aggregated into crumbs. In the tropics, the maintenance of this crumb structure is difficult under fairly continuous arable cropping. Soil crumbs on the surface break up due to the impact of falling raindrops and the fine material formed clogs the larger pores forming a surface skin largely impermeable to water. Cultivation also breaks down aggregates and accelerates the decomposition of organic matter whose presence aids in aggregate formation.

The maintenance of a good crumb structure will therefore be assisted by a protective cover of crop vegetation or a mulch and by minimal and timely cultivation. Periodic vegetative fallows may become necessary to restore structure where it has deteriorated. Stable aggregate formation caused by hydrated iron oxides occurs on certain red soils and fairly continuous cultivation is possible, without complete loss of structure on such soils.

6.2 Soil Aeration

The air content of a soil ranges between 5 and 40 per cent of the pore-space. Larger pores and cracks in the soil are necessary for easy root penetration. For diffusion of oxygen into and of carbon dioxide out of the root zone a continuous system of pores containing air is necessary, extending from the soil surface throughout the root zone. In a well-drained wet soil, pores of a size greater than about 0.06 to 0.03 mm are full of air ; smaller pores down to about 0.003 mm in size hold water available to crops. Still finer pores hold water so tightly that plant roots cannot absorb it. A fertile soil must have a fairly even distribution of larger and smaller pores for holding air and water available to crops respectively.

6.3 Changes in Physical Properties under Tropical Forest and Savanna Grass Fallows

Nye and Greenland (1960) refer to the beneficial physical changes brought about by fallows under shifting cultivation. In the forest fallow, the fine surface roots of the trees mould the soil into soft porous granules or crumbs, worms deposit their casts on the surface and large numbers of drainage channels are created by the constant passage to the surface of termites and other animals. The surface of the soil is thus maintained in an excellent state to permit rapid infiltration of water, and to resist erosion. It has been said that heating promotes an improved structure in the top 1–2 cm of soil.

The physical condition deteriorates rapidly, however, as a result of exposure to sun and rain until an improved first crop forms an effective cover. Under the impact of the early rains, sheet or rill erosion may occur. Soon after harvesting the soil surface may be exposed again but in the period when crops are growing, some protection is given to the soil by crops, weeds and the trash of harvest. Where permanent crops have been planted there is usually a good cover by the end of the first year. Where annuals such as rice and maize are grown the processes of deterioration are repeated.

In the savanna under a grass fallow, the improvement in the physical condition of the top soil is less marked than in a forest fallow. After burning, the grass grows slowly and the cover is incomplete. The intense sun and the early rain cause the structure to deteriorate rapidly, the surface soil becomes more compact and less permeable and surface run-off as well as erosion are greater.

In the savanna the surface of the soil is very much more thoroughly disturbed on clearing than in forest, because of the need to clear it of grasses. Furthermore, if the first crops are planted into loose freshly hoed earth, the structure deteriorates rapidly, a hard surface crust develops and infiltration is reduced.

6.4 Soil Erosion

Geological erosion is natural. It is only when it is accelerated by human activity that erosion becomes a problem. Such erosion may be termed *anthropic erosion.*

6.4.1 Wind Erosion

Wind erosion occurs only when a soil is dry and one of the chief measures for its control is the conservation of water to maintain the soil in a moist condition.

Wind erosion is most serious in dry areas, or at the end of the dry season. Soil blown about by wind has an abrasive action on crops and vegetation. Large sand dunes may form and these can travel covering farms and dwelling places. Windbreaks are essential in areas where wind erosion is likely to occur. Wind strip cropping was effective for moisture conservation and wind erosion control in Jodhpur, India (Mishra *et al.* 1968). Crops were rotated and cultivated in between strips of perennial grasses, 5 m widths of grass strips being interspersed with 30 m of cultivated strips.

Cultivation of the soil should be restricted to a minimum. Tillage should be carried out only when the soil is around field capacity. The inversion type of tillage such as ploughing with a mould board plough has been found unsuitable for the Reddish Brown Earth highland soils of Ceylon (Alles, 1967). Non-inversion tillage methods such as disking have been suggested instead. Practices which encourage a good soil structure, conserve moisture and build up organic matter should be adopted. Crop residues and stubble must be left on the soil surface until the next crop is planted to prevent soil disturbance by wind.

6.4.2 Water Erosion and its Control

Soil erosion by water results from surface runoff of rainfall in excess of the amount that the soil can absorb. The high intensity of tropical showers causes disintegration of the soil aggregates and a clogging up of the pores with fine material. A sealing of the surface soil takes place and infiltration is reduced. Consequently the ability of the soil to absorb the rainfall is decreased and there is more surface runoff.

With the recent rapid development of the tropics, soil erosion by water has become a serious problem because of faulty land use practices. Both shifting cultivation and pastoralism are conducive to accelerated erosion. Soil erosion in plantations has been reduced to some extent by suitable conservation measures but there is considerable erosion in areas recently cleared for cash crops and subsistence agriculture.

The muddiness of tropical rivers and the large amount of suspended matter they deposit when in flood indicates the intensity of erosion in the tropics.

The amount of velocity of water running off the surface of the soil will depend on (1) rainfall; (2) topographical location; (3) soil conditions;

(4) vegetative cover; and (5) land use.

The amount, distribution and intensity of rainfall are important. While the volume and distribution of rainfall contribute to erosion hazards, the most important rainfall characteristic affecting erosion is intensity. In the tropics a large part of the annual erosion loss occurs during relatively infrequent high intensity storms. At Enugu in eastern Nigeria, 66 mm out of a total of 77 mm over a four day period, i.e. 83.4 per cent of the rainfall, was recorded in one day, in a matter of a few hours (Ofomata, 1964).

There is usually more erosion in regions with alternating wet and dry seasons than where the rainfall is more evenly distributed. The first rains after a dry period are liable to cause much erosion both on pastures and arable land. On pastures, the herbage would have been greatly reduced by overgrazing while arable land would have been exposed by preparatory tillage. Shrinkage cracks form in the dry season. Runoff from the early rains attacks those cracks and gradually transforms them into gullies. Also, the surface crusting occurring during the dry season favours runoff rather than infiltration.

There is little relationship between the total annual rainfall and the annual volume of surface runoff and consequent soil loss.

The degree and length of slope as well as the nature of the surface affect surface runoff. Water losses increase with degree and length of slope but not necessarily in proportion. Runoff is less severe over a rough surface than over a smooth one; obstacles present on a rough surface tend to slow down the speed of movement and will increase infiltration.

Those soil characteristics which tend to increase infiltration and drainage will assist in reducing surface runoff and hence erosion. These depend on lithological composition of parent material, texture, structure, porosity, organic matter and clay content, depth of profile and impervious layers (Childs and Collis-George, 1950).

A sandy surface soil is often one that is strongly predisposed to erosion. Erosion is usually accelerated on soils when fertility has declined. Humus content and structure decrease and, consequently, the ability to absorb rainfall is lessened.

A good vegetative cover will protect even poor soils from being eroded whereas the most fertile soil stripped of its vegetation can be readily and destructively washed away.

Closed forest provides effective protection. The canopy of the trees and shrubs, together with herbaceous ground cover, break the impact of rainfall. Raindrop impact is further reduced by the forest litter, which also acts as a filter and allows water to percolate through slowly to the soil without much silt in it. The roots of forest trees open up channels which aid water percolation.

Next in order of effectiveness is a dense grass sward in good condition. Raindrop impact is broken while soil structure and water percolation are improved. However, overgrazing or indiscriminate burning will expose bare soil and excessive trampling by stock will destroy soil structure.

Perennial crops vary in their effect on surface runoff and erosion. Cocoa normally affords good protection. A mature cocoa plantation, with its permanent shade trees provides a complete canopy and abundant leaf litter. Mature rubber affords good protection but the canopy provided by immature rubber is incomplete. Other crops like citrus and coconut that do not form a complete canopy when mature do not provide adequate protection to the soil and require additional conservation measures. Planting on the contour and conservation measures are necessary with short-term perennial crops like sisal and pineapples. With arable crops, the protective effect depends on the type of crop and the cultural methods employed. Spacing of crops is important; a higher density will be more effective in breaking raindrop impact. Cultural practices such as mulching will assist in reducing erosion.

Faulty land use can be an important cause of erosion. Very steep land should either be left under forest, or used for tree crops such as cocoa or for permanent pasture. Faulty agronomic practices such as cultivation up and down slopes, the clean weeding of tree plantations and excessive tillage can also result in much erosion.

In controlling water erosion, the land should only be used for suitable purposes. The steepest slopes, unsuitable for cultivation should be left under forest or permanent pasture. Less steep land may be used for very protective tree crops with a full canopy of foliage, such as cocoa. Lesser slopes may be planted with less protective tree crops, such as citrus and coffee. Only gently sloping and flat land should be used for arable crops.

Appropriate agronomic measures such as proper tillage, contour strip cropping, cover crops and mulching should be used. Mechanical conservation measures such as silt pits or narrow-based ridge terraces can also be used where necessary. In heavy storms contour ridges may not hold water for sufficiently long. A better method is tie-ridging, which involves growing crops on ridges at regular intervals by barriers or ties made from soil, scraped up from the furrows, to form a saddle of earth at right angles to the ridges. The series of basins so made holds the water where it falls and, except during the most intense storms, prevents surface wash. The ridges are generally about three feet wide and four to five yards long. Tie-ridging has been successfully used in Nigeria (Lawes, 1961) and in Tanzania (Le Mare, 1954). But, as Russell (in Moss, 1968) points out, while tie-ridging can be a useful method of water conservation in years of light or erratic rainfall, in years of heavier rainfall, water can accumulate in the basins and stagnate long enough to harm the crop seriously. Lawes (1961), in northern Nigeria, obtained optimum results in both wet and dry years by tying alternate furrows only leaving the untied furrows to act as drains.

6.5 Soil Temperature

More radiant energy reaches the surface of the earth in tropical regions where the sun is high in the sky than in temperate regions. Also, in arid climates there is much less scattering and absorption of radiant energy by water droplets than in humid climates.

Temperatures of surface soils vary. Ramdas and Katti (1934) quoted a surface layer temperature of 74°C during the day falling to as low as 16°C at night. This variation decreases rapidly with depth. Vagelar (1910) in east Africa recorded bare soil surface temperatures of 50-54°C, while at a depth of only 5 cm below the surface the temperature did not exceed 37°C, and at a depth of 10 cm it was scarcely more than 30°C. Conduction of heat into the lower layers depends on the soil material and on its colour. Table 27 summarizes information on soil temperatures for some stations in east Africa (Griffiths in Russell, 1962).

TABLE 27: Soil Temperatures at Selected Stations in East Africa (Russell, 1962)

Station	*Altitude* m	*Mean air temp.* °C	*Mean soil temp.* °C	*Mean temp. range* °C at 15cm	122 cm	*Excess of mean soil temp. over mean air temp.* °C
Dar es Salaam	0	26.1	29.4	7.2	3.9	3.3
Entebbe	1190	21.7	23.9	—	1.1	2.2
Kabete (Nairobi)	1830	17.8	22.2	5	2.2	4.4
Muguga (near Nairobi)	2100	16.2	20.6	6.1	2.2	4.4

Such conduction is rapid in soils containing much quartz sand and is rather slow in black heavy clays. There is a time lag for the attainment of maximum daily temperatures with increasing depth. Mohr and Van Baren (1954) estimate a retardation of 1 to 2 hours for every 5 cm of soil. They have also stated that for many tropical locations, the mean temperature of the soil from the surface to a depth of at least 110 cm exceeds that of the atmosphere by about 3.4°C, this difference increasing with the density of vegetation.

When land in the tropics is cleared for cultivation and planting, deterioration in soil structure is enhanced if the soil is left bare and exposed to the sun for any length of time. In certain parts of east Africa, surface temperatures on dry, bare sandy soil may well exceed 65°C, but where vegetation gives an appreciable shading this would be reduced to about 38°C (Russell, 1962). It is therefore important to provide a vegetative cover as soon as possible after clearing. Annuals or perennials should be planted with the least delay and with tree crops it is desirable to establish a cover crop as quickly as possible.

When shade plants are necessary as with cocoa or tea, they too should be planted as soon as possible after clearing.

Surface layers of soils can be kept both cooler and at a more even temperature by the use of mulches (see 14. 2. 3). Gilbert (1945) found that the temperature of a coffee soil in Tanzania at 5 cm depth might vary by 12°C during the day, whilst under a mulch it varied by 2° to 3°C. A mulched soil is much cooler during the heat of the day and rather warmer during the night.

Shade trees will protect the surface of the soil and low-growing crops from the direct heat of the sun and will also to some extent break the fall of raindrops before they hit the soil.

When vegetation is burnt under shifting cultivation, the direct effects of heat on the soil are confined to local spots where the wood is piled. Surface temperatures may rise to 500°C or more for a few minutes but the soil temperature at 5 cm depth is barely affected (Cook, 1939).

6.6 Soil Colour

The colour of a soil influences its temperature. Where there is no vegetation, the darker the colour the greater is the absorption and transformation of the irradiated energy into heat (Table 28). Lighter soils are therefore generally cooler than darker ones. Soil temperatures will be most in excess over atmospheric temperatures in the darker soils. Ramdas and Dravid (1936) conducted some interesting experiments on the influence of soil colour on the temperature of the soil.

TABLE 28: Influence of Soil Colour on Solar Radiation Absorbed

Soil	*Percentage of total solar radiation absorbed*
Black cotton soil	86
Grey alluvial soil	40
Grass covered soil	60
Charcoal powder	94

Red and yellow colours in soils are derived from iron oxide in one form or another, the yellow colours being caused by hydrated forms while typical red soils are probably characterized by haematite as a colouring compound.

The black colour is commonly attributed to organic matter but the composition rather than the total amount of organic matter is important. Theron and Van Niekerk (1934) suggested that the nature of the clay minerals present and the stage of humification were responsible for the black soil colour and that the black colour is associated with the degree of lime saturation. Ashgar *et al.* (1949) found that the black colour was positively correlated with exchangeable calcium.

Singh (1954) found that a fairly concentrated solution of hydrogen peroxide (20-40 vols) if used in sufficient quantity can completely remove the dark colour of regur soils, provided the carbonate is removed first. A greater decolourization occurred with a greater loss of carbon indicating a very close association of the dark colour of these soils with their organic matter. The lignin content of the black soils was comparatively higher than those of the red clays found in neighbouring soils. The nature and amount of the clays present was also important. The black soils contained 40-60 per cent of clays of the montmorillonitic type while red soils had lower clay contents, with kandites and iron oxides. Presumably when the organic matter is dispersed over a larger surface area it dominates over the iron oxides in imparting colour to the soil.

In a subsequent consideration of the principles of colour production, Singh (1956) found that sodium clays sorbed comparatively more organic matter and gave distinctly darker products than H, Cl or Mg clays. Sorption of the organic complexes took place appreciably at low *p*H values of 3-5 to give very dark products but at higher *p*H values sorption was less and non-existent at *p*H 7. At least periodic anaerobic conditions were therefore necessary for the formation of black cotton soil. Sorption of organic matter and the formation of dark colours were invariably associated with the sorption of such inorganic constituents as Fe, Mn, Ca and Mg indicating that these inorganic constitutents also contribute to the dark colour formation of the clay-organic complex.

Dias *et al.* (1959) stated that the origin of the colour of tropical black clay soils can be due to a number of possible factors, among them, the type of organic and mineral colloids, clay-organic complexes, iron sulphide, and manganese. The action of these factors can be cumulative, but under natural conditions one or more of them is likely to be dominant.

7 Nutrient Supply : Macro- and Secondary Nutrients

7.1 Nutrient Status of the Great Soil Groups

THE WELL-LEACHED soils of the humid tropics are generally poor in plant nutrients. *Latosols* are so low that if no fertilizers are applied they are entirely unsuitable for agriculture. The primary nutrients, nitrogen and phosphorus, and to some extent, potassium are often deficient in latosols. The secondary nutrients, calcium, magnesium and sulphur are also in short supply. A common characteristic of latosols is their ability to hold phosphate ions strongly. Liming, though it may be necessary, is not effective because of the small cation exchange capacity.

The *Red-Yellow Podsolic soils* are also poor in plant nutrients but their cation exchange capacity is somewhat more than of latosols and usually increases with depth. They hold phophate ions more weakly than the latosols. Larger amounts of exchangeable aluminium are present than in most latosols of the same texture.

Reddish Brown Earths have a medium to appreciable base saturation. They are poor in nitrogen and phosphorus but medium to rich in potassium. These soils are well supplied in calcium and magnesium and are agriculturally much more valuable than most other tropical soils. Most of Brazil's coffee plantations are found on the *terra roxa estruturada,* which belong to the Reddish Brown Earth category. The agricultural potential of the Reddish Brown Earths of Sri Lanka, with a cation exchange capacity of about 50me/100g clay and well supplied with calcium and magnesium, is rated very high.

Although *Grumusols* have many undesirable physical features, their chemical fertility is fairly high. They are poor in nitrogen and phosphorus but they are rich in calcium and have a fair magnesium content. The base saturation is medium to high. Their *p*H ranges from 7 in the surface to 8 in depth. Their high *p*H and a relative imbalance of absorbed cations pose special problems in fertilizer application.

The chemical fertility of *Andosols* varies, probably because of the composition of the parent material. The volcanic ash or easily weatherable volcanic glass varies in the amounts of calcium, magnesium and potassium they contain. But they are usually strongly leached and poor in nutrients, except perhaps potassium.

Alluvial soils show great variations in chemical fertility, which is directly related to the sources of the material from which they are formed. Most areas of alluvial soils along large rivers are of mixed origin and are generally moderately well supplied with plant nutrients. In the drier areas, the alluvial deposits may contain much $CaCO_3$ but in the wet tropics many alluvial soils do not contain $CaCO_3$ and they may be quite acid. In the river alluvium of some African rivers, Mg solonetz soils have been observed because much M_g from a region with weathering basic igneous rocks rich in magnesium has been carried down into the river alluvium.

7.2 NITROGEN

7.2.1 NITROGEN CONTENT OF TROPICAL SOILS

Jenny (1930) found that for grassland soils of the United States there was an inverse relationship between nitrogen level and mean annual temperature. Soil nitrogen decreased from north to south and for each 10°C drop in mean annual temperature the average nitrogen content increased two or three fold. Jenny also found that along an annual isotherm the nitrogen content increased with increasing rainfall and humidity.

Extrapolation of Jenny's results to tropical soils would lead to the conclusion that these soils are very poor in both organic matter and nitrogen. But some tropical soils contain much more nitrogen than would be expected on the basis of Jenny's work on temperate zone soils. Indeed, from localities with identical mean annual temperatures and rainfall, the nitrogen and organic matter levels of Colombian, Costa Rican and Puerto Rican soils, for example, were found to be severalfold more than those of United States soils. The nitrogen content of the soils studied by Birch and Friend (1956) varied from 0.05-1.64 per cent and similar ranges have been reported from other tropical countries. Less than about 4 per cent of the total, depending on the crop grown and the cultural practices used, becomes available for plant use during any given season. Nitrogen gains and losses in tropical soils were reviewed by Greenland (1959).

7.2.2 ADDITION OF NITROGEN TO SOIL

Atmospheric sources—Nutrients received by soils from the atmosphere may be derived from the sea, from fine ash from volcanoes or from bush fires, from desert dusts or from industrial installations. In many parts of the world, rain may provide amounts of nutrients that significantly increase the supply available to natural vegetation but the figures quoted show such larger variations that it is unsafe to generalize (Cooke, 1967).

The amounts of nitrogen added to the soil by rainfall varies from 2-20 kg/ha/year, the amounts being greater in the vicinity of industrial areas. Eriksson (1952), from numerous estimates, worked out a median of nitrate N and ammonium N falling in tropical regions to be 8 kg/ha/year. Nye and Stephens (1962) mention a value of 11.2 kg N/ha/year as average for the tropics. Occasional reports of much larger amounts of nitrogen in rain water are probably due to contamination of the samples (Greenland, 1959).

Fixation—In soils there is both symbiotic and non-symbiotic fixation of nitrogen by micro-organisms. In the heavily leached acid soils of the rain forest areas, symbiotic fixation in legumes may be important but there is little evidence to show whether or not such symbiotic fixation does occur. In the drier forest areas and in the savanna zones it is doubtful whether the leguminous trees and shrubs support much active nitrogen fixing bacteria. Non-symbiotic nitrogen fixation is probably more important in these soils.

Non-symbiotic fixation is carried on chiefly by *Azotobacter* and *Beijerinckia spp.*, both of which are widely distributed in the humid tropics while some species of *Clostridium* and blue-green algae, common in rice soils and also probably in forest soils, do fix nitrogen. *Azotobacter* are more abundant in well-drained neutral soils while *Clostridia* are anaerobic and develop best in poorly-drained acid soils.

Meiklejohn (1962) showed that there is little difference in the numbers of nitrogen-fixing organisms between the forest and grassland soils in Ghana; the numbers were generally large, and slightly more in grassland soils. *Clostridia* were always present. *Azotobacter* were dominant in neutral soils and *Beijerinckia* in acid soils. Earlier, Meiklejohn (1954) reported that on east African soils *Azotobacter* was found only in a calcareous, sandy soil : *Beijerinckia sp.* was isolated from an acid sandy soil while *Clostridium spp.* were obtained from all the samples studied. Meiklejohn suggested that the occurrence of *Beijerinckia*, an acid tolerant nitrogen fixer, and of *Clostridia* probably explains why some acid tropical soils are considerably more fertile than temperate zone soils of equal acidity. Florenzano *et al.* (1968) found that *Beijerinckia* in Venezuelan soils were unusual, occurring both in acid as well as alkaline soils.

The absence of *Azotobacter* from neutral soils under east African conditions was ascribed by Meiklejohn to some soil deficiency other than calcium. Jensen (1940) reported earlier that *Azotobacter* were relatively scarce in Australian soils and that *Clostridium spp.* thought common, were not efficient nitrogen fixers. Parker (1954), using improved culture methods, showed that *Clostridium spp.* were really as efficient in fixing nitrogen as was *Azotobacter*.

British Guiana sugarcane soils and many other soils in the Caribbean region showed abnormally small C/N values in the subsoil layers, the C/N ratio diminishing with depth. Clay soils had smaller C/N ratios than sandy soils and the C/N ratio decreased with depth more in clayey than in sandy soils. Rodrigues (1954) showed that 14-78 per cent of the total nitrogen in these soils was fixed ammonium nitrogen. He did not identify the ammonium fixing clay minerals but attributed the fixation to hydrous mica. Native fixed ammonium in Hawaiian soils ranged from 0-585 ppm; volcanic ash soil horizons generally had a lower fixed ammonium content (4 to 178 ppm) than those from basalt (0 to 585 ppm) (Mikami and Kanehiro, 1968).

7.2.3 Mineralization of Organic Nitrogen

The decomposition of organic matter results first in the formation of ammonium; although nitrate and other forms may be released first sometimes. The factors affecting the release of ammonium are those influencing the decomposition of organic matter, namely the nature and age of plant material;

its state of subdivision ; the moisture content, aeration and temperature of the soil ; and climatic factors which affect these soil properties, as well as light.

Nitrification, or nitrate production, is a biochemical process taking place in soils in which ammonium is oxidized first to nitrite and then to nitrate by soil bacteria. The oxidation of ammonium to nitrite is carried out chiefly by bacteria of the genus *Nitrosomonas*. The bacteria producing nitrate from nitrite belong to the genus *Nitrobacter*.

Nitrification is influenced by the composition of the organic matter, its rate of decomposition, as well as by soil conditions. The nutrient status of the soil, especially its calcium, phosphorus, magnesium and iron content, is important (Meiklejohn, 1953). The temperature range for nitrification is 5-55°C, about 35° C being the optimum. A soil moisture content of between 50 to 60 per cent of the waterholding capacity is best. Nitrate accumulates in soils with poor moisture status and high temperatures, while much water and low temperatures are not favourable for its production and accumulation in soils. Nitrifying bacteria are very sensitive to lack of aeration; a good supply of oxygen is necessary for nitrification. The optimum *p*H for nitrifiers is between 7.0 and 8.0 ; extreme acidity and alkalinity retard their activity. Nitrification results in the production of acid and, for continued nitrate production, liming may become necessary.

Maximum nitrate production occurs in the top few centimetres of soil which, when exposed to tropical sunshine, can rise to temperatures above 55° C. Robinson and Gacoka (1962) found an upward movement of nitrate during the dry season in a Kikuyu red loam coffee topsoil (0-15 cm layer), this upward movement contributing towards the build-up of the nitrate in the immediate topsoil (0-5 cm layer) as it dries out. Crops assimilate nitrate and therefore the nitrate content of soils under crops is low, particularly so under a grass cover.

The high temperature and low moisture content of tropical soils during a dry season, in addition to making nutrients more available, also have a partial sterilization effect. Consequently the content of nitrifiable nitrogen rises and, since this is slowly converted to nitrate, nitrate accumulates during the dry season (Simpson, 1960). As the soil moisture content becomes more favourable with the rains, there is an upsurge of bacterial activity. The nitrate level rises rapidly at the onset of the rains (Jones, 1957; Moore and Jaiyebo, 1963) and nitrate values as high as 150 ppm (Griffith, 1949) and 200 ppm (Mills, 1953) have been reported; but these values fall rapidly as the rains continue (Birch, 1958; Semb and Robinson, 1969). Birch (1960) discussed the beneficial effect of soil-drying on soil fertility. He showed that the intensity of drying, the length of the drying period, and the drying temperature are all important in determining the magnitude of organic matter decomposition on remoistening the soil.

Meiklejohn (1962) attributed the lack of available nitrogen in Ghana grassland soils to the absence of nitrate-forming bacteria. Forest and grassland soils differed little in their content of nitrogen fixers but there were large differences between forest and grassland soils in the numbers of nitrifiers.

Under a forest cover, there were many ammonia and nitrite oxidizers, whereas grassland soils contained few ammonia oxidizers and very few or no nitrite oxidizers. None of the grassland samples taken after the start of the rainy season contained any nitrite oxidizers.

7.2.4 Losses of Soluble Nitrogen

Denitrification—Reduction of nitrate to nitrous oxide or nitrogen gas occurs extensively only under anaerobic conditions. Obligate anaerobes as well as facultative anaerobes may be involved. In flooded rice soils, much nitrogen can be lost by denitrification of nitrate leached down through the reduced furrow slice.

Denitrification increases rapidly as soil temperatures increase from 0° to 30°C and attains a maximum at about 60°C. Further, for rapid denitrification, the partial pressure of oxygen in the soil should be low, a sufficient supply of carbonaceous material should be present to provide energy for the denitrifying bacteria, the *p*H should be between 6 and 7, and sufficient nitrate should be present (Greenland, 1962). Cunningham (1962) investigated the extent of denitrification in a range of Ghanaian soils but did not find any serious loss of nitrogen by denitrification in cultivated soils. Under gassland conditions too, losses were negligible, whereas under standing forest denitrification losses were likely to be greater.

Ammonia volatilization:—Jewitt (1942) drew attention to the serious losses of nitrogen by ammonia volatilization when ammonium sulphate was applied to the surface of alkaline soils. Parish and Feillafe (1960) reported better responses to ammonium sulphate than to urea by sugarcane in Mauritius, and attributed this to the greater loss of ammonia from urea. Watson *et al.* (1962) studied the loss of ammonia by volatilization from surface dressings of urea in rubber cultivation. Nye and Greenland (1960) showed that food crops grown on newly cleared land in Ghana did not usually respond to ammonium sulphate while Cunningham and Arnold (1962) obtained the same result with urea and ammonium sulphate applied to shaded cocoa. Cunningham (1962) subsequently showed that lack of response to nitrogen was probably caused by production of much mineral nitrogen in forest soils but reduced responses to added nitrogen may also be the result of large losses of nitrogen by volatilization from moist, near neutral, warm soils such as those under cocoa in Ghana.

Acquaye and Cunningham (1965) studied the losses by ammonia volatilization from nitrogen fertilizers applied to six tropical forest soils in Ghana. They found that volatilization increased when soil *p*H, temperature and drying increased; it was also greatest from soils with the same CEC. Nitrogen losses were decreased by burying the fertilizer in the soil, by sterilizing the soil, and by adjusting soil moisture above and below 25 per cent of the water-holding capacity, at which moisture content the greatest loss was observed.

Leaching—Nitrate nitrogen is very susceptible to loss by leaching whereas ammonium nitrogen is seldom found in drainage water. Nitrate ions are lost principally in association with calcium ions. The amount so removed will

TABLE 29: PHOSPHORUS CONTENTS OF TROPICAL SOILS WITH COMPARATIVE EXAMPLES OF TEMPERATE SOILS

Country	*Soil*	*No. of sites*	*pH*	*Inorganic acid-sol.*	*Inorganic alk-sol.*	*Organic*	*Residual*	*Total*	C/ org. P.	N/ org. P.	*Authority*	*Method of fractionation*
Malaysia	Residual loams and clay loams	26	3.8-5.6	2	10	27	56	94	—	—	Owen (1953)	Williams (1950)
	Residual sands and sandy loams	—	—	7	5	38	23	73	—	—	Owen (1953)	Williams (1950)
	Coastal alluvial clays and clay loams	—	3.8-5.6	13	42	105	92	251	—	—	Owen (1953)	Williams (1950)
Sri Lanka	Lateritic (6) and non-lateritic (5)	11	5.6	26	289	312	309	942	83	—	Kandiah (1948)	Ghani (1943)
Angola	Ferrallitic soils	6	5.1-6.8	4	122	234	456	817	—	—	Almeida and de Miranda (1954)	Ghani (1943) (modified)
Ghana	Forest soils	21	6.3	10	34	99	183	326	233	21.6	Nye & Bertheux	Williams (1950)
	Savanna soils	67	6.2	6	15	27	87	134	247	19.5	(1957)	
India	Rice soils (acid and neutral)	20	5.5	79	86	189	169	516	50	—	Ghani & Aleem (1943)	Ghani (1943)
United Kingdom	Range of mineral soils	18	6.2	162 (0.5N H_2SO_4)	237	359	429	1186	101	—	Dean (1938)	Dean (1938)
United States	Iowa, Texas, Colorado	25	6.3	—	187	246	—	579	182	—	Thompson *et. al.* (1954)	
South Australia	Red-brown earths	15	7.0	11	22	61	126	220	—	15-23	Williams (1950)	Williams (1950)

depend on the nature and extent of crop and the root system as well as on soil texture. Grasses are especially effective in reducing losses. On bare fallow plots in the tropics, nitrates are lost primarily by leaching (Greenland, 1958).

Cropping—Soluble forms of nitrogen are assimilated by crops and the amount so removed from the soil will depend on the nature of the crop and on the extent to which it is removed from the land.

7.3 Phosphorus

7.3.1 Phosphorus content of Tropical Soils

In mineral soils, the total phosphate content varies (Table 29), the variation depending on parent material, degree of weathering and organic matter content. Total phosphorus is, however, of little value in assessing the phosphorus fertility of a soil. 'Available phosphorus' has been determined using a number of extracting solutions. Some of these are: 1 per cent citric acid solution (Dyer, 1894); 0.002N H_2SO_4 (Truog, 1930); dilute acetic acid/sodium acetate, *p*H 4.8 (Morgan, 1937); ammonium fluoride (Bray and Kurtz, 1945); and sodium bicarbonate, *p*H 8.5 (Olsen *et al.*, 1954). Schofield (1955) used a solution of 0.01M $CaCl_2$ to evaluate the phosphate potential of a soil while radioactive phosphorus (P^{32}) has been employed to determine the pool of 'labile' phosphorus. Most of these methods have been widely employed in temperate countries but their validity and usefulness for tropical soils need to be determined by the kind of study reported from Jamaica (Weir, 1962), from Kenya (Robinson and Semb, 1968; Oyot, 1970) and from Nigeria (Bache and Rogers, 1970).

Nye and Bertheux (1957) reported low values of total soil phosphorus in Ghana soils developed over both crystalline and sedimentary quartzose rocks compared with other parts of the world (Table 29). Soils developed over basic rocks contained, on the average, more than twice as much phosphorus as the corresponding soils over quartzose rocks, the higher values being probably due to the presence of more clay in those soils. While a low phosphorus content in the rocks themselves would have partly accounted for the low amounts of phosphorus in the soils, it was more likely that their great age and intense weathering were largely responsible.

No special differences between the Ghana soils and those in temperate countries were observed in the distribution of the different chemical fractions of phosphorus. Forest soils contained much more inorganic phosphorus than savanna soils, but there was little difference between the relative proportions in the dilute acid-soluble, the dilute alkali-soluble and the residual forms. Forest soils contained more organic phosphorus than savanna soils, corresponding to increased amounts of organic matter in them. Organic phosphorus was closely correlated with organic carbon content. Commenting on the relatively high C/P and N/P ratios for both forest and savanna soils, Nye and Bertheux (1957) suggested that high C/P and N/P ratios were characteristic of phosphorus-deficient regions.

A comparison between the N/P ratio in the fresh litter of forest (about 20) and tall grass savanna vegetation (about 4), and in the organic matter of the corresponding soils (22 for forest soils and 14 for savanna soils), showed that as forest litter decomposed, nitrogen and phosphorus were released at roughly the same relative rates, whereas in savanna the grass litter decomposed releasing mineral phosphorus more rapidly than mineral nitrogen. This was in accord with the fact that tall grass savanna sites were more acutely deficient in nitrogen but showed little need for phosphorus in the first year after clearing, while freshly cleared forest soils showed little need for either nitrogen or phosphorus in the first year.

It is commonly believed that one of the more important effects of allowing land to revert to natural fallow is the enrichment of the surface soil in phosphorus extracted by deep-rooting weeds from lower levels (Russell, 1968). Nye and Bertheux (1957) observed an accumulation of total phosphorus in surface horizons of leached forest profiles; the total and acid-soluble phosphorus decreased sharply, and the alkali-soluble inorganic phosphorus rather more slowly with depth. There was no evidence of such accumulation for savanna soils. The forest vegetation was more effective than the savanna in transferring nutrients from the subsoil to the surface (Nye and Stephens, 1962). Vine (in Moss, 1968), however, felt that 'pumping-up' from the depths was a more important source of potassium than of phosphorus.

In incompletely leached profiles, such as those of the lower members of a catena or soils on the semi-arid coastal Accra plains, Nye and Bertheux (1957) found a sharp rise in the acid-soluble and total phosphorus in the subsoil. This seemed analogous to the customary rise found in temperate subsoils. An increase of dilute acid-soluble phosphorus with depth is a feature of temperate soils which are not so leached, are much less weathered at comparatively shallow depths and in which the *p*H usually rises in the C horizon.

Organic phosphorus decreased with depth but more slowly than would be expected from the fall in total organic matter. The slow fall in organic phosphorus was associated with a narrowing of the C/P ratio similar to a decrease in the C/N ratio, with depth. This suggested that organic phosphorus compounds, like organic nitrogen compounds, were more readily leached than other organic compounds and were also more stable in the subsoil.

In the surface horizons, usually free from concretionary gravel, most of the inorganic phosphorus was associated with the clay fraction, though in sandy soils some phosphorus was found in the iron oxide coating the grains. In the subsoil, much phosphorus was associated with concretionary iron oxide gravel or pan, if present.

7.3.2 Phosphate Fixation

Phosphate fixation in tropical soils is, to a large extent, caused by the formation of insoluble compounds of iron and aluminium. On soils rich in sesquioxides, like the latosols (ferrallitic) and fersiallitic soils, it is therefore difficult for plants to obtain phosphorus from insoluble phosphorus compounds. The *Terra Roxa legitima* soils of Brazil, derived from basalt and containing much iron sesquioxides cause considerable reversion of superphosphate fertilizer to unavailable forms.

There are instances where residual effects of phosphate fertilizers have been shown (e.g., Kamprath, 1967 ; Fox *et al.*, 1968). Under a diversity of conditions in Kenya, residual effects were consistent and persistent (Boswinkle 1961). Experiments indicate that organic matter can play an important role in making phosphorus available to crops. Decomposing organic matter increases phosphate availability, but not so much perhaps in soils rich in ferric, aluminium or titanium phosphates (Dhar, 1957).

7.4 Potassium

Of all nutrient elements, potassium is usually the most abundant, to the extent of 0.05-3.0 per cent with an average content of about 1.5 per cent. Potassium is derived from potassium-bearing minerals, especially the felspars and micas, present in parent rocks. Potassium in soils ranges from the water-soluble to forms that are increasingly unavailable to plants.

Potash deficiency in tropical soils is more localized than that of nitrogen and phosphorus, and there are fewer reported instances of crops responding to potassium fertilizers in the tropics than in temperate regions. Potash deficiency occurs more commonly on heavily leached lateritic soils in the humid parts of the tropics and also on the light, sandy soils of the drier regions.

Potassium is retained in large amounts by the soil colloids but it can move down the profile more readily than phosphate. Much of the potassium leached to a depth of several feet can be recovered by deep-rooted plants and returned to the surface soil. Vine (in Moss, 1968) considered this 'pumping-up' from the depths to be a more important source of potassium than of phosphorus.

Considerable quantities of potassium are accumulated by both forest and grass fallows in shifting cultivation and returned to the soil in the ash when the leaves and litter are burned.

7.5 Calcium

Well-leached acid soils of the humid tropics, like the latosols and Red-Yellow Podsolic soils, are poor in exchangeable calcium while other soils, like the Reddish Brown Earths and the Grumusols, have relatively high contents of exchangeable calcium and magnesium.

Signs of calcium deficiency have sometimes been recognised in crops susceptible to it, like cocoa, groundnuts, and sisal, especially in the humid tropical forest zones. Calcium is supplied in a number of fertilizer and liming materials.

7.5.1 Liming in the Tropics

Responses to liming are not as marked on tropical as on temperate zone soils. The addition of calcium to soils with a low CEC can possibly cause nutrient imbalances and lower availabilities, particularly of micronutrients (Ignatieff, 1949). There have, however, been reports of benefits from applying lime. In southern Nigeria, much benefit has been reported from small amounts of lime applied to sandy latosols with a *p*H less than 5.0, mainly caused by increased production of nitrate.

Ignatieff and Page (1958) reported that for wheat and pyrethrum in east Africa, *p*H 5 seems to be the critical level of soil acidity below which lime may benefit the crop. With Kenya soils, crop responses to lime were unlikely when the *p*H was over 5.2, and even in more acidic soils such responses were not always shown. Both in Tanzania and in Uganda economic responses from lime applications were not generally obtained with a number of food crops. Responses to lime have been obtained on the more acidic soils of Tanzania with sweet potatoes and rice as well as with sisal in Kenya and Tanzania. In the Union of South Africa, the quantity of lime recommended on acid soils is 500-1000 kg/ha depending on *p*H and the kind of soil.

Foster (1970) recorded 18 significant responses to lime in 84 factorial trials on continuously cultivated soils in Uganda. Groundnuts responded to lime one year after its application, if exchangeable Ca^{++} was below 6 me/100 g. Cotton, sweet potatoes and beans responded only if the exchangeable Ca^{++} was below 6 me/100 g and the soil *p*H was also below 5.25. Cereals responded only if the *p*H was below 5.25 and exchangeable Ca^{++} was very low indeed. The beneficial effect of lime on groundnuts was probably because of the increased supply of calcium while for other crops the raising of soil *p*H appeared to have removed toxic aluminium.

It has been suggested that liming may have more to recommend it than any other low-cost soil amendment in tropical America (Miller, in Turk and Crowder, 1967); the amelioration of aluminium toxicity in typically leached soils has been considered to be one possible advantage (Abruna *et al.* 1964).

Tropical legumes do not require a high base status for fixing large amounts of nitrogen, although such fixation may be adversely affected by low availability of molybdenum at *p*H values below 5.0. Non-symbiotic nitrogen fixers of the genus *Beijerinckia*, which are widespread in acid tropical soils, especially latosols, are active fixers of nitrogen.

7.6 Magnesium

Nye and Greenland (1960) estimated a return of 71 kg Mg/ha/year from vegetation in a high forest fallow at Kade, Ghana.

Magnesium deficiency symptoms have been observed on oil palm and cocoa on soils in west Africa (Bull, 1954), on tea, rubber and coconut in Sri Lanka, and on commercial citrus orchards in the Caribbean area (Weir, 1971). In all these cases, symptoms of magnesium deficiency have been induced by high potassium applications.

In oil palm, the deficiency symptom 'orange frond' (Bull, 1954) is common on west African soils with low levels of exchangeable magnesium and the symptom can be readily induced in both seedling and adult palms by potash manuring. Over a wide range of soils in Nigeria and Congo Brazzaville the 'orange frond' symptom occurred where the Mg/K ratio fell below about 2.0 to 2.5 (Tinker and Ziboh, 1959). With young palms, deficiency symptoms can appear at a higher level of exchangeable Mg/K ratio of about 3.5 (Tinker and Smilde, 1963). The 'sickle-leaf' disease of cocoa has been observed in Ghana and is probably induced by excess potassium.

In Sri Lanka, dolomite is invariably used as an insurance against magnesium deficiency which is widespread in all tea growing areas. Crushed dolomite is used in all fertilizer mixtures for rubber. Magnesium deficiency has been observed in all the major coconut growing areas in Sri Lanka, especially on the Red-Yellow Latosols and on the Red-Yellow Podsolic soils. Soil applications of magnesium have restored a healthy green colour to yellowed leaves after three years, magnesium sulphate giving better results than dolomite.

Weir (1971) reported fertilizer trials conducted in commercial citrus orchards in the West Indies to study the occurrence of widespread magnesium deficiency. Soil applications of Kieserite ($MgSO_4.H_2O$) were effective in controlling the deficiency on light soils. On heavier clay soils, where Mg uptake was slow, foliar sprays of 1 per cent $Mg(NO_3)_2$ solutions were necessary. Application of additional N fertilizers increased the uptake of magnesium on calcareous soils, where control of Mg deficiency was difficult.

7.7 Sulphur

Sulphur deficiency is commonly reported in the tropics, especially on sandy soils and in heavily leached areas, and increasingly so where S-free nitrogenous fertilizers are used.

Beauchamp (1953) drew attention to a deficiency of sulphur in rivers and lakes in east Africa and to a possible deficiency in soils. The occurrence of excess sulphur in acid sulphate soils in west Africa and in Uganda was referred to by Chenery (1954 a) and other authors.

Bolle-Jones (1964) reviewed the widespread incidence of sulphur deficiency in various parts of Africa. He found that sulphur deficiency occurred on well-leached ferrallitic soils on the remnants of an old erosion surface of Tertiary age. A minimum rainfall of 600 mm per annum seemed to be one prerequisite for the occurrence of sulphur deficiency. However, under conditions of drought the absorption of soil sulphur by plants is reduced (Dutt, 1962). Legumes, such as lucerne, and certain oil-bearing plants, such as groundnuts and cotton were good indicators of the deficiency.

Storey and Leach (1933) reported the 'tea-yellow disease' of the tea bush in Nyasaland (now Malawi), found also in the south and north-west of Tanganyika (now Tanzania). There were indications of sulphur deficiency in some

Ugandan tea soils and in Kitale (Child, 1957).

Sulphur deficiency in groundnuts has been shown in northern Nigeria (Greenwood, 1951, 1954; Goldsworthy and Heathcote, 1963), in Senegal (Bolle-Jones, 1964), and in Ghana (Nye, 1952; Stephens, 1960). Other studies in western Africa (Goldsworthy and Heathcote, 1963; Prevot and Ollagnier, 1964) have confirmed these findings. Oil crops have relatively high sulphur requirements and sulphur deficiency may be more widespread than hitherto believed (Stanford and Jordan, 1966). The prevalence of sulphur deficiency in coconut plantations in Papua and New Guinea was responsible for low yields and a high incidence of defective or 'rubbery' copra in those countries (Southern, 1967).

Among other crops which have shown responses to applications of sulphur are cotton in several African countries (Bolle-Jones, 1964), coffee in northern Rhodesia (now Zambia) (Bolle-Jones, 1964) and pasture legumes in Kitale in Kenya (Chenery, 1958).

Aiyar (1945) observed sulphur deficiency in rice in Burma. Sulphur deficiency in sugarcane has been reported by Venema (1959) and Dutt (1962). The relative sulphur status of a number of sugarcane soils in Hawaii was reviewed by Ayres (Stanford and Jordan, 1966); values ranged from almost nil to several thousand parts per million.

Pineapple yields in Puerto Rico were reduced when sulphur was deficient (Cibes and Samuels, 1958), while experimental plots of pineapples fertilized with potassium sulphate gave a higher yield per acre and a larger mean weight per fruit than those which received potassium chloride (Samuels and Diaz, 1960). Potassium sulphate was more effective and profitable than potassium chloride for pineapple culture in Taiwan (Su and Li, 1962).

8 Nutrient Supply: Micronutrients

IN ADDTION to the macro-and secondary elements, plants require the so-called minor elements or micronutrients for their nutrition. These include, so far as is known at present, iron, copper, zinc, manganese, molybdenum, boron, chlorine and cobalt. These elements are usually needed by the crop only in very small amounts. An average crop of rice removes only about 1 kg/ha of iron and manganese compared with as much as 20 kg/ha of phosphorus. Apart from the quantities removed by plants, there is no sharp distinction between the macro-and the micro-elements. Magnesium and sulphur are examples of intermediate elements for many crops. Table 30 and Figure 3 give the range of micronutrient content of soils.

TABLE 30: RANGE OF MICRONUTRIENT CONTENT OF SOILS

Nutrient	*Normal Rnage*	
	Percentage	ppm*
Iron	.500 — 5.000	5000 — 50,000
Manganese	.020 — 1.000	200 — 10,000
Zinc	.001 — .025	10 — 250
Boron	.0005 — .015	5 — 150
Copper	.0005 — .015	5 — 150
Chlorine	.001 — 0.1	10 — 1,000
Cobalt	.0001 — .005	1 — 50
Molybdenum	.00002 — .0005	0.2 — 5

*ppm.—parts per million. These estimates are based on published data from a number of sources, especially R. L. Mitchell, "Trace Elements" in F.E. Bear, *Chemistry of the Soil* (New York: Reinhold, 1964), chap. 8.

8.1 MICRONUTRIENT PROBLEMS IN TROPICAL SOILS

In the present state of our knowledge of the micronutrient content of tropical soils it is difficult to state the extent to which micronutrient deficiencies or toxicities occur. But wherever soils, especially the more sandy ones, have been strongly leached deficiencies can be expected. Dry soil conditions, on the other hand, may immobilize nutrients (Schütte and Amdurer, 1960). Also, acid conditions prevailing in the soils of the humid tropics are conducive to the presence, in soluble form, of excessive amounts of some micronutrients, such as manganese, and to deficient amounts of others, such as molybdenum.

In flooded rice soils, iron and manganese are more soluble and can sometimes be present in excessive amounts. Organic soils are generally poor in available copper and manganese while alluvial soils vary greatly in their micronutrient content.

Schütte (1955) has reported that symptoms of micronutrient deficiencies are very widespread in Africa. About 80 per cent of the agricultural land of South Africa had crops showing trace element deficiencies, which were observed on all crops except the hardy sugarcane. Zinc and copper deficiencies were found on citrus. Deciduous fruits showed zinc, manganese, copper and iron deficiencies. Maize was widely deficient in manganese and, in certain areas, also in iron. Other cereals showed manganese and, probably, copper deficiency. In pastures, molybdenum, copper, manganese and cobalt deficiency symptoms were observed. Micronutrient deficiencies were widely scattered throughout the rest of Africa. The Kalahari sands have low levels of copper which prevent cattle production in the Kwango region of the Congo. But, as Nye and Greenland (1960) have pointed out, deficiencies have rarely been demonstrated under the conditions of local subsistence farming even on inherently poor soils that are intensively cropped. Responses in the field to micronutrient applications have been small.

In other parts of the tropics, iron deficiency has been reported with pineapples in Hawaii, iron and zinc deficiencies in coffee, citrus and sugarcane, boron deficiency in coffee, cocoa (Cunningham, 1964; Mestanza and Lainez, 1970) and oil palm, and manganese deficiency in coffee, rubber and citrus. Soil analysis data in India and Pakistan show that manganese and copper may be limiting crop yields in certain areas (Tamhane *et al.*, 1964). Raychaudhuri and Biswas (1964), in a review of the trace element status of Indian soils, referred to reports of the widespread occurrence of trace element deficiency in soils and crops, while Datta (1964) stressed the need for a comprehensive survey of the trace element status of cultivated and pasture lands.

TABLE : 31 DISTRIBUTION OF MICRONUTRIENTS IN SOILS AND ROCKS (Buckman and Brady, 1969)

in ppm

Element	*Earth's crust*[a]	*Earth's crust*[b]	*Basic rocks*[c]	*Acid rocks*[c]	*Sedimentary rocks*[c]	*Soils*[c]	*Soils*[d]
B	10	3	10	15	12	10	10
Mn	1,000	1,000	2,000	600	670	850	1,000
Fe	50,000	50,000	86,000	27,000	33,000	38,000	—
Co	40	23	45	5	23	8	8
Cu	70	45	140	30	57	20	20
Zn	80	65	130	60	80	58	40
Mo	2.3	1	1.4	1.9	2	2	1

a Goldschmidt (1954) b Mason (1960) c Vinogradov (1959)
d Swaine (1955)

8.2 Micronutrient Supplies of Tropical Soils

Table 31 shows the distribution of micronutrients in soils and in rocks.

Since the micronutrient reserves in uncultivated soils come originally from

TABLE 32 : Major Natural Sources of the Micronutrients and their Contents in the Surface Layers of Tropical Soils

Nutrient	*Major natural sources*	*Country*	*Total Content ppm*	*Reference*
Iron	oxides, sulphides, silicates		5000-50,000*	
Manganese	oxides, silicates, sulphides	Sri Lanka	<15-889	Kalpagé (1967)
		Colombia	850	Barshad and Rojas-Cruz (1950)
		Hawaiian Islands	5750, 9450	Matsusaka & Sherman (1950)
		India	391-1656	Raychaudhuri & Biswas (1964)
		Indonesia	450	Koenigs (1950)
		Taiwan	76-1028	Chiu (1950)
Zinc	sulphides, oxides, silicates	Cook Islands	38.5	Healy (1952)
		India (Madhya Pradesh)	9.6-127.6	Sharma & Motiramani (1969)
Copper	sulphides, hydroxy carbonates	Borneo	0.2. 2.2	Vermaat & van der Bie (1950)
		India (Madhya Pradesh)	8.9-134.1	Agrawal & Motiramani (1966)
Boron	borosilicates, borates	India	8.5-57.5	Raychaudhuri & Biswas (1964)
Molybdenum	sulphides, molybdates	Hawaiian Islands	0.002-0.13	Fujimoto & Sherman (1951)
			1.8, 13.9	Robinson & Alexander (1953)
		India (Gujarat)	1.10-3.01	Raychaudhuri & Biswas (1964)
Chlorine	chlorides		10-1000*	
Cobalt	silicates	Hawaiian Islands	10.9-127.9	Fujimoto & Sherman (1950)
		India (Gujarat)	8.0-47.0	Raychaudhuri & Biswas (1964)

*normal range for surface soils

the rocks that weathered to form the soil, soil micronutrient content depends on geological composition. Basic rocks contain much micronutrients; granites and acid igneous rocks are poorer. Sedimentary rocks, such as shales and slates, are richer in micronutrients than sandstones and limestones. Hodgson (1963) reviewed the chemistry of micronutrient elements in soils and discussed their distribution as well as availability in soils.

As Oertel (1961) has pointed out, the relation between the concentration of a micronutrient in a soil and its concentration in the parent material is often not close enough for satisfactory estimation of the former from the latter. Pedogenetic factors other than parent material also affect the concentration of micronutrients in the solum. During weathering the concentrations of a micronutrient may be greatly increased, and later the range may be reduced to its original value. No more than a rough indication of the average concentration of a micronutrient in a soil can usually be obtained from its concentration in the original rock.

Table 32 gives values for the total contents of micronutrients reported in the surface layers of some tropical soils.

8.3 Micronutrient Availability

Micronutrient availability in soils depends on soil reaction, organic matter content, microbial activity, redox potential, climate and the nature of the rhizosphere.

The effect of *p*H is indicated in a generalized way in Figure 4. All the micronutrients, except molybdenum, are more available in acid than in neutral or alkaline soils. Molybdenum availability increases with increasing *p*H.

The presence of organic matter may promote the availability of certain micronutrients, presumably forming organo-metallic complexes and thus preventing their fixation. The effect of organic matter has been studied on the availability of copper (Stenberg *et al.*, 1948), of manganese (Heintze, 1957), of zinc (Miller and Ohlragge, 1958), of boron (Berger and Truog, 1945) and of molybdenum (Davies, 1956).

According to Alexander (1962), micro-organisms may affect micronutrient availability in the soil in the following ways:

1. the release of inorganic ions and soluble organic complexes during organic matter decomposition;
2. immobilization of ions by incorporation into microbial tissue;
3. oxidation of an element, generally to a less available form;
4. reduction under anaerobic conditions;
5. by indirectly affecting *p*H or redox potential;
6. changing the amount of a micronutrient e.g. in nitrogen fixation.

Redox potential influences micronutrient availability of those elements such as Mn, Fe and Co which occur in different valence states. The state of oxi-

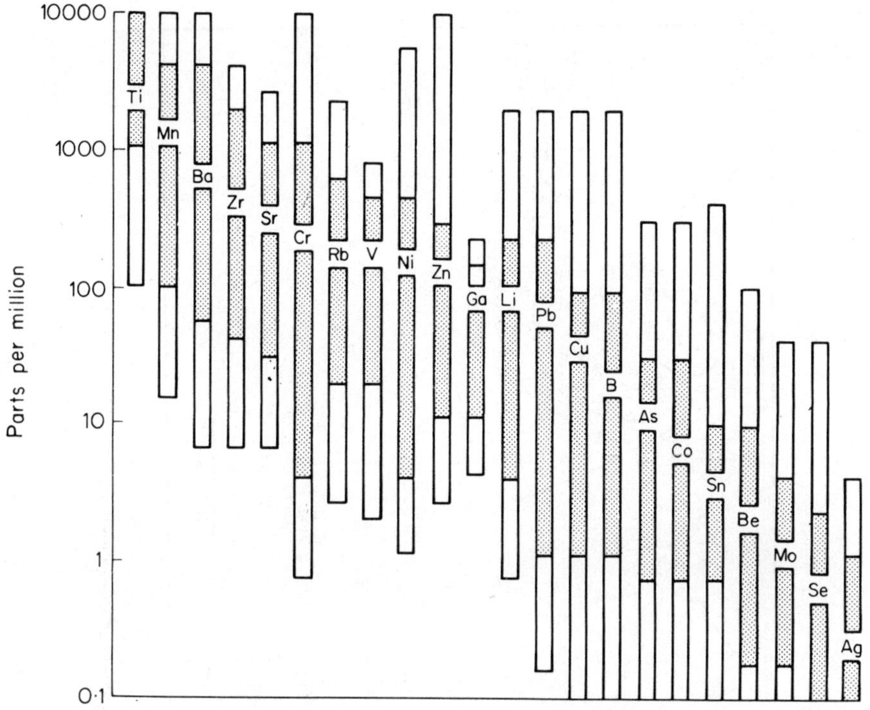

Fig. 3 : Micronutrient content of mineral soils

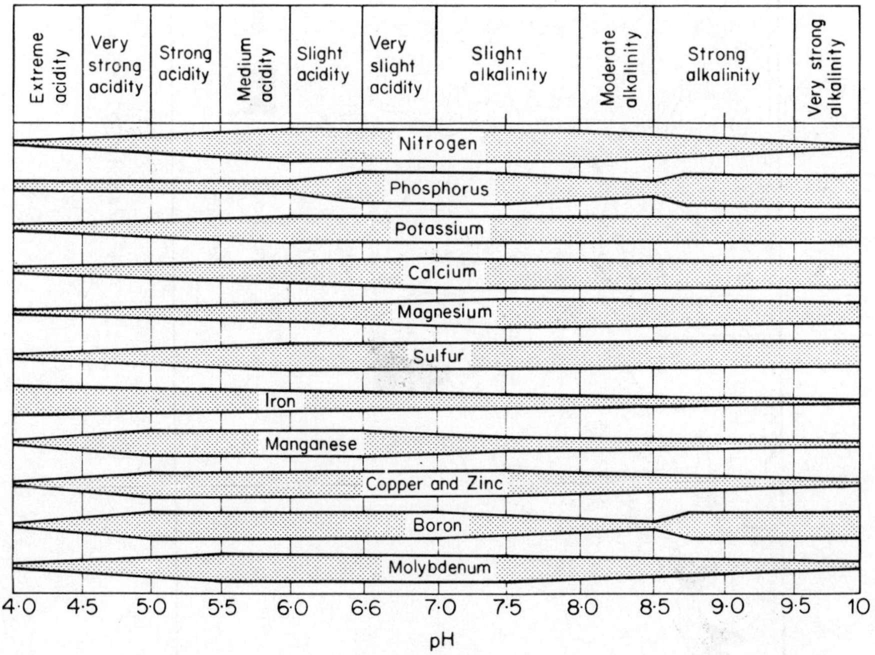

Fig. 4 : Nutrient availability as influenced by soil reaction

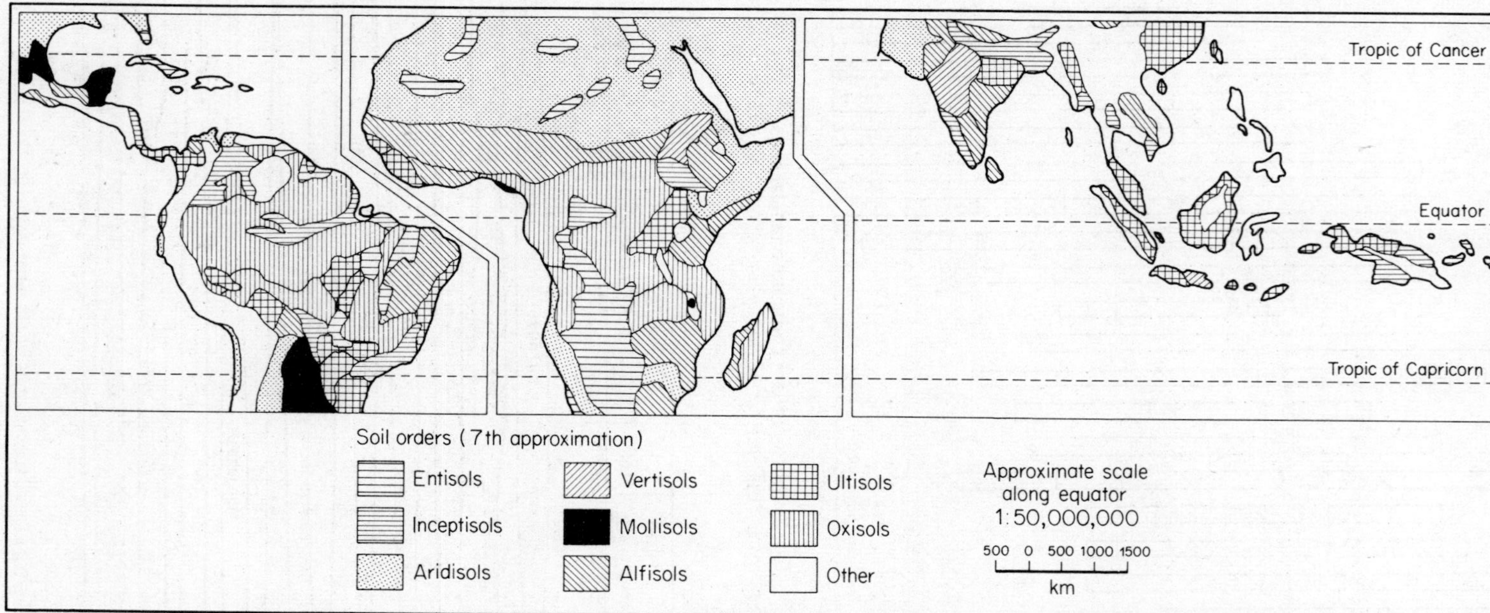

Map 8 : Soil orders of the tropics (Seventh Approximation)

dation is governed by the redox potential and this in turn is affected by drainage conditions.

There is a seasonal variation in trace element availability brought about by seasonal fluctuations of plant composition. Manganese exhibits the most pronounced seasonal variation in availability, probably because of microbially induced oxidations and reductions.

The chemical environment of plant roots is altered by root exudates which may influence micro-organic activity or cause direct interaction with soil constituents (Reuszer, 1962). Rhizosphere bacteria are known to reduce the availability of manganese to oats (Timonin, 1946).

8.4 Micronutrients in Plant Nutrition

The function of micronutrients in the physiology of tropical crops is the same as in temperate crops. Certain elements have attracted more attention in tropical than in temperate agriculture; for example, sulphur and copper in the tea estates of Nyasaland (now Malawi). Micronutrients take part in many vital activities in the plant. Most of them are either plant constituents or serve as prosthetic groups in enzymes or as integral components of enzyme systems, in the synthesis of chlorophyll, amino acids, proteins and vitamins. They are directly involved in the nutrition of the plant quite apart from their possible effects in correcting some unfavourable microbiological or chemical condition of the soil.

Plants suffer when these elements are in short supply, either through absence in the soil or, even if present, through their being unavailable. A shortage of one or more of these elements usually, though not always, affects the appearance of the plant, giving the leaves a chlorotic, bronzed or mottled colour, altering their habit of growth, or causing the death of growing points. These symptoms may not be shown at all or they may be displayed for a short period in the growing season. Roach (quoted by Schütte, 1955) has estimated that in southern Africa 50 per cent decrease in crop yields occurs before deficiency symptoms are visible. The symptoms may be sufficiently characteristic for the deficiency to be diagnosed visually; often, however, they are so indefinite or even suppressed altogether that chemical tests may have to be used for diagnosis. Such tests involve either the analysis of leaves or stems for the elements concerned or the application of an element suspected of being deficient to selected parts of the plant. T. Wallace (1951) has dealt at length with mineral deficiencies in plants.

Schütte (1955) has observed a relationship between micronutrient deficiencies and susceptibility to disease. In areas with sub-optimal concentrations of micronutrients, fungal infections of plants were more severe and also more complex than in plants grown on normal soils.

Micronutrients influence the water economy of plants, especially in hot dry regions prone to drought. Biebl (1952) showed that plants with trace elements in the fertilizers applied to them transpired throughout hot days, while the controls did not. The latter did not assimilate carbon while not transpiring. Schütte (1957, 1959) observed that wilting occurs more rapidly in deficient plants.

The vitamin content of plants is decreased by trace element deficiencies; manganese deficiency lowers the vitamin A content of spinach and the vitamin C content of tomatoes and black currants (Schütte and Amdurer, 1960). These authors also reported (1960) that boron deficiency lessens the carotene content of carrots, and of tomatoes and other fruits. They also found an adverse effect of trace element deficiency on the amino acid content and hence on the nature of plant protein.

Once a micronutrient deficiency has been detected it is not always necessary to add this element to the soil. It may be more economical to spray it on the leaves of the plant or, if it is a tree, to insert pellets containing it under the bark in the trunk. Soils may contain adequate quantities of a micronutrient but in forms unavailable to the plant. Adding a micronutrient element to the soil under these conditions will not be of much use. In general, increasing the acidity of a soil, by placing sulphur near the plant, for example, will increase the availability of all the micronutrients except molybdenum. In the presence of much organic matter, Zn deficiency disappeared and Mn concentration increased when the soil *p*H was lowered on a cocoa soil in Ghana (Cunningham, 1964).

On many soils, plants suffer from a deficiency of a number of elements simultaneously. Treatment with all the elements concerned will then be necessary to improve crop yields. Again, the vitality of growth may be affected without any symptoms being obvious.

Micronutrients, though essential in small amounts, are usually very poisonous if present in the soil in an available form in larger quantities. To apply them indiscriminately, except perhaps in very light dressings, is therefore very dangerous. Cauliflower grown on sandy soil may need 2-5 kg of borax per hectare but 10 kg per hectare would injure the crop.

Many elements invariably found in plant tissue have not been shown to be essential for plant growth. These accessory elements for which the term *trace elememts* has been used by some authors (Hodgson, 1963) and which have been found in plants include the following: aluminium, silicon, sodium, selenium, arsenic, barium, bromine, chromium, cobalt, gold, iodine, lead, nickel, silver and titanium. The first three of these are found in relatively large quantities in plants. The presence of silicon does not appear to exert any injurious effects. Silicon appears to be an essential element for rice. Aluminium and sodium both appear to be toxic when absorbed in critical amounts. Some sodium can replace a small amount of potassium in plant nutrition. Selenium is harmful to animals when ingested from herbage.

Aluminium may accumulate in the roots of plants growing on acid soils to the extent of reducing the power of the plant to absorb and translocate phosphate ions. Excess aluminium may therefore cause the plant to suffer from phosphorus starvation which applications of phosphatic fertilizers will not correct. Soluble aluminium (180-480 ppm) in the acid 'lunyu' Uganda soils cause problems for plants other than those accumulating aluminium, such as tea, which do well under such conditions (Chenery, 1954a).

8.4.1 IRON

Iron in soils—Iron is present in quantity in most soils, to the extent of 0.5 to 5.0 per cent iron, mainly as oxides, which are largely responsible for the red and brown colours in soils, and as phosphates. Among the oxides reported are: magnetite, ilmenite, limonite, haematite, goethite and lepidocrocite (Brown, 1953).

Where drainage and aeration are good, ferric compounds predominate and where waterlogging and bad aeration occur, ferrous compounds are formed. Organic forms of iron, such as iron humate, appear to be mobile. The availability of iron to plants increases with acidity and is depressed by phosphates. Most soils contain much more available iron than is needed by plants. Iron deficiency because of an actual lack of the element in the soil seldom occurs. A high *p*H or nutrient imbalance are frequent causes of chlorosis.

Iron deficiency is likely on alkaline soils and at high redox potential values. Under such conditions, phosphorus may play an important role in preventing the absorption of iron. Iron deficiency is not common on acid soils. The only outstanding examples are the pineapple soils in Puerto Rico and Hawaii. In both cases, the deficiency probably results from a high Mn/Fe ratio in the soil. Calcium, zinc and copper can similarly induce a physiological deficiency of iron. 'Pineapple yellows' or chlorosis of pineapples is common on Hawaiian soils containing high amounts of manganese dioxide and calcium carbonate.

On fruit trees on calcareous soils, iron deficiency may be aggravated by potassium deficiency. There is a close relationship between iron and potassium. In maize plants which are poorly supplied with potassium, iron accumulates in an ionic form at the nodes of stems. Potassium contributes to the mobility of iron within the plant and hence alleviates iron deficiency. Conversely, symptoms of potassium deficiency can also, in certain circumstances, be decreased by supplying iron. The presence of actively decomposing organic matter in soil decreases the possibility of iron deficiency, probably by increasing the carbon dioxide content of soil and hence decreasing the *p*H. In Ghana, iron deficiencies have been found around old termite mounds (Nye and Stephens, 1962), where a calcareous horizon is often formed in the subsoil, and on areas with much wood ash consisting largely of alkali salts.

Correction of iron deficiency–The application of iron salts to soils, especially if the *p*H is high, is usually ineffective (Cunningham, 1964). The most economical method is the application of foliar sprays. These sprays are often damaging to the foliage and must be used with caution. A solution of 0.2 to 1.0 per cent ferrous sulphate may be used. The injection of solid iron compounds, ferrous sulphate, ferric citrate or tartrate, into stems is also used for fruit trees. The *p*H may be lowered and iron made more available by dressings of sulphur. Organic iron complex salts, the chelates such as Fe EDTA, have shown promise and are being widely used. Chelates can be applied as foliar sprays as well as to the soil. Growing cover crops, such as grasses and clovers, over the roots of affected trees is also successful. The carbon dioxide produced lowers the *p*H and stimulates the formation of ferrous iron.

8.4.2 Manganese

Manganese in soils—The manganese content of soils is very variable. In the USA it varies from less than 0.001 per cent to 1.27 per cent. Some tropical soils contain up to 15 per cent manganese. Most of the element is found in the surface soil; it is least in the B horizon while there are increased concentrations in the C horizon.

As with iron, the oxides are the important forms. Both iron and manganese play an important role in soil redox reactions. Both form concretions in imperfectly drained soils. The higher oxides are of low availability to plants.

The solubility of soil manganese increases with increasing acidity. In many soils, manganese is largely unavailable to plants above *p*H 6.5 and deficiencies occur in such soils. In strongly acid soils manganese is often present in toxic amounts. Chenery (1954a) reported that the exchangeable manganese in the acid 'lunyu' (salty) soils of Uganda was about 200-400 ppm compared with 1-20 ppm in a normal fertile soil, even when the soils were only moderately acid, *p*H 5.3–6.5. In temperate climate soils such high exchangeable Mn figures are usually associated with high acidity. Manganese toxicity in the lunyu soils was cured by long continued mulching with banana trash and elephant grass, the benefits being due at least in part to an interaction between soil moisture and exchangeable Mn. Moist fresh soil samples contained less exchangeable manganese than air dried soil samples.

The availability of manganese is greatly affected by organic matter content and drainage. Deficiency is common on calcareous peats and high organic matter content soils.

Under waterlogged conditions, e.g. in rice fields, where reducing conditions prevail, manganese compounds are reduced to the manganous state and availability is increased (see 13.3).

Certain soil organisms oxidize Mn^{++} to Mn^{+++} at *p*H values above 6.5 and thus render manganese unavailable to plants (Mann and Quastel, 1946). The absorption of manganese is promoted by potassium.

Correction of manganese deficiency—The application of manganese salts to soils may be satisfactory in some cases, not so in others. Sprays are more economical and very effective. Spray damage is negligible. Manganese deficiency, if not too severe, can be prevented by keeping the soil acidic by addition of sulphur or by use of farmyard manure and compost.

Manganese deficiency symptoms can be controlled by a single treatment of 45 to 65 kg of manganese sulphate per hectare or by spraying with a 0.5 to 2.0 per cent solution of manganese sulphate. The injection of manganese salts into fruit trees has proved effective in many instances.

8.4.3 Molybdenum

Molybdenum in soils—The total molybdenum content in soils usually varies from 1 to 10 ppm but soils containing 25 to 75 ppm molybdenum are found in Hawaii (Davies, 1956). Molybdenum may be present in soils either (*a*) as part of the mineral structures of soils as a compound of magnesium and cobalt; (*b*) as anions adsorbed by soil clay minerals; and (*c*) bound with organic matter (Dobritskaya, 1961). The anionic form is affected by soil conditions and influences plant uptake of molybdenum. Molybdenum differs from most of the other trace elements in that its availability is highest at soil *p*H values between 6 and 8. Under acid conditions molybdenum combines very strongly with sesquioxides and clay minerals.

Molybdenum deficiencies in the field may rise in two ways. The total amount may be small because the soil has been formed of parent material poor in molybdenum and under conditions in which the molybdenum is leached away. On the other hand, the soil may be well supplied with total molybdenum which is fixed in forms unavailable to plants.

Acidic soils rich in iron and aluminium are most likely to be deficient in molybdenum. Liming an acid soil on which molybdenum deficiency symptoms are shown by plants will release the molybdenum fixed by such soils.

Excess molybdenum is found in copper deficient peats (Davies, 1956). No relationship has been established between molybdenum and climate, parent rock, soil texture or organic matter in mineral soils.

Correction of molybdenum deficiency—The use of molybdenum as a fertilizer was reviewed by Anderson (1956). For many soils that have too little molybdenum, fertilization with 0.07 kg/ha (1 oz/ac) of molybdenum is ample and there is no need to repeat the application perhaps for several years. Anderson (1942) found that 4 g of molybdenum per hectare was sufficient to overcome molybdenum deficiency in pastures. Excess molybdenum is harmful, 3-4 kg/ha being toxic to most plants.

Usually, a soluble molybdenum compound such as sodium molybdate or molybdic acid is mixed with superphosphate fertilizer. 'Molybdic superphosphate' contains 400 to 600 g of sodium molybdate per ton. When the phosphate is spread on the field, the molybdenum is distributed with it. In practice, soils poor in molybdenum have usually been low in phosphorus as well, so that small amounts of molybdenum could best be spread through the use of molybdenized phosphatic fertilizers. Treating seed by dusting with a molybdenum salt is adopted in large plantations of legumes. Molybdenum may be sprayed on to the plants in solution as is done to correct yellow spot of citrus (Stewart and Leonard, 1952).

8.4.4 Boron

Boron in soils—Boron derived from igneous rocks is present in soils to the extent of 45 to 225 kg/ha in the plough depth and originates mainly from the highly resistant mineral, tourmaline. It is readily leached out when tourmaline weathers. Boron deficiency is most common in light soils and in organic soils. Less than 5 per cent of the total boron content in soils may be available for use by plants. Boron deficiency for plant growth is manifested when boron soluble in hot water is less than 0.35 ppm (Bear, 1953). Martin (1969) found that oil palms growing in soils containing less than 0.1 - 0.2 ppm available boron may show signs of boron deficiency.

Correction of boron deficiency—The use of 10-45 kg/ha of borax, mixed with superphosphate whenever possible, will correct symptoms of boron deficiency. Mestanza and Lainez (1970) corrected boron deficiency in cocoa in Ecuador with a foliar spray of a soluble boron compound. Excess of boron is toxic and should be avoided. Toxicity symptoms first occur in the older leaves.

8.4.5 Copper

Copper in soils—Copper occurs to the extent of 5-12 kg/ha in the plough depth of most soils. The copper content of organic soils is small and their copper fixing capacity is high (Nikitin, 1954). On such soils, somewhat larger applications of copper can be used without toxic effects. Yet, excessive applications of either copper or manganese may lead to a deficiency of the other.

Pinkerton (1967) found that copper deficiency in wheat crops in the Rift valley, Kenya, was associated with soils derived from pumice and ash containing less than 3 ppm Cu available to *Aspergillus niger*. Broadcast applications of copper sulphate and copper oxychloride were both effective in eliminating the deficiency in wheat and resulted in a marked increase in the copper content of the grain.

Correction of copper deficiency—Copper deficiency symptoms can be cured by the application of 10 kg/ha of copper sulphate or by spraying a 1 per cent solution of copper sulphate. Excess of copper can be toxic.

8.4.6 Zinc

Zinc in soils—Zinc occurs to the extent of 20-550 kg/ha in the plough depth of soils. There is usually more zinc on the surface than in the subsoil. Its content is low in strongly weathered coarse-textured soils. In soil solutions Zn^{++} cations are present. Maximum availabiliry of zinc is between *p*H values of 5.5 and 7.0.

Where soil *p*H is less than 6, a zinc content of 100 mg/kg of soil is considered harmless. As the reaction rises above *p*H 7, Zn^{++} ions may be converted into negatively charged zincate ions, and calcium containing minerals will also lessen the availability of zinc. Thus zinc deficiency is acute in arid and semi-arid areas. Soils with a high organic matter content also exhibit a deficiency of zinc (Nikitin, 1954).

Citrus has a high zinc requirement, and zinc deficiency is widespread in central and southern Africa. In the citrus growing areas of the Deccan plateau, the total zinc content varies from 8 to 14 ppm and the exchangeable zinc from 0.01-0.10 ppm indicating a deficiency.

The 'sickle-leaf' disease in cocoa on alkaline surface soils in west Africa has been attributed to zinc deficiency and has been cured by zinc application (see also 16.4). Zinc deficiency on coffee is widespread in Brazil and is usually remedied by a zinc sulphate spray.

Correction of zinc deficiency—Zinc deficiency can be controlled by spraying with 0.5 to 1.5 per cent solutions of zinc sulphate or by soil application of 10-15 kg/ha of zinc sulphate.

8.4.7 Chlorine

In 1954 Broyer *et al.* demonstrated that chlorine was essential for tomatoes. Chlorine deficiency is characterized in the early stages by a wilting of the leaflet blade tips which is followed by chlorosis, bronzing and necrosis in areas close to the wilting. In more advanced stages fruit production is inhibited. *In vitro* studies have shown that the chloride ion is capable of accelerating the photosynthetic process. The chloride ion may function as the essential co-enzyme for photochemical reactions in photosynthesis.

Studies in Africa, Indonesia and South America show the importance of chlorine in oil palm nutrition (Ollagnier and Ochs, 1971). In Colombia, correcting chlorine deficiency increased oil yield by 700 kg/ha. Low levels of chloride in the leaf were found in west Africa, Colombia, Peru, Indonesia and the Congo. The optimum leaf concentration was 0.5-0.6 per cent of the dry matter.

8.4.8 Cobalt

The essential role of cobalt in animal nutrition has been known for some time but it is only recently that it has been recognized as an essential element for plant growth. In general, a cobalt content of less than 5 ppm in the soil or 0.08 ppm in forage may lead to deficiency diseases in animals.

8.4.9 Silicon

Silicon has not been shown to be essential for the growth of all plants. But it plays an important, perhaps a critical, role in the nutrition of the rice plant (Jones and Handreck, 1967) and in sugarcane (Fox *et al.*, 1969). The rice plant absorbs more silicon than other cereals, the stems and leaves of the rice plant containing 10-20 per cent of silicon. Rice plants showed retardation of their vegetative growth and decrease of the degree of seed setting when their silicon content was extremely small (Mitsui and Takatoh, 1963).

9 Soil Organic Matter and Soil Organisms

The luxuriant forest vegetation of the humid tropics led to the early belief that the soils underneath were inherently fertile. However, when the forest was cleared and the soil cultivated according to methods in vogue in temperate countries, fertility decreased rapidly as a result of erosion, leaching of soluble nutrients, and loss of organic matter. This in turn led to doubts as to whether the fertility of tropical soils could ever be raised adequately to meet the needs of continuous cropping. Opinions were even expressed (Stamp, 1953; Gourou, 1953) exaggerating the lack of fertility of tropical soils. Later studies (e.g. Vine, 1954) have made a less pessimistic view possible.

9.1 Organic Matter Content of Tropical Soils

The high temperatures characteristic of lowland tropical regions are more conducive to the mineralization of organic materials added to soils rather than to their humification. On the basis of Jenny's finding (1930) that the organic matter and nitrogen contents of soils in the United States decrease as the annual temperature increases, it was expected that tropical soils would, as a rule, be very low in organic matter and nitrogen. This belief was, in fact, held by early workers like Mohr (1922) and Corbet (1935), the latter attributing low organic matter contents in tropical soils to high insolation as well as to high temperatures.

Dean (1930), Hardon (1936), Ancizar-Sordo (1941) and Jenny *et al.* (1948) questioned the general validity in tropical regions of the organic matter/temperature relationships derived for the United States. In a study of soils collected from 100 different localities in Colombia, South America, Jenny *et al.* (1948) observed that at high annual temperatures, organic matter and nitrogen increased as precipitation became higher. At higher altitudes, a marked increase in organic matter and nitrogen contents was noticeable. In belts of constant rainfall and NSQ, the nitrogen content decreased exponentially as temperature increased. Comparing Colombian and North American soils having equal annual temperatures and annual moisture values, these workers found that the Colombian soils had much higher organic matter and nitrogen contents than the North American soils.

Jenny *et al.* (1949) found that, like the Colombian soils, Costa Rican soils were also richer in organic matter and nitrogen than Californian soils of the Sierra Nevada mountains. They estimated that the annual production of organic matter in tropical forests (8500-12000 kg/ha) was much greater than

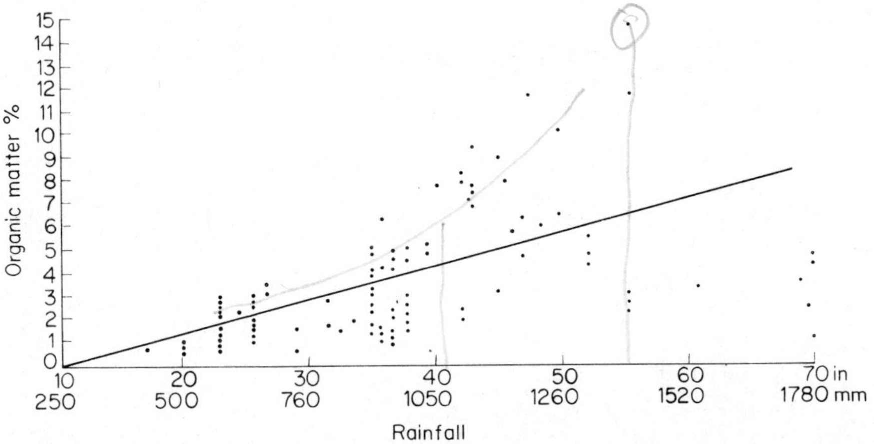

Fig. 5 : Organic matter content and rainfall

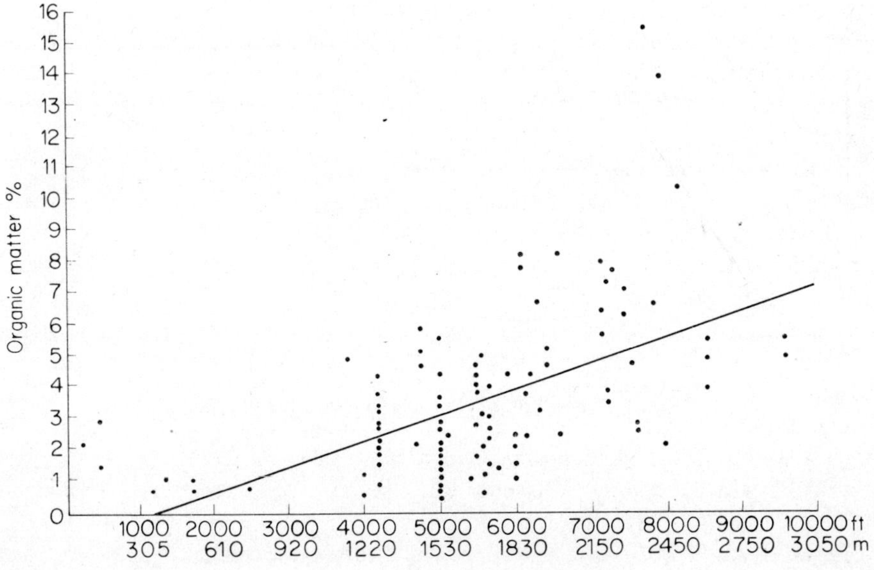

Fig. 6 : Organic matter content and altitude

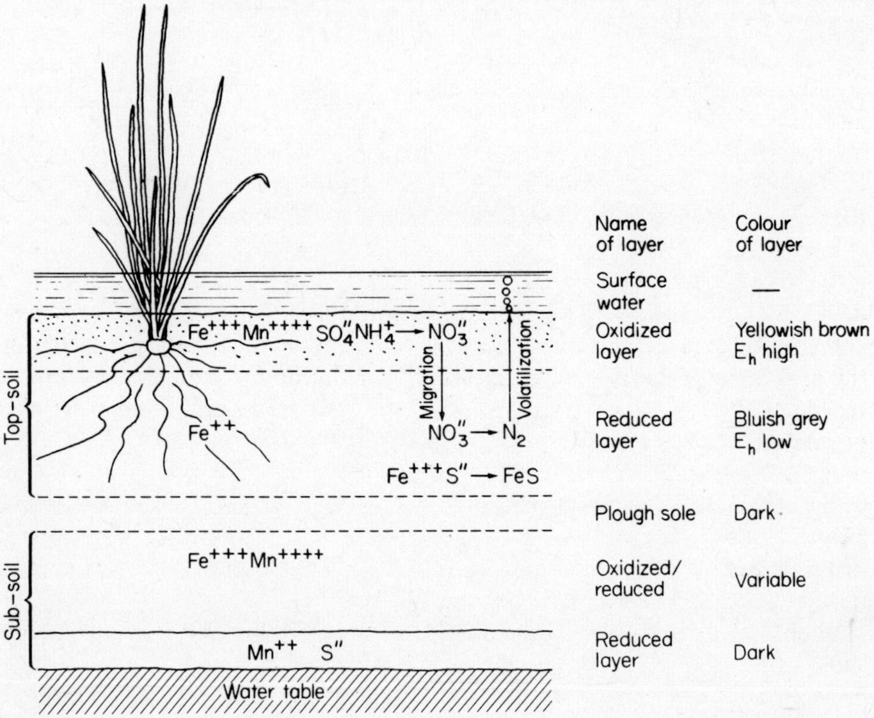

Fig. 7 : Redox potential profile in a flooded rice soil

in Sierran forests (920-3100 kg/ha). At the same time they also found that alfalfa leaves placed in tropical soils decomposed more rapidly than in temperate soils. With these seemingly contradictory results, they could not provide a direct clue as to the cause of the high organic matter levels in some tropical soils.

Jenny (1950) attempted to elucidate the factors responsible for organic matter accumulation in tropical soils and attributed this to (*a*) favourable climatic conditions and (*b*) high annual rates of nitrogen fixation by legumes, leading to a high production of organic matter. Further, he thought that a considerable portion of organic matter decomposition products infiltrated into the mineral subsoil where they decomposed slowly and hence tended to accumulate.

Smith *et al.* (1951) showed that organic matter contents of 2 to 3 per cent were common to depths of 30 cm or more in a number of Puerto Rican soils and that, where erosion was controlled, even on steep slopes, there was no clear evidence that organic matter reserves were declining. Under good fertilization and proper management, on sugarcane and coffee soils, it was possible to maintain high organic matter levels.

The influence of climatic and topographic factors on the organic matter status of east African soils was investigated by Birch and Friend (1956) in a detailed study of soil samples collected from sea-level to 3,000 metres in Kenya, covering a range of climatic conditions. The amounts of organic matter in the soils were fairly high and compared very favourably with the amounts in temperate soils. Nearly 47 per cent of the soils contained over 4 per cent organic matter even though many of them from the Kenya Highlands had been under cultivation for anything up to thirty-five years. Organic matter values below 2 per cent were generally found on sandy soils and in soils under low rainfall. They reported a general range of about 1.2 to 3.2 per cent organic matter in east African topsoils (to 15 cm depth), depending primarily on rainfall but also affected by the percentage of clay.

Birch and Friend (1956) concluded that the main factor governing the organic matter content of soils was rainfall. Except on very sandy soils, temperature and clay content of the soils were minor factors. The regression of the percentage organic matter on rainfall was very significant (Fig. 5) and was similar for east Africa and North America in spite of big seasonal temperature differences. Increasing rainfall worked almost entirely in favouring vegetative growth and organic matter production. The direct effect of temperature on organic matter status was, however, small compared with that of rainfall. Increasing temperature affects both the synthesis and breakdown of organic matter; the production of organic material with increased temperature, particularly in frost-free areas, is counterbalanced by increasing microbiological activity and organic matter breakdown. The regression of organic matter percentage on altitude was also very significant (Fig. 6); on an average, each 300 metres increase in altitude in east Africa was associated with a 0.8 per cent increase in organic matter. The organic matter contents of high altitude soils were generally high (Table 33). This was to be expected since increasing

TABLE 33: ORGANIC MATTER CONTENTS OF SOME HIGH ALTITUDE SOILS

Altitude (metres)	*Depth* (cm)	% O.M.	% N	C/N	*Remarks*
2750	0-15.2	38.2	1.64	15.5	Evergreen forest
	15.2-45.6	18.4	0.56	21.9	Evergreen forest
2750	0-10.2	33.8	1.10	20.4	Moorland
	10.2-20.4	20.8	0.71	19.5	Moorland
	20.4-38.0	14.4	0.38	25.2	Moorland
3650	0-10.2	7.0	0.18	25.9	Moorland
	10.2+	1.2	0.05	16.0	Moorland
3820	0-10.2	10.8	0.28	25.7	Moorland
	10.2+	6.4	0.17	25.0	Moorland
3650	0-30.4	12.4	—	—	Old moraine
4450	surface	3.4	—	—	Foot of glacier

altitude meant increasing rainfall and decreasing temperature both of which contributed to accumulation and slower decomposition of organic materials.

Jenny and Raychaudhuri (1960) in a study of over 500 soil samples from cultivated fields and forests throughout India in accordance with a climatic grid in which mean annual temperature and mean annual precipitation occurred as independent variables, found that climatic effects were very pronounced on soil carbon and nitrogen contents. Soil carbon and nitrogen rose with increasing mean annual precipitation, both in the hot coastal areas and in the cool Himalayan ranges. Strong negative correlations were observed between carbon, nitrogen and temperature. At high elevations the soils were much richer in organic matter than at sea level.

These workers found that the organic matter reserves of both virgin as well as cultivated Indian soils were higher than those of the United States, provided sites having equal annual values of temperature and precipitation were being compared, but that they were lower than those of Central America and the equatorial regions of South America. They concluded that the low level of organic matter in many Indian soils was primarily caused by environment, and only secondarily by cultural practices.

Carbon and nitrogen losses were most pronounced in areas having drier climates and long histories of human occupation. In such places, soil organic matter had reached a steady state at about 30-40 per cent of the original forest soil. Rice soils contained either equal or higher quantities of organic matter, seldom less, than related non-rice, cultivated soils; also they tended to be more acid. Calcareous soils were not richer in N than non-calcareous soils.

Besides high rainfall and low temperatures, there are other conditions under which organic matter accumulates in tropical soils. Soils derived from rhyolite and andesite ash are known to contain much organic matter and this has been attributed to the presence of amorphous clay minerals (Jackman, 1955.)

Volcanic ash soils which have large amounts of humus in the surface layers also contain much aluminium and this may be responsible for suppressing the microbial decomposition of the humus. Further, organic matter accumulates under anaerobic conditions caused by waterlogging as, for example, in flooded rice soils (Table 36).

Freshly drained soils in the tropics seldom, if ever, have a surface layer of decomposed organic material or vegetable mould. In spite of a large litter fall, the proportion of carbon on the forest floor to the total carbon in the profile is usually low, being about 1 per cent, as compared with 11 per cent for a temperate oak forest (Jenny, 1950). The leaf litter, woody residues and roots are rapidly attacked by ants and termites, fungi and other organisms and are converted partly into carbon dioxide and nutrients and partly to humus which is incorporated in the topsoils. Generally, the surface soil contains about 2-5 per cent organic matter and the next 15 cm about 0.5-2 per cent. The bulk of the organic matter is usually concentrated in a thin superficial layer of soil. The levels of organic matter in the soils of Ghana are given in Table 34.

TABLE 34: ORGANIC CARBON (C) AND NITROGEN (N) IN GHANAIAN SOILS (PER CENT)

	Forest			*Savanna*	
Depth cm	*Organic* C	N	*Depth* cm	*Organic* C	N
0-7	6.26	0.44	0-15	0.51	0.04
7-20	1.25	0.11	15-30	0.32	0.03
20-43	0.72	0.07			

In forest soils, the level in the surface horizon is fairly high, but falls rapidly with depth. In savanna, the level in the 0-15 cm layer is much lower than in forest, but the fall in the 15-30 cm, layer is relatively less steep.

The rate of decomposition of humus is much higher in the forest but in the savanna the supply of fresh material is greatly reduced by annual burning (Greenland and Nye, 1959). Nye (1959) found that about 2.5 per cent of the humic carbon was oxidized each year in forest soils compared with only 0.5-1.0 per cent in savanna soils. Russell (1960) estimated that the rate of loss in arable soils is nearer that in forest than in savanna soils. Nye and Greenland (1960) suggested that 1/5th to 1/10th of the carbon content of fresh organic material is converted to humus while Laudelot (quoted by Vine in Moss, 1968) estimated that about 50 per cent of straw and root residues of maize and rice was converted to humus.

The amount of humus in a soil depends on a balance between its rates of formation and decomposition. Its formation takes place at a fairly steady rate by conversion of a proportion of added plant material. Loss occurs through the action of micro-organisms on the humus so formed and on microbial residues. The rate of humus decomposition is dependent on its amount and ultimately an equilibrium level is reached. When the forest is cleared and crops are grown, the rate of addition of plant residues changes from about 11,200 kg dry matter per hectare each year to about 1,120-2,250 kg per hectare per annum. The rate of decomposition is high at first but decreases later so that after some years a new equilibrium level is attained. When cropped land is reverted to a natural fallow, humus decomposition is relatively slow at first but as the amount of humus present increases so does its rate of decomposition and a new equilibrium is reached.

Mohr (1922) considered 25°C to be the critical temperature below which organic matter tended to accumulate in well-drained tropical soils and above which it tended to disappear. Corbet (1935) reported that sunlight increased organic matter losses and a number of subsequent studies show that exposure to sunlight increases the rate of destruction of organic matter.

The organic matter contents in the different Great Soil Groups found in Sri Lanka are shown in Table 35. Soils of the wet zone are generally higher in organic matter than those of the dry zone and there is more soil humus at the higher elevations of the wet zone. Bog soils occupy lowland depressions along the west and south-west coasts. Elsewhere in the tropics, there is a similar range of values for soil organic matter (Table 36).

TABLE 35: ORGANIC MATTER IN THE GREAT SOIL GROUPS OF SRI LANKA
(Moorman and Panabokke, 1961)

Great Soil Group	% C	C/N
Dry Zone Soils		
Reddish Brown Earth	1.2	9
Noncalcic Brown Soil	1.3	10
Red-Yellow Latosol	0.4	7
Red-Yellow Latosol (calcic red subgroup)	1.1	8
Immature Brown Loam (dry zone)	1.7	12
Rendzina	1.7	9
Grumusol	2.0	12
Solodized Solonetz	1.2	15
Wet Zone Soils		
Reddish Brown Latosolic	4.1	10
*Red-Yellow Podsolic	1.6	10
Red-Yellow Podsolic (with a prominent A_1)	1.8	18
Red-Yellow Podsolic (with dark horizon)	9.4	41
*Immature Brown Loam (wet zone)	1.0	10
Meadow Podsolic	4.5	18
*Bog	14.5-17.5	

*Soils at higher elevations

TABLE 36 : ORGANIC MATTER AND NITROGEN CONTENTS OF TROPICAL SOILS

Location	*Depth* cm	*pH*	OM %	N %	C/N	*Reference*
1 Horton Plains,						
Sri Lanka	0-25	5.1	26.1	1.15	13.2	
2200 m ; 2130 mm,						
16°C ; granites and	25-50	5.1	18.1	0.80	13.2	Kalpage and
gneisses ; montane	50-63	5.2	4.6	0.22	12.2	Thenabadu
grassland ; midslope	63-89	5.2	0.9	0.11	5.0	(1969)
2 Masaiti, Malawi.						
1220 m ; 1300 mm,	0-6	6.3	1.50	0.064	23.5	
20°C ; woodland ;	15-20	5.8	0.64	0.035	18.2	
biotite gneiss ;	30-35	5.8	0.38	0.024	15.8	Webster (1965)
midslope.	54-60	5.6	0.52	0.028	18.6	
3 Usun Apau plateaux,						
Sarawak, Borneo.						
6600 mm 18.4°C	0-63	4.2*	1.9	0.13	8.4	Beckett and
shales and sand	63-130	2.7	1.4	0.13	6.1	Hopkinson
stones ; evergreen						(1961)
montane forest.	0-5	4.2	11.2	0.38	16.9	Wood and
690 m, steep slope	5-86	4.2	1.8	0.06	16.7	Beckett (1961)
1000 m.	86-130	4.3	1.5	0.05	18.6	
4 Ibadan, Nigeria.	worm casts	7.1	11.9	0.44	15.7	
1220 mm 26.6°C ;	0-5	7.7	6.8	0.29	13.6	
lowland rain forest ;	5-18	6.4	0.66	0.047	8.1	
coarse granites and	18-30	6.7	0.52	0.043	7.0	
gneisses ; mid-slope.	30-51	6.7	0.38	0.032	6.9	Bates (1960)
	51-76	6.7	0.36	0.026	8.1	
	76-102	—	—	—	—	
	102-127	6.5	0.36	0.030	7.0	
	152-183	5.6	0.08	0.016	3.1	
5 Hawaii Island, Hawaii.	0-5	6.2	22.5		11	
980 m ; 1040 mm,	5-13	7.5	8.8		8	
18.4°C ; grass ; basic	13-21	7.6	8.3		9	USDA (1960)
volcanic ash	21-58	7.6	7.4		11	
	58-89	7.4	5.5		12	
	89-120+	7.4	5.0		11	
6 Matara, Ceylon.	0-15	6.7	13.1	0.47	16.1	
2196 mm, 26.7°C ;	15-25	6.8	17.8	0.54	19.1	Takutome (1970)
peaty muck rice soil.	25-61	6.4	35.3	0.87	23.5	
	61-71+	6.0	33.1	0.49	39.5	
7 Montserrat series,	0-15	6.8	4.1	0.27	8.9	
Trinidad. 2300 mm,	15-30	6.4	2.4	0.16	8.7	
25°C ; calcareous	30-61	6.0	2.1	0.12	10.0	Ahmad *et al.*
glauconitic sand	61-91	6.7	2.4	0.11	12.9	(1968 a)
stone ; 5-15° slopes,	91-122	6.7	3.4	0.14	14.2	
cultivated with cocoa.						

*In M/100 $CaCl_2$

9.2 Contribution of Organic matter to Soil Properties

The effectiveness of humus in influencing soil properties and hence soil fertlity depends on humus quality as well as on its amount. Humus quality is not easy to determine or to understand. Such properties as physical nature, CEC, *p*H, percentage base saturation, humic acid/fulvic acid ratio, carbon/nitrogen ratio, have been used in describing humus. Some of these characteristics for the high organic matter content soils of Sri Lanka are given in Table 37

9.2.1 Plant Nutrients

The significance of soil organic matter as a source of inorganic plant nutrients has been well established by the classical experiments at Rothamsted and confirmed by numerous experiments elsewhere, including the tropics. Both fresh organic materials as well as soil organic matter (humus) release plant nutrients as they decompose, especially nitrogen, phosphorus and sulphur. In most soils 98 per cent of all nitrogen, 80 per cent of all sulphur and 60 per cent of all phosphorus are in organic forms. The greater rapidity with which organic materials generally decompose in tropical soils must imply a greater release of available nitrogen. Bradfield (1968) estimated that when a soil was planted to intertilled crops in temperate regions, about 2 per cent of the total store of humus was decomposed each year releasing about 2 per cent of the nitrogen and organic phosphorus stored in the soil. Nye and Stephens (1962) reported minimum values of 3 per cent of the total per annum in forest and 4 per cent in savanna on the long term fertilizer trial sites in Ghana. In most tropical soils higher rates are likely to prevail. Thus Cunningham (1962) found that crops on newly cleared forest land in Ghana failed to respond to nitrogen because the soil produced much mineral nitrogen and not because added nitrogen was rapidly leached.

Organic matter exerts a protective function in phosphorus fixation phenomena in acid tropical soils. Bhat and Bouyer (1968) reported that in a ferruginous soil from Senegal, deficient in organic matter and with weak fixing capacity, the addition of organic matter forced a larger amount of phosphorus into solution than in the case of a soil that had not been enriched in organic matter. The amount of isotopically exchangeable phosphorus was also greater. When a soluble phosphate was added to the soil, the quantity of phosphorus fixed was lower when organic matter was present. Above a certain dose of phosphate, corresponding to the maximum fixation capacity, all the phosphorus added remained capable of isotopic dilution. In contrast, in alluvial hydromorphic soils deficient in organic matter and with a high fixing capacity, the organic matter added led to an equal increase in the phosphorus content of the soil and in the amount of isotopically dilutable phosphorus but fixation was quite considerable and a portion of the fixed phosphorus was no longer capable of isotope dilution.

9.2.2 Cation Exchange Capacity

Organic matter makes an impotant contribution to the cation exchange capa-

TABLE 37: CHARACTERISTICS OF ORGANIC MATTER IN SOME HIGH ORGANIC MATTER CONTENT SOILS OF SRI LANKA (Handawela *et al.*, 1970)

Location			*Great Soil Group*	% OM	% C	% N	C/N	% *Humus*	*Humic C* / *Fulvic C*	pH	% *Humus C* *in org.* C	% *Humus* N in *org.* N
1 Bopatalawa	Wet Montane	1500 m	Red-Yellow Podsolic	14.4	8.0	0.66	12.2	7.3	0.72	4.5	53.0	72.0
2 Ambawela	,,	1800 m	,,	8.7	5.1	0.47	11.1	3.8	0.51	5.0	43.8	58.9
3 Horton Plains	,,	2100 m	,,	23.2	13.5	1.01	14.6	9.1	1.89	5.5	37.6	38.6
4 Muthurajawela	Coastal swamp		Bog soil	60.0	34.8	1.08	35.7	16.2	3.49	6.0	33.0	42.8
5 Kotte	River valley swamp		,,	52.0	30.2	0.95	34.9	10.6	4.94	5.4	32.7	39.7
6 Bombuwela	Valley swamp		,,	23.6	13.7	0.56	26.2	7.3	3.93	5.4	31.4	39.4

city of all soils. While values between 150-250 me/100 g organic carbon are common, values ranging from 30-60 me/100 g have been reported (Cooke, 1967.) For equal weights, the cation exchange capacity of organic matter is between 2 and 20 times that of clay.

The contribution of organic matter to cation exchange capacity is perhaps of greater significance in tropical than in temperate regions. In well-leached, intensively weathered tropical soils with clay minerals of the kandite group, organic matter can make all the difference to soil fertility. For example, Tinker and Ziboh (1959) found that CEC values of some Nigerian soils ranged from 4-23 me/100 g soil. The figures were directly proportional to organic matter contents and the CEC of the organic matter was equivalent to 500 me/100 g of organic carbon. For forest ochrosols and oxysols in Ghana, Nye and Greenland (1960) estimated CEC values of about 350 me/100 g organic carbon.

Cation exchange capacities varying from 10.0 to 67.7 me/100 g soil and directly proportional to organic matter content were reported by Kalpagé and Thenabadu (1969) for some montane soils in the wet zone of Sri Lanka. Sombroek (1966), studying the soils of the Amazon basin, found that CEC increased by about 3.9 me per 1 per cent carbon and was also positively correlated with the percentage of clay in the topsoil but not in the subsoil.

The dependence of cation exchange capacity on organic matter in tropical soils such as latosols stresses the importance of maintaining a satisfactory level of organic matter in these soils. When organic matter reserves are low, retention of plant nutrient cations is poor and fertility will decline. Bramao (1968) was of the opinion that one reason why soil fertility problems in South America are somewhat different from those in Africa and tropical Asia was because many South American tropical soils have a very low exchange capacity and low base saturation.

9.2.3 Carbon/Nitrogen Ratio

The C/N ratio generally indicates the availability of nitrogen in the organic matter and a C/N ratio of around 10 is considered to suggest satisfactory conditions for normal microbial activity and humus decomposition. This was not so in the long-term fertilizer experiments conducted by Djokoto and Stephens (1961*b*) in Ghana.

But not all crops are probably influenced in the same manner. Trees, in general, grow well with C/N ratios up to 35. Above this, nitrogen deficiencies may occur (Bramao and Riquier, 1968).

Nye and Greenland (1960) found that the C/N ratio was characteristic of the major soil groups they studied. In the forest ochrosols (Reddish Brown Earths) in which bacterial activity was high, it had a value of 8-12 similar to that of temperate regions. In the acid oxysols (Red-Yellow Podsolic soils) under forest, the C/N ratios were higher, 14-17. High C/N values (15-20) were also obtained on the neutral or slightly acid ochrosols under long-established high-grass (*Andropogoneae*) savanna. Similar results were reported earlier by Hardon (1936) for old secondary or virgin forest soils in the lowland areas of Sumatra.

In accordance with the expectation that soils of low C/N value mineralize rapidly, the humus of the forest ochrosols was shown to undergo rapid decomposition. While the general level of microbiological activity in the oxysols was much lower than in the ochrosols there appeared to be no reason for believing that in most oxysols the higher C/N ratios were necessarily associated with low mineralization rates. Under cropping, however, nitrification in these oxysols was rapid and thus their C/N ratio was no guide to their ability to supply crops with available nitrogen. The high C/N ratio in the high-grass savanna ochrosols was however associated with a low rate of mineralization (0.5-1.2 per cent per annum) and this persisted to some extent after the vegetation was cleared and the land cultivated. Odu and Vine (1968) refer to the possibility that the high C/N ratios in these savanna soils indicate some peculiarity in the conditions of 'turnover' of elements, especially nitrogen.

9.2.4 Soil Structure

Soil structure is a parameter of soil fertility favourably affected by organic matter in temperate regions. But it is somewhat of an unknown quantity in soil fertility in the tropics. Structure determines plant-soil-water relationships, soil aeration and root proliferation.

Structure is characterized by soil aggregation and pore space relations. Baver (1968) has discussed the effect of climate on soil aggregation. The extent to which the silt and clay fractions are aggregated was found to increase with rainfall from a minimum of about 15 per cent in desert soils to a maximum of about 50 per cent in the chernozems, with a decrease thereafter to approximately 25 per cent in podsols. The increase reflects an augmentation of the clay and organic matter contents of the soil as a result of the higher rainfall. The decrease from the maximum corresponds to the eluviation of colloidal particles and a lower organic matter content in the surface layers of podsols as there is a change from grass to forest vegetation. Aggregation decreases with increasing temperature corresponding to a lower organic matter content.

The distribution of pore size has an important bearing on soil structure especially in its influence on soil-air-water relationships. Non-capillary porosities below 10 per cent are restrictive to adequate aeration while values between 20 and 30 per cent are optimum.

Organic material, without biological transformation has little, if any, effect on soil structure. Micro-organisms without organic materials as sources of energy are ineffective in producing soil aggregation. The intense microbial activity consequent to the incorporation of organic materials in the soil is associated with metabolic processes of degradation and synthesis. The sum total of these effects is the production of stable soil aggregates. This stability can result from the mechanical binding action of the cells and filaments of organisms, the cementation effects of the products of microbial synthesis, or the stabilizing action of the products of decomposition, acting individually or in combination.

It would seem as if much of the humus of tropical soils has relatively little effect in stabilizing soil structural aggregates. The cementing gums produced by various bacteria are probably fairly readily decomposed by the soil microbes. Steady or frequent addition of fresh organic material to support vigorous activity of the gum-forming bacteria would therefore be necessary to maintain surface soil structure.

In east Africa, grass fallows were thought to improve soil structure but the investigations of Pereira, Chenery and Mills (1954) and Pereira (1955) have shown that the physical effects of grasses do not last long, even on clayey soils, but disappear largely during the first year of cultivation. On the other hand, mulches (see 12.3.2 & 14.2.3), were found to be beneficial in improving soil structure. A stable surface structure was built up under a mulch by microbial activity aided by the mechanical protection of the soil from the beating rain.

9.2.5 Soil Moisture Availability

Soil organic matter has a favourable effect on soil moisture availability. Much of the tropics has a dry season of varying duration, in which the rainfall is too small and too irregular to provide for uninhibited plant growth. In such periods productive cropping depends on effective exploitation of soil moisture. For this, the maintenance or increase of the soil organic matter level is an important management measure.

It is clear that soil organic matter contributes just as much, if not more, to the fertility of tropical soils as to that of temperate soils. Maintenance of an adequate level of organic matter is however not as easy under the high temperature conditions prevalent in the tropics. In view of the rapid rate of organic matter turnover, large additions would merely lead to faster decomposition. The equilibrium level of soil organic matter can only be raised by minimizing soil disturbance, by growing crops that leave large residues and by returning all possible residues to the land (Jones, 1971). Continuous applications are needed since the residual effects, if any, are much less than in temperate regions. Green manures, composts, farmyard manure, cover crops, shade trees and mulches should all be employed according to availability and local conditions for increasing equilibrium levels of soil organic matter.

9.3 Soil Organisms

9.3.1 Soil Fauna

Of the larger soil animals, ants and termites are probably more important in tropical soils than even earthworms whose importance in temperate soils is well-known. Nye (1955-6) working in an area of rather small termite mounds, estimated that 1.2 tons of subsoil per hectare per year were transported to the surface by termites, as compared with an estimated 50 tons per hectare of earthworm casts produced during the six-month rainy period by the worm *Hippopera nigeriae* on a neighbouring area under bush. The mound of *Macrotermes nigeriensis* studied by Nye (1955c) was about 4 ft (1.2 m) high and 8 ft (2.4 m) in diameter whereas Russell (in Nye, 1955-6) considers that in the 'miombo' areas of equatorial and southern Africa the mounds of *Macrotermes* Spp. can be up to 25 ft (7.6 m) high and 40-60 ft (12-18 m) in diameter. Such mounds may contain hundreds of tons of earth above soil level and commonly occur at a spacing of about one per 1.5 acres (0.6 ha).

9.3.1.1 ANTS AND TERMITES: Ants and termites have many features in common. They are both social insects living in nests, which are in the soil for many generations. Some are mound builders whereas others are not. Mound builders bring subsoil to the surface to build their mounds but do not mix this with humic matter as earthworms do.

Ants differ from termites in that they are not confined to tropical and subtropical regions, but are common inhabitants of temperate zone soils as well. Their feeding habits are also different. Ants are predaceous and feed on soil fauna as well as on insects feeding on plants. Some genera of ants cultivate aphids or coccids in their nests and feed on their exudates. Termites, on the other hand, use wood as the base on which to grow their fungal gardens.

Black ants of all sizes are also active in depositing loose sandy loam on the surface. In some areas the superficial earthmoving activities of ants are overshadowed by the earthworms while, in other areas, where there are fewer worms, the topmost layers of the soil are formed largely of earth deposited by ants.

In many tropical soils, termites are the dominant soil animals. There are many different kinds of termites such as the dry-wood termites, the wood-feeding termites, the fungus-growing termites and the humus-feeding termites. The last two categories have most effect on the soil (Harris, 1954). Some species of termites have much ability to penetrate through hard pans and possibly into laterite crusts. The characteristic structure of vesicular laterite has been attributed to the activity of the termites removing the finer soil particles from lateritic layers in the subsoil (Sys, 1955).

Nye (1955c) has described the activities of *Macrotermes nigeriensis* in the area around Ibadan, Nigeria. Watson (1962) described the soil below a termite mound in southern Rhodesia.

From the mechanical composition of the mounds it was evident that the termites appeared reluctant to carry particles larger than 2 mm and were unable to carry those larger than 4 mm. The earth particles that had passed through the bodies of the termites were not larger than 0.5 mm. The material of the mounds was on the average brought up from between 30 and 75 cm depth. The termites did not seem to use the humic surface, 15 cm deep, as building material.

In chemical composition, the casing of the mounds differed little from the nearby soil except that the organic carbon contents were more, due possibly to the secretions used to stick the pellets together. In the nest, however, the organic carbon, *p*H, exchangeable Ca and Mg, as well as the base saturation were all more than in the neighbouring soil. Harris (1961) considered the presence of lime-rich termite mounds in east Africa a purely local phenomenon. Mounds were merely heaps of subsoil with an altered soil structure and in cases where termite mounds were more fertile than the surrounding soil it was because the subsoil was more fertile than the topsoil.

In Ibadan, the standing mound did not support vegetation, probably because the casing was too compact and dried out rapidly. The collapsed mound had a topsoil poorer in nutrients and more compact than the surrounding topsoil but it was possible that the subsoil below the mound was richer. Crops such as maize grew relatively poorly on collapsed mounds.

The most important activity of the termites on the Ibadan soil was the formation of a gravel-free topsoil by movement of the finer soil particles from depth to the surface.

Robinson (1958) investigated the activities of the termite *Odontotermes badius* (Hav.) in the soils of the drier coffee producing areas of Kenya, where termites hasten the decomposition and destruction of grass mulches used foı coffee. They found that the termite soil *in situ* was more closely related to the topsoil than to the subsoil samples. Moreover, the termite soil material had a higher percentage content of calcium plus magnesium in the total exchangeable bases, and a higher *p*H value than either the surrounding topsoil or the subsoil.

9.3.1.2 EARTHWORMS : Earthworms were found to be important in forming a surface horizon of fine sandy loam. The species found in the Ibadan area was *Hippopera nigeriae* which occurred in all sizes up to or exceeding 25 cm long and 8 mm in diameter. Their casts were irregular cylinders up to 6 cm long and 1.5 cm in diameter, standing vertically on the surface of the soil and dark brown in colour. There was a small hole running through the centre of the cast almost to the top, which was sealed. These worms were extremely active during the rainy season but there was no sign of casting during the dry season except near streams. The earth from the decomposing casts was sufficient to form a well-defined layer 1.25 to 2.5 cm thick of fine sandy loam above the coarse sandy loam below, both under secondary forest and under tall grass.

The casts contained virtually no grains greater than 0.5 mm and a low proportion between 0.2 and 0.5 mm. The maximum depth from which the earth in the casts was being obtained was around 30 cm, the earthworms operating mainly in the surface soil. The casts contained more organic matter than the topsoil, had a greater CEC value, base saturation, and *p*H. The exchangeable Ca, Mg and Mn contents were greater in the casts, while the exchangeable Na and K were about the same. The casts were in fact an intimate mixture of digested earth and plant remains.

The physical and chemical composition of casts by *Pheretima tumulifaciens* Lee, a large earthworm found on lowland soils bordering the Sepik river floodplain in New Guinea showed that the cast material was from the upper and not from the underlying soil horizons (Lee, 1967).

9.3.1.3 NEMATODES : A number of plant-parasitic and soil-inhabiting nematodes cause much damage to a variety of tropical crops. Their distribution in tropical soils and in plant roots was reviewed by Peachey (1969). Soil fumigants are used in controlling a number of these nematodes. Crop rotations including fallows are effective against others (e.g. Scotto La Masses, 1969; Edmunds, 1970). Winchester (1968) reported that nematode populations in tropical rain forest soils of Costa Rica, Puerto Rico and Rio de Janeiro decreased during dry periods and increased with the rains ; also, that they were generally more abundant in cultivated than in forest soils. Dao (1970) found that the nematode faunas of tropical Venezuela and the Netherlands differed conspicuously in species and that the distribution range of nematodes was determined largely by climate, and especially by temperature.

9.3.2 SOIL MICROFLORA

9.3.2.1 BACTERIA : Bacteria play an important role in the nitrogen, phosphorus and sulphur cycles in soils.

Nitrogen fixation by symbiotic and non-symbiotic organisms has already been discussed (pp.89). The role of bacteria in other aspects of the nitrogen cycle has also been referred to earlier (pp.91).

9.3.2.2 ALGAE : Fuller *et al.* (1960) found that many blue-green algae growing on desert crusts in Arizona were autotrophic both with respect to nitrogen as well as carbon. Algal and lichen crusts had four to five times as much nitrogen as the soil below. The algae fixed nitrogen which was then available to plants.

Most of the nitrogen in flooded rice fields is fixed by blue-green algae growing on the soil surface or floating on the surface of the irrigation water. De (1939) showed that algae fix nitrogen in rice soils, and Watanabe *et al.* (1951) demonstrated that thirteen species of blue-green algae, all common in south-east Asia, were capable of fixing nitrogen.

Azotobacter may be capable of fixing some nitrogen in flooded soils if they are in association with living rice roots from which oxygen diffuses outwards but algae were the main nitrogen fixers (De, 1939)

Estimates of the amounts of nitrogen fixed by algae in flooded rice fields have varied. Watanabe *et al.* (1951) estimated fixation at the rate of 22.5 kg of N per hectare in an unspecified period during the growth of one rice crop. De and Sulaiman (1950) found that the nitrogen content of a soil increased from 895 to 1014 ppm in the presence of algae but declined to 814 ppm in the absence of algae.

In a study on the distribution of micro-organisms in samples of soils from **Malyasia, Philippines and Thailand, Kobayshi *et al.*, (1967) recognized the presence of nitrogen fixing algae *Tolypothrix* and *Nostoc*.**

Kobayashi *et al.* (1967) found Azotobacter and Clostridium in rice soils from four south-east Asian countries. Photosynthetic bacteria were also widely distributed in these soils and there was evidence of some symbiotic association between *Athiorhodaceae* and *Azotobacter*, after a few weeks of flooding, at the tillering stage of the rice plant.

9.3.2.3 FUNGI : Fungi causing diseases in tropical agriculture were outlined by Wrigley (1969). Soil fungi must depend for their energy and carbon on the soil organic matter and they play an important role in humus formation and decomposition as well as in soil aggregate stabilization. Associations between some fungi and the roots of higher plants, *mycorrhizae*, are widespread and of practical significance. They are especially pronounced in soils low in phosphorus and nitrogen, and high nutrient levels are correlated with poor mycorrhizal development.

Work on fungi in tropical soil has been limited (Robinson, 1970). Farrow (1954) listed fungi in soils from Panama and Costa Rica. Goos (1960, 1963) and Goos and Timonin (1962) studied the fungal flora of the banana rhizosphere in Latin America. Mills and Vlitos (1965, 1967) investigated the soil and rhizosphere flora of sugarcane in Trinidad. Khan and Siddiqi (1962) reported on the fungal population of soil and roots of sugarcane in India.

10 Soil Use in the Tropics

10.1 Land Use Patterns in the Tropics

The 'developing countries', classified as Zone C in the Provisional Indicative World Plan (IWP) for Agricultural Development (FAO, 1970b), are for the most part in the tropical and subtropical region. It was estimated that only a little over a tenth of the land surface in these countries in 1961-63 was being used for cultivation including both annual and permanent crops (Table 38), about a fifth of the land was used for permanent pasture (mainly natural grassland) and almost a third was under forest, mostly of very poor quality. Thus about 60 per cent of the land surface was being used for some agriculture and forestry. The remainder was too arid or too mountainous for such use or was put to non-agricultural uses.

TABLE 38: Pattern of Land Use in the Tropics and Subtropics

	Million Hectares	*Per cent of Total Area*
Arable land and permanent crops	563	11
Permanent pasture	1106	21
Forest	1447	28
Used for some agricultural purpose	3116	60
Not used for any agricultural or forestry purpose	2126	40
Total	5242	100

The broad regional contrasts are shown in Table 39. In Asia and the Far East, the large proportion of land under crops and the very small proportion under permanent pasture reflect the fertility of soils as well as population pressures on land resources. The figures for the Near East indicate the large proportion of arid land in this region. The low figures for South America reflect poor land utilization rather than poor natural resources. To some extent, this is true of Africa south of the Sahara.

Within the tropics, arable land is only 18 per cent of the area actually used for agriculture and yet much of the production comes from this land. A poor intensity of cropping is characteristic of Latin America, Africa and the Near

TABLE 39: REGIONAL DISTRIBUTION OF LAND USE IN THE TROPICS
(per cent of total area)

Region	*Arable*[1]	*Permanent Pasture*	*Forestry*	*Not used for Agriculture*
Africa south of the Sahara	10	34	25	31
Asia and Far East	39	3	26	32
Latin America[2]	7	20	40	33
Near East and N.W. Africa	6	13	11	70

[1]Including tree and other permanent crops.

[2]Crops and pasture cover 20% of the land area in most countries of tropical America e.g., Peru and Brazil: crops 1.4 and 2.2%, forest 56 and 71%. In Ecuador and Panama 60% is forested (Grobman, in Turk and Crowder, 1967).

East (Table 40). In these countries, and even in Asia where cropping intensities are higher, much attention is being given to improving this. Cropping intensity is naturally more on irrigated land but in many areas the intensity of use of

TABLE 40: CROPPING INTENSITIES[1] (1961-1963)

Region	*Arable Land Million ha.*[2]	*Area Harvested Annually Million ha.*[2]	*Cropping Intensity Percentage*
Africa south of the Sahara	152	64	42
Asia and Far East	211	211	100
Latin America	130	71	54
Near East and N.W. Africa	70	39	56
Total or average	563	385	68

[1]Cropping intensity $= \dfrac{\text{Harvested area}}{\text{Arable area}} \times 100$

[2]Includes perennial crops

irrigated land is less than the desirable level (Table 41).

Table 40 shows that fallows are a part of the land use pattern in most areas. These are generally of two types: grass or bare fallows in poor rainfall areas of the Near East and Latin America, and long term fallows of shifting cultivation in tropical forest zones in Africa south of the Sahara and in parts of South America.

The proposals for land and water development and use made by the IWP

TABLE 41 : CROPPING INTENSITIES WITH AND WITHOUT IRRIGATION

	Arable Land Million hectares	*Area Harvested Annually Million hectares*	*Cropping Intensity Percentage*
Irrigated	73	72	99
Non-irrigated	490	313	64
Total	563	385	68

mostly favours intensifying rather than adding new land in most of the tropical and subtropical regions, mainly for economic reasons. To achieve this, the present limitations need to be eliminated or mitigated, e.g., water shortage, poor soil fertility.

Annual (food) crops—Table 42 shows the wide range of agricultural products of countries within the tropical zone. A few of the crops, like wheat and potatoes, are grown mainly outside the tropics; tobacco and groundnuts are also found to a large extent in subtropical and warm temperate regions. But world supplies of commodities like tea, rubber, coconut, oil palm and sugarcane come exclusively from the tropics.

TABLE 42 : AGRICULTURAL PRODUCTS OF COUNTRIES WITHIN THE TROPICS

America entirely within the tropics	
Barbados	Sugarcane
British Guinea	Rice, Sugarcane, Citrus, Timber
British West Indies	Sugarcane, Spices, Bananas, Cocoa, Cotton
Colombia	Coffee, Cocoa, Bananas, Cotton, Sugarcane, Soyabean, Oil Palm, Beans
Costa Rica	Bananas, Coffee, Cotton, Cocoa, Sugarcane, Rice, Maize
Cuba	Sugarcane, Tobacco, Coffee, Cocoa, Potatoes, Fibre, Timber
Dominican Republic	Sugarcane, Bananas, Rice, Tobacco
Guatemala	Coffee, Cotton, Bananas, Sugarcane
Guyana	Sugarcane, Rice
Haiti	Coffee, Bananas, Sugarcane, Sisal, Cotton, Tobacco
Honduras	Bananas, Maize, Rice, Cotton, Coffee
Jamaica	Sugarcane, Sisal
Nicaragua	Cotton, Coffee, Sugarcane
Panama	Bananas, Sugarcane, Coffee, Rice, Maize
Peru	Maize, Wheat, Barley, Sugarcane, Cotton, Rice

Puerto Rico	Sugarcane, Citrus, Pineapples
Surinam	Rice, Sugarcane, Citrus, Bananas
Trinidad and Tobago	Tobacco, Sugarcane, Cocoa, Citrus
Venezuela	Rice, Sugarcane, Cotton, Cocoa, Bananas, Citrus
Virgin Islands	Fruit, Vegetables

partly within the tropics

Argentina	Sugarcane, Maize, Alfalfa, Wheat
Bolivia	Potatoes, Wheat, Rice, Cotton, Coeffe, Maize
Brazil	Coffee, Cocoa, Sugarcane, Rubber, Oil Palm, Sisal, Bananas, Citrus, Tobacco, Soyabean, Beef cattle
Chile	Wheat, Barley, Oats, Maize, Rice, Potatoes, Pulses
Mexico	Cotton, Coffee, Maize, Rice, Citrus, Sugarcane Alfalfa, Tomatoes
Paraguay	Cotton, Wheat, Soyabean, Tobacco, Rice, Sugarcane

Africa
entirely within the tropics

Angola	Coffee, Sisal, Maize
Burundi	Coffee, Cotton
Cameroon	Cocoa, Oil Palm, Bananas, Coffee, Cotton
Central African Republic	Cotton, Coffee, Groundnuts
Chad	Cotton
Congo (Brazzaville)	Oil Palm, Tobacco, Bananas
Congo (Democratic Republic of)	Oil Palm, Coffee, Rubber, Tea
Dahomey	Oil Palm, Coffee, Groundnuts, Cotton
Equatorial Guinea	Cocoa, Coffee
Ethiopia	Coffee, Groundnuts
Gabon	Groundnuts, Oil Palm
Gambia	Groundnuts, Millets, Rice, Oil Palm
Ghana	Cocoa
Guinea	Bananas, Oil Palm, Coffee
Ivory Coast	Coffee, Cocoa, Bananas
Kenya	Coffee, Sisal, Tea
Liberia	Rubber, Oil Palm
Malawi	Tobacco, Tea, Groudnuts, Cotton, Maize
Mali	Groundnuts, Cotton
Niger	Groundnuts
Nigeria	Groundnuts, Cocoa, Oil Palm, Rubber, Cotton
Portuguese Guinea	Groundnuts, Oil Palm, Cocoa, Coffee, Rice
Ruanda	Coffee, Cotton
Rhodesia	Tobacco, Tea, Groundnuts
Senegal	Groundnuts, Millets, Rice
Sierra Leone	Oil Palm, Coffee, Cocoa, Rice, Millets, Sorghum, Maize, Cassava
Somalia	Bananas
Sudan	Cotton, Millets, Groundnuts
Tanzania	Sisal, Coffee, Cotton, Spices (Cloves), Cashew
Togo	Oil Palm, Coffee, Cocoa

Uganda	Coffee, Cotton, Tea, Maize, Sweet potatoes, Groundnuts
Upper Volta	Groundnuts, Cotton
Zambia	Tobacco, Groundnuts
partly within the tropics	
Algeria	Grapes
Botswana	Pulses, Sorghum, Cattle
Libya	Groundnuts, Olives, Esparto grass
Madagascar	Rice, Coffee, Vanilla, Sugarcane, Sisal, Tobacco
Mauritania	Sugarcane
Mauritius	Groundnuts
Mozambique	Cashew, Cotton, Sisal, Sugar, Coconut, Tea
South Africa	Sugarcane, Bananas, Citrus
South-west Africa	Dairy cattle
United Arab Republic	Cotton, Rice
Asia/Australia	
Bangladesh	Rice, Cotton, Pulses, Sugarcane, Jute
Borneo[1]	Rubber, Rice, Coconuts, Sago, Tapioca
Cambodia	Rice, Maize, Rubber, Sugarcane, Groundnuts
Fiji	Sugarcane, Coconuts, Bananas
Hawaii	Sugarcane
Hong Kong	Vegetables, Sweet potatoes, Groundnuts, Fruits
Indonesia	Rice, Cassava, Maize, Sweet potatoes, Soya beans, Coffee
Laos	Rice, Maize, Tobacco, Coffee
Malaysia	Rubber, Rice, Coconuts, Oil palm
Maldives	Millets, Coconuts
New Guinea[2]	Coconuts, Coffee, Cocoa
Pacific Islands	Coconuts, Bananas, Yams
Philippines	Sugarcane, Rice, Maize, Coconuts
Portuguese Timor	Maize, Rice, Sweet potatoes
Singapore	Rubber, Coconuts, Vegetables
Sri Lanka	Rice, Tea, Rubber, Coconuts
Thailand	Rice, Rubber, Maize, Kenaf, Sugarcane, Tobacco, Cotton
Vietnam	Rice, Rubber, Maize, Tobacco, Sugarcane, Sweet potatoes
Yemen (Arab Republic)	Dates, Coffee, Tobacco
Yemen (Peoples' Democratic Republic)	Dates, Coffee, Tobacco
Australia	Sugarcane, Maize
Burma	Rice, Sugarcane, Pulses, Groundnuts, Tobacco
China	Rice, Sugarcane, Tea, Soyabeans, Cotton
India	Rice, Pulses, Groundnuts, Sugarcane, Tea, Chillies
Oma	Dates
Saudi Arabia	Dates, Coffee, Tobacco
Taiwan	Rice, Sugarcane, Sweet potatoes, Groundnuts, Vegetables, Fruits

[1]including Sabah and Sarawak, which are a part of Malaysia.

[2]a portion belongs to Indonesia while the remainder is a protectorate of Australia.

The areas of land under food crops is shown in Table 43.

TABLE 43 : LAND UNDER FOOD CROPS IN 1962 (thousand hectares)

Region	*Starchy roots*	*Pulses dry*	*Ground-nuts*	*Other oil crops*	*Vegetables*	*Fruits*
Asia	1763	26018	6962	8975	6509	2937
Africa south of Sahara	7784	7339	4414	—	1228	2679
Latin America	3215	5775	801	1019	858	2616
Near East and north Africa	152	1472	1	2220	1037	1761
Total	12914	40604	12178	12214	9432	9993
World	—	—	17000	27000	—	—

Almost all the cassava and yams produced come from the tropics, mainly from Africa. Starchy roots also include about 4/5ths of the total world production of white potatoes, which is the staple diet in parts of Latin America where, in 1962, about 7 million tons were produced. White potatoes constitute the most important root crop in India and Pakistan.

Chickpeas are the main pulses grown in the Near East and in south Asia; cowpeas are popular in tropical latitudes, principally in Africa. Haricot beans are also grown widely.

Groundnuts constitute the main edible oil crop in the tropics. But, out of a total world acreage of 44 million hectares under groundnuts and other oil crops, such as soya beans, only about 24 million hectares are in the tropics. The remainder are in the southern United States, China and Japan.

Perennial crops— About half the earnings from the agricultural exports of tropical and subtropical countries in 1961-63 came from perennial crops; in Asia and Africa south of the Sahara, their contribution was 70 and 55 per cent respectively. The areas of land under perennial crops is given in Table 44.

TABLE 44 : LAND UNDER PERENNIAL CROPS IN 1962 (thousand hectares)

Region	*Oil Palm*	*Coconut*	*Cocoa*	*Coffee*	*Tea*	*Rubber*
Asia			21	159	647	
Near East				42	22	
Africa south of the Sahara			3071	1934	49	
South America			829	5794	21	
Central America			—	977	—	
Total			3921	8907	739	

Perennial crops are grown mostly on a regional basis, between 70 and 85 per cent of each crop being produced in a particular region (Table 45).

TABLE 45: REGIONAL DISTRIBUTION OF MAJOR TREE CROPS IN 1962 (as per cent of total production)

Region	*Palm-oil*	*Coconuts*	*Cocoa*	*Coffee*	*Tea*	*Rubber*
Africa south of the Sahara	85.9	6.7	74.5	16.5	6.4	8.7
Asia	10.8	74.9	0.6	2.6	90.1	88.7
Central America	2.1	12.7	3.5	11.8	—	—
Near East	—	—	—	0.5	1.9	—
South America	1.2	5.9	21.4	68.6	1.6	2.6

Again, only a few countries produce most of the crops. Thus only three supply 65-85 per cent of all the crops, and two produce 80 and 75 per cent respectively of the tea and rubber from tropical countries (Table 46).

TABLE 46: MAJOR PRODUCERS OF THE PRINCIPAL PERENNIAL CROPS (IN 1962)

Palm oil	Nigeria, Congo-Kinshasa, Malaysia
Palm-kernels	Nigeria, Congo-Kinshasa, Dahomey
Coconuts	Philippines, India, Sri Lanka
Cocoa	Ghana. Nigeria, Brazil
Coffee	Brazil, Colombia, Ivory Coast
Tea	India, Sri Lanka, Pakistan
Rubber	Malaysia, Thailand, Sri Lanka

10.2 THE FERTILITY OF TROPICAL SOILS

Any one of the classifications in Table 47 may be adopted in an exposition of tropical land use.

A discussion of fertility problems arising from the different uses to which tropical soils are put can, however, be more conveinently made under the following headings:

Soils under shifting cultivation
Soils under alternative forms of upland arable farming
Soils under rice culture
Soils under perennial crops in plantations
Soils under tropical grassland

TABLE 47: CULTIVATION SYSTEMS IN THE TROPICS
(Ruthenberg, 1970)

According to:

1. Type of rotation
   e.g. Fallow
   Ley
   Field crops
   Permanent crops

2. Intensity of rotation
   e.g. Shifting cultivation
   Semi-permanent cultivation
   Permanent farming

3. Water supply
   e.g. Irrigation farming
   Dry farming or Rainfed farming

4. Cropping pattern and animal activities

5. Implements used for cultivation

6. Degree of commercialization
   e.g. Subsistence farming
   Partly-commercialized farms
   Semi-commercialized farms
   Highly-commercialized farms

The luxuriant and vigorous growth of tropical vegetation led to the early belief that tropical soils had much inherent fertility. After some striking and sometimes disastrous experiences that showed that this was not so, soil fertility in the humid tropics was considered poor and incapable of producing food crops over a period of time.

Stamp (1953) discussed two main types of tropical soils; rain forest soils and savanna soils. He referred to the incompletely decomposed organic matter in rain forest soils and their rapid exhaustion by excessive leaching when the forest cover is removed. In savannas, Stamp mentions the formation of surface hardpans, the abrasive effects of angular quartz grains on agricultural machinery and the formation of alkali in irrigated land. Pierre Gourou (1953) thought that tropical soils were nearly always very infertile, poor in plant available bases and phosphorus, in humus because of rapid oxidation, and losing nitrate N and bases rapidly by leaching. He described 'laterization' as a 'pedological leprosy which occurs only in hot, wet climates with an alternation of dry and rainy seasons' and thought that 'a good deal of the hot, wet regions is covered with soils which are highly lateritic or of pure laterite' — the laterite being 'utterly infertile'. He thought that, more than half of the land in the tropics was worthless, that sandy soils formed from sandstones were of little use even when not developing laterite, and that the very severe soil erosion that can occur in the wet tropics was unavoidable.

The wide diversity of climatic regimes in the tropics has been referred to earlier. A general statement regarding the fertility of soils in all parts of the tropics is clearly not possible. What applies to the hot wet equatorial forest regions, is not likely to be valid in the semi-dry savanna grasslands. Stamp (1953) drew a distinction between tropical rain forest soils and tropical savanna soils while Gourou (1953) was inclined to consider all tropical soils as being worthless.

Three broad ecological regions can be identified : the hot wet regions, the hot dry and semi-dry regions, and the high altitude regions. The problems of soil fertility in each of these areas are clearly different. The soils of the hot wet regions are acid, leached of the essential plant nutrients but contain substantial amounts of humus. The dry region soils are less acid, even neutral or alkaline, and are poor in humus. Salts may accumulate and irrigation is necessary to supplement an inadequate, irregular or unreliable rainfall. Some of the most fertile soils and best managed farms and plantations are found in the wetter and higher regions. For instance, the Kenya Highlands of the eastern rift, mainly more than 1,500 m high, have deep volcanic soils with rich natural reserves of plant nutrients under a rainfall of 1020-1520 mm. These regions, within which alluvial and volcanic soils are usually very fertile, are now discussed in turn.

10.2.1 Hot Wet Tropical Regions

Vine (1954) produced evidence on the forest zone of Nigeria, to show that Stamp's and Gourou's disparaging descriptions of tropical soils were not universally applicable. Dennison (1959), reviewing work in the southern Guinea zone of nothern Nigeria, found that fallows, cattle manure and artificial fertilizers were effective in maintaining soil fertility in this area; only short-term green manure crops proved a failure.

Analyses of forest soils (Table 48) from moderate-(1320 mm), intermediate -(1780 mm), and high-(3300 mm) rainfall regions showed that examined profiles had much humus. The soils in the high-rainfall sites had lower *p*H values and less exchangeable bases than soils in moderate rainfall sites. In all profiles, *p*H and exchangeable bases were most in the surface layer, but more so in the moderate rainfall sites. C/N ratios were small in the soils in the moderate and intermediate sites but larger in the wetter areas. A ratio of 10-12 in the drier regions indicated optimum conditions for micro-organic activity and nitrification.

Long-term experiments at Ibadan (1220 mm rainfall) demonstrated that these soils could be very productive under intensive cropping. Continuous rotational cropping using only leguminous green manure— chiefly *Mucuna deeringiana* — showed a definite but gradual decrease in yields. Yields eventually declined seriously in some rotations, especially where the green manure crops in the rotation had been reduced. The main factor responsible for this

TABLE 48: Analyses of Some Nigerian Forest Soils (Vine, 1954)

Depth cm	% Stones and Gravel	% Coarse sand	% Fine sand	% Silt plus Clay	pH	Base-Exchange Capacity me per 100g	Total Exchange-able Bases me per 100g	% Carbon	% Nitrogen	C/N Ratio
A—MODERATE-RAINFALL LOCALITY										
A (1). Clayey Upper-slope Soil (Egbeda Series)										
0-10	1	34	35	31	6.3	15.0	13.3	2.64	0.257	10.3
10-23	9	31	32	37	5.8	7.5	5.7	0.76	0.082	9.3
23-58	47	33	20	47	5.7	6.4	4.4	0.34	0.042	8.1
58-84	20	20	14	66	5.4	6.9	4.4			
84-	6	14	14	72	5.2	6.8	4.5			
A (2). Sandy Lower-slope Hill-wash Soil (Apomu Series)										
0-5	0	26	59	15	6.4	7.7	6.3	1.58	0.137	11.5
5-23	6	40	42	18	5.4	3.8	2.1	0.40	0.046	8.7
23-56	1	33	46	21	4.7	3.3	1.5	0.18	0.026	6.9
56-91	4	31	25	34	4.7	5.1	2.3			
91-127	15	30	28	42	4.7	6.4	3.3			
B—HIGH-RAINFALL LOCALITY										
B (1). Clayey Upper-slope Soil										
0-5	17	60	25	15	5.0	9.9	2.6	2.63	0.147	17.9
5-15	62	50	28	22	4.3	6.7	1.4	1.54	0.095	16.2
15-46	65	42	25	33	4.3	6.2	1.5			
46-91	55	34	18	48	4.4	7.2	1.9			
91-122	56	28	16	56	4.3	7.8	1.3			

TABLE 48 (*Contd.*)

Depth cm	*% Stones and*	*% Coarse sand Gravel*	*% Fine sand*	*% Silt plus Clay*	*p*H	*Baes-Exchange Capacity* me per 100g	*Total Exchange-able Bases* me per 100g	*% Carbon*	*% Nitrogen*	*C/N Ratio*
B (2). Sandy Lower-slope Hill-wash Soil										
0-3.8	1	37	50	13	4.2	7.8	1.1	1.64	0.133	12.3
3.8-15	3	36	50	14	4.1	4.9	0.3	0.81	0.070	11.6
15-30	14	34	48	18	4.2	4.2	0.7			
30-84	54	33	43	24	4.4	4.3	0.8			
84-122	53	35	38	27	4.4	4.1	0.8			
C—INTERMEDIATE LOCALITY C (1). Upper-slope Soil (Orlu Series)										
0-5	0	71	12	17	6.7	8.2	6.9	1.83	0.160	11.4
5-15	0	74	12	14	5.9	4.0	2.9	0.64	0.061	10.5
15-46	0	69	14	17	5.1	3.6	1.6	0.40	0.031	12.9
46-76	0	65	11	24	4.6	3.7	1.3			
76-122	0	56	8	36	4.6	4.2	1.6			

decline was presumably nitrate deficiency. Nitrogen supply increased after green manuring and yields were better. After incorporating *Mucuna*, the nitrate content of the soil increased considerably. An experiment on land newly cleared from secondary forest, showed that the humus in the soil provided a nearly adequate supply of nitrate for about eight to ten years. If erosion was controlled, the soils were suitable for mechanical cultivation and it was not difficult to maintain fertility.

Applying green or organic manures to these soils was not sufficient to increase their humus content, but their regular use showed that the soils could be very productive provided the soil was not intensively cultivated. Shade was evidently necessary for increasing humus in these soils. In fact, they were better than soils of temperate regions in that they needed less nitrogenous fertilizer because vegetation deposited on or added to the soil was more rapidly nitrified.

10.2.2 Hot, Dry and Semi-dry Regions

In these areas, water rather than temperature is the critical factor. Water is perhaps the most important limiting factor in the physical environment. Even though evapotranspiration is large, total rainfall is adequate for crop growth except in arid and desert areas, but unreliable annual rainfall and the length of the dry season are the real problems. These areas occur in Mexico, extending into Colombia and Brazil, in the Sudan, and northern Australia and have some of the best soils in the tropics. However, these soils require careful management if humus reserves are not to be rapidly depleted and erosion by wind and water are to be minimized. With adequate irrigation and careful management, the fertility of soil in the drier tropical areas can be maintained and even improved.

10.2.3 High Altitude Regions

These are fertile soils of great agricultural value in high-altitude tropical regions with moderate rainfall. One such group, the Black Andean, or High Paramo, soils, is found in the Andean Plateau at altitudes varying from 3,000 to 4,000 metres above sea level and rainfall from 1,000 to 2,000 mm with no dry season. Temperature varies little seasonally but may vary markedly between day and night. The natural vegetation consists of grasses and shrubs (Bramao and Lemos, 1961).

Black Andean soils are derived from volcanic ash materials. They are nearly uniform in colour throughout the profile, the horizon boundaries being diffuse. The profiles are very dark coloured and deep, from 180 to 330 cm (6-12 ft). The soils are very friable, non-sticky, non-plastic, very permeable, with a massive to fine weak sub-angular structure, without clay films and very porous. They withstand erosion but are deficient in nitrogen and phosphorus and require liming.

Black Andean soils are very productive, being excellent for pasture and cattle production. They are also suitable for potatoes, wheat, rye, barley, carrots and cabbage.

11 Soils under Shifting Cultivation

11.1 SHIFTING CULTIVATION IN THE TROPICS

SHIFTING cultivation is the traditional way of maintaining soil fertility in the tropics. It involves the clearing of forest or grassland (eg. savanna), burning of the debris, growing crops for a period which may vary from one to ten years, but is usually two to four years, and then reverting the land to its natural vegetation so as to restore soil fertility, the length of the fallow period usually extending from less than ten to twenty or more in grass.

Although it may be defined in such simple terms, shifting cultivation is extremely complex and varied. There is great variation in the crops grown, cultural methods adopted and the intensity of cultivation influenced by climates, natural vegetation, soils, culture and customs of the people, land tenure systems, and settlement patterns adopted by the people. Even in one region such as the Congo Basin, cultivation methods range from the simple slash-burn-plant technique to more intensive systems involving hoeing, and still more intensive ones characterized by use of specially prepared ash fertilizers, composting, application of animal manures, irrigation, or terracing. In addition to variations in the systems of cultivation, there is also diversity in crop associations and sequences as well as in the combination of agricultural enterprises found. All these variations have one common factor; in shifting cultivation reliance is placed on nature to restore soil fertility through the long fallow.

Shifting cultivation has been adopted at some time or another by people all over the world. It was practised in Europe and by the early European settlers of the seventeenth century in the forests of the eastern United States. Because of the faster vegetative growth it is easier to adopt in the tropics. Today, it is being widely practised in many parts of the tropics and sub-tropics, especially in the less developed regions of tropical America, Africa, south-east Asia, and Oceania. It has been referred to by a number of different names. The early Europeans called it 'slash and burn'; in Britain, 'swidden' meant a burnt clearing; other names common in literature are: 'milpa' in South America, 'kaingin' in the Philippines, 'ladang' in Indonesia, 'rai' in Thailand and Veitnam, 'taungya' in Burma, 'jhum' in Assam, and 'chena' in Sri Lanka. The term 'shifting cultivation' indicates appropriately that this is a kind of temporary land use.

As a method of crop growth, soil management and land use, shifting cultivation has been severely criticized, but constructive possible alternatives have seldom been suggested. Some detailed studies (e.g. Nye and Greenland, 1960; Kellogg, 1948, 1963) were made of the problems posed by shifting cultivation and these have contributed much to our present knowledge. About a third of the agricultural acreage in tropical Asia was estimated to be still farmed by shifting cultivation. When one considers that some 200 million people (Kellogg and Orvedal, 1969) in the tropics and sub-tropics, scattered over an area of nearly thirty-six million square kilometres (fourteen million square miles) rely for their sustenance on systems of shifting cultivation, it becomes apparent that solutions to the problems that arise are urgently needed.

Shifting cultivation is a most extravagant use of the land. It presupposes the existence of cultivated to uncultivated land at least in the ratio 1:4 and, accordingly, about 40-50 people per square kilometre can be supported by this form of agriculture. As population densities increase and the land available diminishes, the cropping period is prolonged and the length of fallow reduced. The land is not rested sufficiently for adequate regeneration of fertility and becomes progressively less productive.

In the nomadic 'Bantu system', the forest was cleared and burnt, and interplanted crops were grown until yields decreased. The field was then abandoned to a long bush fallow. The Research Centre of INEAC at Yangambi modified the Bantu system and evolved the Corridor System (Coene, 1956; Kellogg, 1963). The forest is cleared in alternate blocks so that the cleared blocks have their long axes east-west, so as to provide maximum light for the crops. The length of the strip is anything beyond a minimum of 200-300 metres, but the optimal breadth is about 100 metres. The proximity of the forest at the north and south edges promotes rapid regrowth of forest fallow on the abandoned field. This occurs by the maintenance in the clearing of a forest micro-climate and by the presence of mother-trees. There is some loss of yield—10 per cent in heavy forest clearings falling to 3 per cent in cleared forest fallows. But these losses are compensated in drought periods by the protective action of the border on the field. A minimum of fifteen years' fallow is adopted. A number of possible rotations have been worked out and perennial crops have also been introduced into the settlement.

11.2 Land Preparation

In the forest zone, the land is cleared during the dry season. Small and medium sized trees are cut using hand implements and burnt. Some of the larger trees may be left standing and provide light, patchy shade. A wind during burning helps the fire to spread easily over the area selected but such fires may also be difficult to control. In the savanna, the grasses are fired during the dry season and excess woody vegetation is cut and burned. Because the grass is usually burned every year during the dry season, the bare soil tends to be exposed to the early rains. Weeding in savanna is troublesome, particularly if *Imperata cylindrica* is present and the soil has to be more disturbed than in forest to get rid of the roots of the grasses.

Shifting cultivation usually involves clearing by burning, and in grassland this distinguishes it from ley farming. Burning is essential in both forest and savanna. It is only by burning that the cut mass of vegetation can be easily cleared in a forest. Although organic carbon, nitrogen and sulphur are lost in the burn, there is no other practical way of incorporating the felled material in the soil. Similarly, it is equally impractical to bury grassland vegetation in the soil with simple implements. Although burning is usual and, in the 'citemene' system practised along the Congo-Zambesi watershed, branches lopped from trees on surrounding land are added to the piled debris and burned, yet there are places in Malawi and in southern Tanzania, in predominantly grass savannas, where the grass is not burned after clearing but is hoed up and buried in mounds towards the end of the rainy season.

Burning causes changes in the nutrient content of the soil, heats the soil surface and may result in a different microflora being established. Results from forest fallows indicate a substantial increase in the exchangeable cations in the soil and a corresponding rise in the *p*H. On very acid forest soils particularly this is one of the important effects of burning. Nye and Greenland (1960) have pointed out that if it were possible to incorporate the organic matter of the fallow into the soil without burning, the effect would be similar to a continued addition of litter, the humus level would rise and with it the exchange capacity, but the *p*H and the availability of the cations would change little.

The carbon, nitrogen and sulphur in the fallow and litter are lost in the burn, but not the amounts in the soil humus. This is important, namely that while the litter layer is largely destroyed in the burn, humified organic matter in the soil is not affected.

Burning of forest vegetation seems to have little effect on the bulk of the soil, except locally where logs and piles of wood burn for long and the surface soil may change profoundly. The thick layer of ash in these patches remains hot for some hours. Heating promotes an improved structure, making the soil more friable, leaves a weed-free seedbed, and discourages the activities of reddish seed-storing ants. Heating also sterilizes the soil partially resulting in an initial decrease in the microbiological population followed by increased activity of a modified microflora. In savanna grassland, however, the change in the soil population following the burn is relatively small. In Kenya, Meiklejohn (1955) found that the nitrifiers were virtually absent for the first month after the burn. One of the important effects of burning may be to kill soil-borne pathogens, but very little is known about this.

11.3 The Cropping Period

11.3.1 Crops and Cropping Sequences

A number of factors influence the crops grown, the cropping sequences, and the cultural methods adopted. Among the more important of these are the climate, soil, crop characteristics, the need to control weeds, the requirements and availability of labour, and the customs and dietary habits of the people.

Mixed cropping is generally practised probably because this would enable maximum returns with minimum effort. It also ensures a more complete cover and checks weed growth. It is moreover an insurance measure in case some crops fail because of adverse weather conditions, pest or disease attack.

Little preparatory cultivation is practised in forest clearings, the seeds are merely scattered on the surface or dibbled into holes with a digging stick or hoe. Sometimes when roots are planted as a first crop, ridges or mounds are made.

A cereal, such as maize or upland rice, is usually the first crop to be planted with a mixture of pulses and vegetables. Annual or semi-perennial roots are then interplanted and, subsequently, perennials such as bananas. The cereals and other annual crops can be harvested in the first year, the roots mainly in the second year while the harvesting of the perennials starts in the second year and may continue for several years after the clearing has been abandoned.

Maize or rice, or both, are the cereals first planted but this may be mixed with millet, sorghum and other crops such as okra, melon, cucumber and pigeon pea. Either during the growth of the maize or soon after it has been harvested, root crops such as cassava, sweet potatoes, yams, coco-yams, tannia, and bananas are planted.

In most places only one cereal crop is taken, followed by roots and perennials, over a cropping period of about three years. In other places, another annual crop, not necessarily a cereal, may be grown and this often necessitates a second clearing and burning.

In the grassland clearings, more preparatory tillage is needed, especially to get rid of the grass root-stocks. The land is dug with a hoe to a depth of about four inches before planting the first crop and this procedure is repeated each season before replanting at the beginning of the rains. Further hoeing may be necessary during the growth of the crop to control weeds. Unlike in the forest clearings, cereals are not the first crop to be grown. The nitrate content of the soil under grass fallows may not be sufficient for the quick-growing cereals, especially if the grass has been buried instead of being burnt. But if predominantly woody vegetation has been felled and burnt, a cereal may be grown.

At the end of the cropping period, pigeon pea or cassava is planted and weeding is discontinued after they have been established. Growing perennial crops and harvesting them after the clearing has been abandoned is not possible because annual fires are common.

11.3.2 Decline in Soil Fertility

Yields in successive seasons decrease because the fertility of the soil gradually diminishes, and not so much because pests, diseases or weeds increase, though in some areas these do contribute.

Until the first crop is fully established, rapid organic matter decomposition and the battering effect on the soil of raindrops will cause soil structure to degrade somewhat (Cunningham, 1963). Cultivation, weeding and the growing of annual crops in succession can accelerate this deterioration most markedly in savanna clearings, where soil structure under grass fallows recovers less than under forest and the soil is tilled and cultivated more initially to prepare seedbeds and control weeds. Also, the soil is probably eroded more in the savanna zone than in lowland forests, and increasingly on the steeper slopes. Preparatory cultivation and hoeing to control weeds tends to destroy soil aggregates and to seal the soil surface.

Contrary to popular belief, humus in tropical soils does not decrease to very low values rapidly under cultivation. An equilibrium value is reached eventually and humus content declines more gradually after the first year or two of cultivation. For forest soils, cropped by traditional methods, Nye and Greenland (1960) estimate the rate of humus decomposition to be about 3 per cent per year of the total humus content. This is roughly equivalent to the rate of increase under fallow, estimated at between 2 and 5 per cent per annum. In cropped grassland soils, the rate of humus decomposition is more, about 4 per cent per annum, greatly exceeding the estimated 0.5 to 1.2 per cent per annum increase under fallow. Humus in forest clearings will not be appreciably depleted if the cropping periods are short and the fallows relatively long; whereas in grassland clearings, unless the fallows are lengthened considerably, humus content will fall markedly.

Because soil *p*H increases on burning, the *p*H of the surface soil in the forest zones remains above 5.0 and there is no significant decrease in the rate of mineralization of organic matter. Total nitrogen contents after fallowing are large and, even after several years of cropping, responses to nitrogen are small when the fallows are sufficiently long. Soil nitrate levels in savanna land are less than in forest. Leaching losses are also less than in forest soils, but may be appreciable at the beginning of the rainy season when there is a flush of nitrate production.

After burning, phosphate availability is increased and the decline in soil phosphate during cropping is caused by uptake in crops and a decrease in availability; leaching losses are negligible. Nye and Greenland (1960) estimate that 27 kg P per hectare is added by a ten-year forest fallow, and about 9 kg P per hectare annually from humus mineralization during a cropping period. In the savanna, the ash provides only 4.5-9 kg P and humus decomposition 3.4 kg P per hectare annually.

Nutrients are lost when part of the ash is washed away by the first rains. The extent of leaching losses during cropping can often be considerable and may exceed the amounts removed by crops. But these losses will be compensated to some extent by nutrients released from non-exchangeable forms. Nye and Greenland (1960) suggest that, except for potassium, no serious nutrient deficiences usually occur during a single cropping period of 1 to 3 years. This is also true for savanna soils, although the total nutrient content added by the fallow vegetation is less than in forest areas.

11.4 The Fallow Period

11.4.1 Nutrient Accumulation in Vegetation

When a clearing is abandoned after cultivation, regeneration of forest vegetation is very rapid. The rate of accumulation of nutrients in the forest vegetation is more in the early years of the fallow when most of the vegetative growth takes place (Table 49).

TABLE 49 : MEAN ANNUAL INCREASES OF NUTRIENT STORAGE IN FALLOW VEGETATION* (kg/ha)

Forest Fallow Period		N	P	K	Ca	Mg
Yangambi	over first 5 years	114	6.3	91		84
Yangambi	over first 18 years	39	5.9	34		46
Kade	over first 40 years	40	2.8	17	54	7

*including roots.

The rate of increase in nutrient storage in grassland cannot be estimated easily because of the annual firing and the variable extent of woody growth. But the amounts are certainly much less than in forests. The annual production of dry matter is limited by lack of moisture because of comparatively less rainfall and a marked dry season. Jaiyebo and Moore (1964) studied nutrient storage under different kinds of fallow.

11.4.2 Humus and Nutrient Buildup in Topsoil

Nutrient changes in the soil occur because organic matter is added from vegetation, nutrients are recycled from the subsoil, are removed by cropping and leaching and react with the soil.

Litter production in moist tropical forests amounts to about 12 tons of dry matter per hectare, compared with an average of a little over 2.5 tons per hectare in temperate hardwood forests and 7-10 tons per hectare in high-grass savanna. In older forest fallows, substantial amounts are added to this by the annual timber fall, especially as regards the calcium content. Large quantities of potassium and phosphorus, but only small amounts of nitrogen, calcium and magnesium are washed out of the leaves into the soil.

The maximum level of litter is built up fairly rapidly on the forest floor, between one and three years for most tropical forests ; and when this has happened, the rate of mineralization equals the rate of addition.

Soil humus increases under fallow. Under moist lowland evergreen or semi-deciduous forest, the maximum attained after very long periods of fallow averages about 67,000 kg/ha C and 6,150 kg/ha N in the 0-30 cm. layer. On free-draining soils, more humus occurs in highland areas, when the soils are clayey and when the clay fraction contains allophane-like clay minerals. Under high-grass savanna, the levels attained are 56,000 kg C and 3,900 kg N in the 0-30 cm layer. In drier areas, the amounts are less. After cultivation, less humus accumulates in soils under fallow, depending on the intensity of the previous cultivation. The C/N ratio is a rough guide to the rate of mineralization of the humus. It is around 8-12 in the forest ochrosols, 14-17 in the forest oxysols and 15-20 in the high grass savanna soils.

Organic matter is added to the soil as litter and simultaneously from dead roots and root products. In forest, about 16,800 kg/ha are added but in savanna, where the litter is burnt annually, only roots and root products are available for increasing soil humus by about 3,000 kg annually. Of the fresh organic material added, only between 1/10th and 1/5th contributes to the soil humus. The humus carbon decomposes at a rate directly proportional to its amount and ultimately an equilibrium level is attained. For forest fallow soils, this equilibrium is at about 75 percent of the maximum level (Nye, 1959; Greenland and Nye, 1959). The rate of increase of humus carbon ranges between 280 and 670 kg/ha annually in the forest zone, but is only between 78 and 190 kg/ha/year in the high-grass savanna zone, because the vegetation is much less. Relatively much more humus is fouud ín forest fallow soils that have been cultivated and fallowed over many cycles.

Assuming a constant C/N ratio of 12 for both forest and savanna soils, Nye and Greenland (1960) estimated that the average rate of increase of soil nitrogen in the forest zone will be around 39 kg/ha/year and that the principal source for the nitrogen increment is the atmosphere. But, whereas symbiotic fixation may be important in the heavily leached acid soils of the rain forest areas, the main contribution to increase in soil N in the moist semi-deciduous forest zones comes from non-symbiotic sources. *Azotobacter*, *Beijerinckia* and *Clostridia spp.* are likely to be the organisms responsible together with nitrogen-fixing blue-green algae in the forest soils.

The increase in soil phosphorus under a ten year fallow at the 75 per cent equilibrium level is around 13-34 kg/ha in the forest and 3-10 kg/ha in the high-grass savanna. The phosphorus accumulated in the vegetation is about 34-45 kg/ha in the foiest and 11 kg/ha in the savanna. Increasing soil organic phosphorus is thus one of the most useful functions of fallow.

The exchange capacity of kaolin-containing tropical soils depends largely on their organic matter content. Assuming an average increase of 1800 kg organic C in a ten year fallow period, the corresponding increase in exchange capacity will be around 6.3 kg equivalents using an average value of 350 me per 100 g C for the exchange capacity of the organic matter. In savanna grassland, with an average increase of organic C of about 1340 kg/ha, the grassland, capacity would increase by about 1.9 kg equivalents.

11.4.3 Cycling of Nutrients

Soils under forest fallows contain more total phosphorus and calcium in the surface horizon (0-10 cm) than in the subsoil. Exchangeable Ca and exchangeable K contents are also not appreciably altered despite losses by leaching and to standing vegetation. This indicates that nutrients are brought up from the subsoil and returned through the litter to the topsoil where most of the feeding roots occur.

Vine (in Moss, 1968) considers it likely that 'pumping-up' from the depths is the more important source of potassium, and release from the insoluble forms in the upper layers of soil is the more important source of phosphorus. Potassium can be moved down the profile more readily than phosphate, and much of the potassium involved in the nutrient cycling is leached to a depth of several feet from where it is recovered by deep-rooted plants in fallows.

This clearly suggests that in the warm humid tropics, unlike in the temperate zone, trees generally build up soils much more effectively than grasses. In the heavily leached soils of the humid tropics plant nutrients are almost entirely in the vegetation. Woody plants, especially large forest trees, can gather and retain more nutrients per hectare than grasses. Furthermore, because grasses get burnt annually, nutrients in the ash are washed away by the rains. Nye and Greenland (1960) report that nutrients are added to soil under forest trees by dust falling, or washed down from the trees. Trees benefit soil structure. They shade the soil from sun and beating rain and provide a favourable temperature, humidity and food supply for the micro-organisms. Trees open up channels which improve aeration and water movement. The expanding root systems of trees help to consolidate loose soil and to open up soils that have become hard. In badly-drained soils, trees remove excess water and thus reduce waterlogging and salt accumulation. Replacing forest by savanna grassland thus leads to soil fertility decreasing permanently because of short fallows and excessive cropping.

11.5 Soil Fertility under Shifting Cultivation

11.5.1 Forest Regions

Nye and Greenland (1960) suggest that shifting cultivation in the forest is perhaps the best system that could have been devised; the labour of clearing, planting and weeding in relation to the harvest being small.

The main objections to shifting cultivation are that forests are destroyed, organic matter and humus is lost by burning, and nutrient loss is considerable.

Agriculture will always involve some destructions of forests but this is harmful only if the forest is more useful for other purposes, such as producing timber or conserving soil and water resources. Small scattered peasant plots do not cause any extra surface runoff and possible adverse climatic effects are negligible.

Organic matter can be maintained at a satisfactory level if the cropping periods are short and the fallow sufficiently long. Early work (Beinaert, 1941) has emphasized the destructive effects of cultivation on humus reserves in tropical soils. Nye and Greenland (1960) attribute some of these reports of humus oxidation to 'exaggerated estimates of litter return under forest.' Instead of the 250 tons per hectare in previous reports, measurements indicated an average litter fall in tropical forests of about 12.5 tons oven-dry weight per hectare annually, plus about half this amount as roots.

The rate of mineralization of humus is more than in soils of the temperate zone but the rate of loss is not so great as to deplete the soil completely. Moreover in shifting cultivation, the soil is not much disturbed, weed residues are returned to the soil while some trees remain on the plots and help to maintain soil humus by their litter, shade, and roots. In fact, there is much evidence to indicate that humus in areas of soils cultivated for many years by shifting cultivation is still relatively large (Coulter, 1950; Birch and Friend, 1956). Only in savanna areas subjected to repeated grass burning, is soil humus content very small.

The mineralization of humus and the release of available soil nitrogen was found to vary considerably between the soils of the moist evergreen forest zone and soils of the moist semi-deciduous forest; also between the forest and savanna soils.

In the evergreen forest zone, nitrification improves on clearing and burning but may thereafter decline slowly with cultivation. In the moist semi-deciduous forest, nitrification is slightly less under cultivation than in the forest soil but soil nitrate levels do not decrease significantly after several years of cropping. In the savanna zone, mineralization improves steadily during the first year or two after clearing from grass and thereafter changes little for several years.

TABLE 50: EFFECT OF THE FALLOW IN SHIFTING CULTIVATION

	Forest		*Savanna*	
	lb/ac	kg/ha	lb/ac	kg/ha
1. Litter production per year	10,000	11,200	8,000	9,000
			(lost in burn)	
2. Roots added per year	5,000	5,600	2,700	3,050
3. Organic materials added to soil per year	15,000	16,800	2,700	3,050
4. Maximum humus level after 10 years (0–30 cm layer) C	60,000	67,000	50,000	56,000
N	5,500	6,200	3,500	3,900
5. Rate of increase of humus C per year	250–600	280–670	70–170	79–191
6. Rate of increase of soil N per year	35	39	10	11
7. Rate of increase of N in fallow vegetation per year	60	67	25	28
	lost on clearing and burning			
8. N added in rain per year	7	7.9	7	7.9
9. C/N ratio	ochrosols	8–12	ochrosols	15–20
	oxysols	14–17		
10. P in soil after 10 years	12–30	13.4–33.7	3–9	3.4–10.1
11. P in fallow vegetation after 10 years	30–40	33.7–45	10	11
12. Increase in CEC (equivalents)	14	15.8	4.2	4.7
13. Erosion losses	less		more	
14. Leaching losses	more		less	

Soil nitrate studies also clearly indicate that early planting helps the crop to make use of the rapid nitrification after the first rains. The crop can benefit from organic phosphorus mineralizing similarly, establish its root system early and thus compete more favourably with weeds. Nutrient status decreases during cropping by crop uptake, loss by soil erosion and leaching. Much of the potassium and less of phosphorus is removed by the crop. Carbohydrate-rich crops, such as cassava, sweet potato and plantain, rich in K, remove more potassium.

In forest soils, normal erosion cannot cause the rapid decline in fertility under cropping. The physical condition of the top soil after clearing and burning, such as its porous, granular nature, is sufficiently stable to withstand the shattering action even of the heavy rains. Within a few weeks, the early planted crops form a protective cover, especially effective when a mixture of perennial crops is grown.

The ash from forest fallow contains cations and anions. NO_3^- and HCO_3^-, little absorbed by the soil, principally compensate the cations Ca^{2+}, Mg^{2+} and K^+ in the soil solution. At pH values < 6, HCO_3^- concentrations are negligibly small and, therefore, in acid top soils, NO_3^- is most important. NO_3^- in soil must be adequate for crop growth but excess may cause the loss of associated cations especially K^+. Little or no phosphorus is lost by leaching. During cropping, soil phosphorus status decreases by crop uptake and by decreasing solubility in the soil.

11.5.2 Grassland Regions

All the evidence indicates that shifting cultivation in savanna exhausts soil nutrient reserves. Annual burning exposes the soil to beating rain, encouraging erosion on all but gentle slopes. Soil structure and stability, developed under savanna grasses, is much poorer than under a forest fallow. A thorough hoeing is necessary to clear the grasses for cultivation and the loose soil, heaped into mounds or ridges, is exposed and easily eroded.

The natural grass savanna is, moreover, a very poor means of restoring soil fertility (Table 50). Nutrient reserves in the soil are small and plant-available nitrogen is scarce because less organic matter accumulates in the top soil under savanna than under forest. Phosphorus is adequate in the first year or two of cropping but crops respond to fertilizer phosphorus in subsequent years.

Potassium in savanna soils, even with intensive cropping, is seldom insufficient for good crop growth because, although these soils have little exchangeable K, their exchange capacity is also small, so that the proportion of potassium in the exchange complex is adequate. Secondly, because soil nitrate is small, less potassium is leached out. Potassium, exchangeable to M ammonium acetate, is greater than in the more weathered forest soils. This compensates somewhat for potassium removed by crops.

12 Soils under Intensive Upland Arable Cropping

12.1 Farming Systems

Rainfall, especially its monthly distribution, determines the vegetation and agricultural system of each tropical region. Although more than half of the world's annual rainfall falls in the tropics the rainfall pattern differs from one tropical region to another; very wet, extremely dry and even desert areas occur. Rainfall varies considerably from year to year because of cyclonic depressions and monsoon failures.

Evapotranspiration is generally more in the tropical lowlands than in temperate regions and therefore rainfall, considerable by temperate climate standards, may be quite inadequate for productive farming in the tropics. Russell (in Moss, 1968) estimates that in Africa, on the equator, the annual potential transpiration at sea level is about 2,286 mm decreasing to 1,270 mm at 2,500 metres above sea level. The mean annual rainfall over much of this area is less than 1,000 mm, exceeding 1,500 mm only in the coastal areas of west Africa and in the highland areas of the east. In a FAO report (1967) it is estimated that in the savanna belt of Africa, evapotranspiration can be as much as 8 mm per day on a sunny day, before the rainy season, decreasing to 4-5 mm during the rainy season and 2-3 mm on a rainy day.

No part of Sri Lanka gets an annual rainfall less than 890 mm, yet nearly two-thirds of the island receiving less than 1,915 mm annually is referred to as the Dry Zone. This area is dominated by a seasonal drought extending from about June to September, the rainfall during this period being inadequate for agriculture. Some of the most fertile soils in the tropics are found in these drier areas. They are not as intensively leached as those in the wetter areas and their chemical fertility is quite high. The Reddish Brown Earths of Sri Lanka and the Vertisols of India and Africa are examples of such chemically fertile soils. Although there are other problems associated with cultivating such soils, productive permanent systems of continuous annual cropping would be possible on them if irrigation water is available. Irrigated agriculture contributed to past civilizations which flourished in these regions.

12.1.1 Irrigated Farming

Efficient systems of irrigation have existed from very early times. In ancient Mesopotamia, irrigation was practised as long ago as 4000 B.C. Today irrigation is available throughout Asia mainly for rice cultivation, but increasingly for upland arable farming as well. Irrigation was not common in Africa south of the Sahara, except the Kilimanjaro region, but irrigation projects have been developed in a number of African countries and more are planned. The Gezira, started in 1925, is the most famous irrigation scheme in Africa. The Volta Project in Ghana and the Kariba Dam Project in Zambia are multipurpose river schemes, providing irrigation water as one of the objectives. Wherever rainfall is inadequate, the considerable cumulative radiation input in the tropics can be effectively utilized for crop growth if irrigation water is available.

Traditionally, irrigation in the tropics was confined to rice cultivation, but it has spread to other field crops and vegetables grown in rotation with rice, and even to perennial crops. For example, irrigation of the tea crop in Malawi during the dry months gave increased production (Shaxson, 1969 as quoted by Willatt, 1970). Sprinkler irrigation has been successful on latosols of the old coffee areas of Brazil (Ignatieff and Lemos, 1963). Irrigation systems involve high capital expenditure and, consequently, better standards of farming involving high-yielding varieties, adequate fertilizer and agrochemicals to combat insects and disease, better cultural practices and efficient use of irrigation water are necessary to make capital investment worthwhile. Various methods for storing irrigation water and irrigating crops are used (Wrigley, 1969).

Profitable irrigation requires a knowledge not only of the total water requirements of the crop from planting to harvest but also how much water is needed at different stages of growth. Additional water may be necessary for land preparation. Pereira (1957) reported on field measurements of water use for irrigation control in *Arabica* coffee under high altitude tropical conditions in Kenya and proposed a method for the control of supplementary irrigation under conditions of limited water supply. Techniques for the measurement of soil moisture storage under east African conditions were discussed by Pereira (1954).

The quality of the irrigation water is important. The total salt content, and particularly the proportions of exchangeable sodium and possibly magnesium, must be kept less than critical levels. A sodium ion concentration of less than 600 ppm is usually safe, more than 1,000 ppm is dangerous. A high Na/Ca ratio is not desirable. More than 12 to 15 per cent of exchangeable sodium ions decreases the permeability of the soil while more than 40 to 50 per cent sodium may restrict the absorption of calcium. The effect of magnesium is similar though less. However, Robinson (1971) reported that on the Vertisols of Gezira the Sudan Republic, an exchangeable sodium percentage of up to 35 per cent did not seriously limit yields of cotton, dura or groundnuts. Irrigation water should not contain trace elements at toxic levels; boron, selenium and lithium are especially harmful. A high concentration of salts in irrigation water will increase the salt content and, therefore, the osmotic pressure of the soil solution, retarding crop growth. Crop growth retardation varies directly with increase in osmotic pressure. Saline water is particularly unsuitable for sprinkler irrigation because leaf absorption of salts is more rapid than root absorption, and leaves get scorched.

Where saline or alkaline water is used for irrigation, the soil may deteriorate by gradually becoming saline and alkaline. Salt accumulation is a major threat to all irrigation projects, particularly in arid areas. For example, four-acre feet (39-ha cm) of irrigation water taken from the White Nile at Khartoum would add 8 cwt (406 kg) of sodium salts and slightly more of soluble salts including calcium chloride and magnesium salts (Wrigley, 1969). Many sources of irrigation water have even more salts. Once salts have started to accumulate, they build up rapidly and so this must be avoided.

To prevent salt accumulation, irrigation water with high salt concentration should preferably be avoided. Irrigation water should be applied in excess of the field capacity so that the surplus will carry the unwanted salt into the drainage water. This will inevitably lower the efficiency of water use and may cause crop damage, if practised on heavy soils where waterlogging can occur. Good drainage is essential to a productive irrigation system. The water table should be kept low, about 2 m below the soil surface, to prevent the upward movement of salts in arid areas. Flood irrigation should be done on flat land so as to prevent the water from flowing into, accumulating and forming concentrated salt solutions in hollows. Water channels should be kept free from weeds as these can cause waterlogging and salinity. This has happened in different parts of the tropics; for example, in a large irrigation scheme in India (Holm, 1966) and in Ghana (Little, 1967).

Soil salinity and alkalinity have been investigated by workers in both temperate and tropical countries. Kelley (1951) dealt at length with the formation, properties and reclamation of saline and alkali soils. The main characteristics of such soils and the remedial measures usually adopted are summarized in Table 51.

TABLE 51: CHARACTERISTICS OF SALINE AND ALKALI SOILS (from Kelley, 1951)

	Saline soil	*Saline-Alkali soil*	*Alakli soil*
Electrical conductivity (m. mhos per cm. at 25°C)	>4	>4	<4
Exchangeable-Na per cent)	<15	>15	>15
*p*H	<8.5	Variable, but usually <8.5	8.5 to 10.0
Structure and Permeability	Variable	Variable	Poor
Remedial measures	Irrigate and drain to leach out soluble salts	Add gypsum, irrigate and drain	Add gypsum or sulphur; irrigate and drain

12.1.2 RAINFED FARMING

Conventional irrigation schemes cannot possibly serve all the agricultural land in dry and semi-dry tropical regions. In much of this area semi-irrigated farming or pure rainfed farming must necessarily be practised. In India, about a third of the arable land is farmed by dry farming (Arakeri *et al.*, 1959). Soil and water conservation as well as decreasing of soil erosion become major programmes in rainfed farming systems. Splash erosion and surface sealing (page 81) are manifest on upland arable soils and soil erosion control, principally, achieves a settled system of arable farming.

In rainfed farming, tillage operations must be timed according to the rainfall pattern. This inevitably leads to a rush in operations, avoidable if supplementary irrigation is available.

For rainfed farming, soils with a permeable surface and a large capacity for holding available water should preferably be used. Such soils are usually deep, have a medium-textured surface, and medium to moderately fine-textured subsoil.

12.2 Cropping Patterns

12.2.1 Rotational Cropping

The crops and their sequences on a peasant's farm in the tropics vary and are governed by factors such as magnitude, frequency and distribution of rainfall, the needs of farmers and the local population, prices and markets for cash crops and surplus food, tradition and social customs and the farmers' credit and debit balance. The rotation is adjusted according to the nature and productivity of the soil. Cassava, maize and upland rice are common. National priorities such as self-sufficiency in food and import substitution have sometimes made governmental intervention necessary. For example, the need for producing chillies and onions locally made it necessary for the Government of Sri Lanka to stipulate that land under lift irrigation in certain parts of the dry zone of Sri Lanka should be planted with chillies from January to June and with red onions from July to September, and that during the north-east monsoon, rainfed legumes like cowpea, green gram and sunnhemp be grown for ploughing in (Anghie in Peries, 1967).

Experiments have been conducted to evaluate the merits of different kinds of resting periods to maintain and restore soil fertility (e.g. Peat and Brown, 1962; Clarke, 1962; Watson and Goldsworthy, 1964). Bennison and Evans (1968) reported an investigation of the effect of crop sequences on soil fertility and crop yield at the Katumani Research Station, Kenya. Jackson and Burnham (1968) studied the interactions between nitrogen fertilizer and crop rotations in their effects on the yield of cotton in the Sudan Gezira, and demonstrated the role of nitrogen as a major factor in the differential effects of rotations. Cotton was the poorest precursor to cotton; lubia (*Dolichos lablab* L.) was a better precursor to cotton than was dura (*Sorghum vulgare* Pers.) or fallow; fallows between cotton crops or between dura and cotton benefited unfertilized cotton, less so when nitrogen fertilizer was given.

Crop rotation alone will not maintain soil fertility but a good rotation will slow the decline in fertility caused by soil exhausting crops grown consecutively (e.g., Cutting *et al.*, 1959). The rotation of crops of different root habit requiring different types of tillage and cultivation, and the alternation of legumes and non-legumes can be used to minimize soil exposure and erosion, reduce weeds, harmful insects and plant disease.

12.2.2 Mixed Cropping

Mixed cropping or intercropping, characteristic of tropical farming, is widespread throughout the tropics and has been adopted over the years for several reasons. It is an insurance against crop failure under the hazards of peasant farming. When several crops, requiring most labour at different times are grown, available labour can be distributed more evenly throughout the year. Spreading the harvest overcomes storage problems. Planting of early—and late—maturing varieties ensures supplies over extended periods. Crops required in small quantities such as vegetables, spices and medicinal herbs can be fitted in on soil areas occupied by termite mounds and around tree stumps. Crops following, or associated with legumes derive nitrogen from them (Agboola and Fayemi, 1971). Roots take up nutrients from various soil depths, exploring a large soil volume. Interplanting of a number of crops also makes them less susceptible to pests and disease than monocultures. Above all, the soil is better and more continuously covered from beating rain, desiccating sunshine and wind and, therefore, protected from erosion.

Various mixed crops and different cropping sequences are practised (Miracle, 1967). Mixed cropping is not confined to annual crops. In east Africa, bananas are interplanted with coffee and provide shade as well as food and material for mulching.

12.2.3 Multiple cropping

Multiple cropping enables maximum utilization of radiation energy for crop growth annually. It consists of an almost continuous cycle of crops on the land for a comparatively short time and uses quick-maturing varieties, interplanting or even ratooning. The aim is to step up total output per annum by harvesting as many crops as possible in a year. In tropical and subtropical regions the potentialities for multiple cropping are great because growing seasons are either long or continuous and several crops may be produced on a given field.

The benefits derived from multiple cropping have been demonstrated by work in Malaysia, Hongkong, India, the Philippines and Taiwan. Research workers in several Asian countries got up to 20 tons dry matter equivalent per hectare using multiple cropping techniques with a well-balanced rotation of soya or mung beans, sweet potatoes, some vegetables, and a cereal. A normal double-crop rotation would only give about 10 tons.

Favourable natural conditions and good irrigation facilities in Taiwan make it possible to grow a number of crops in a year. Double cropping of rice is prevalent all over the island while a short season crop is grown in central Taiwan in the 3-4 months transitional period between the rice crops. Quadruple and even quintuple cropping are practised in some places (Kung, 1969).

However, multiple cropping demands efficient planning and good cultural techniques, an assured and well-controlled water supply, relatively heavy applications of fertilizer and agrochemicals, and a more sustained evenly spread workload. Some mechanical assistance during 'peak labour' periods would also help. Multiple cropping is therefore still a long way from being widespread although its potential for maximizing returns, especially in high-investment irrigation projects, is undoubtedly great.

12.3 Soil Fertility Conservation

12.3.1 Green Manures, Cover Crops

Opinions vary on the methods and value of *green manuring* in the tropics. The rapid decomposition of added organic materials at high temperatures and humidities does not let soil humus increase appreciably. Experiments in India, Nigeria, Trinidad and elsewhere seem to indicate that the chief benefit from green manure crops may be from their plant nutrient content, particularly nitrate ions (Vine, 1953; Joffe, 1955), rather than on any lasting effect from soil organic matter. Joachim (1931) thought that in the tropics the effects of green manuring last only for about six months.

Singh and Agarwal (1953) compared the effects of two green manures, mung and sunnhemp, on wheat, grown on soils of initially poor and rich fertility. Green manuring alone seemed to maintain soil fertility on the poor soil, not on the more fertile soil. Eventually, the fertility of both soils stabilized in about four years.

Singh (1963) in a critical evaluation of green manuring experiments on sugarcane in north India found that green manuring with sunnhemp left little or no residual effect on the soil as measured by the amount of organic C and N in the soil after the harvesting of the following crop. A leguminous crop grown during the monsoon-fallow period, either for fodder or for seed, was equally as effective as green manuring for raising the yield of the subsequent sugarcane crop. The limited effectiveness of the aboveground portion of the green manure in sugarcane cultivation was attributed to the interval of 5 months, from September to February, between the turning-in of sunnhemp green manure and the planting of the subsequent sugarcane crop.

Broadbent's (1948) work showed that immature green plant tissues, when decomposing, stimulate microbial activity and cause soil humus to be lost. Therefore it is likely that the repeated use of green manures may even result in decreasing soil humus (Russell, 1952). Acharya and Rajagopalan (1956) found that FYM given annually from 4,500 kg to 9,000 kg per hectare per year increased the carbon content of the soil by 50 to 40 per cent more than the controls; green manures produced only a statistically insignificant increase. Farmyard manure, as a result of its previous decomposition, contains more resistant organic matter than green manure and thus contributes to humus build-up in the soil.

Green manuring has not been successful in the Kenya High lands (Webster, 1954), being ineffective in the drier areas, and increasing maize yields appreciably only in the first crop. Nor were they effective in the 'Sudan Zone' of west Africa (Touré, 1964).

However, even though they do not significantly increase soil humus, green manures benefit the soil by supplying available nitrogen on decomposition and by improving its physical properties, especially its water-holding capacity.

Experiments in Uganda have shown that green manure crops do not appreciably improve soil structure or maintain soil fertility (Jameson, 1970). On the other hand, growing a perennial grass such as elephant grass (*Pennisetum purpureum*) for two or three years maintained satisfactory crop yields on experimental farms. In Trinidad, growing certain leguminous climbing or twining plants with tall grasses, such as elephant grass, has also been successful.

Water availability and the economics of green manuring are important points to be considered in the use of green manures in the tropics. Sufficient moisture is necessary for growing a green manure crop, its subsequent decomposition after digging in, and for the succeeding crop. The loss of a season's cropping while the green manure crop is in the field and the effort involved in planting and turning it in have also contributed to green manures not being more widely used. Green manure crops are sometimes intersown with the main crop, or material for green manuring brought from elsewhere.

Cover crops in the form of trees, bushes or herbaceous plants sometimes provide material for green manuring, shade for plants and protection against erosion for the soil. The use of shade trees in cocoa, coffee and tea plantations is discussed elsewhere. Green manuring crops such as sunnhemp (*Crotalaria juncea*), cowpea (*Vigna catiang*) and mung (*Phaseolus mungo-radiatum*), have been successfully used for rice, sugarcane and other crops in India. Agarwal (1965) discussed at length the nature and value of the principal green manure crops used in India and their use for the green manuring of rice, wheat and sugarcane.

12.3.2 Crop Residues, Mulches

Crop residues of all kinds are valuable for maintaining soil fertility. When large amounts of high-carbon residues are ploughed in adding a soluble nitrogenous fertilizer it facilitates their decomposition and prevents any temporary shortage of nitrogen for a growing crop. Crop residues can also be used as mulches, as composts and, indirectly, by feeding to stock and returning the manure to the land.

Mulches are especially useful in the tropics and may be more beneficial than ploughing in crop residues. They favourably affect soil moisture, soil temperature, organic matter content and organisms, nutrient content, and soil structure. They protect the soil from splash erosion and surface sealing, promote infiltration and minimize surface runoff and erosion, and they help in suppressing weeds.

The value of a surface grass mulch has been demonstrated in the sandy soils of northern Uganda (Mills, 1954), and it has been suggested that its effect may not only be one of increasing soil moisture but also of depressing toxic concentrations of aluminium and manganese (See p.106).

A groundnut-shell mulch improved the rainfall acceptance properties and increased the soil organic matter in savanna soils under continuous cultivation at Samusai, Nigeria (Jones, 1971).

Perforated plastic sheets placed at 5 cm depth in sandy soils conserved soil moisture sufficient to enable growth and fruiting of egg plants (*Solanum melongena*) dependent solely on 431 mm rainfall in central Sudan (Gerakís and Tsangarakis, 1969).

The use of mulches, valuable though they are, has not been practised as much for annual crops as for perennial plantation crops. Their use in tropical plantations is discussed later.

12.3.3 Animal Manures

Many annual crops benefit from dressings of animal manure but its use is very restricted because supplies are scarce. Experimental evidence from many countries show that cattle manure maintains soil fertility well in continuous upland arable cropping. Farmyard manure may indeed be more beneficial than can be inferred from its N, P and K content (Djokoto and Stephens, 1961 *a*).

Soil fertility experiments in Uganda demonstrated these effects (Jameson, 1970). At Serere, a five-year rotation started in 1936 incorporates various kinds and durations of 'rest' periods and also FYM given at 0, 2.5 and 5 tons per acre (0, 6.2 and 12.4 tons/ha) every five years. Farmyard manure increases yields of all crops, except sweet potatoes, more than do 'rest' periods. Residues from FYM increased crop yields even twenty years or so later. Resting the land three years out of five without FYM maintained soil fertility only marginally, but 2.5 tons/acre of FYM (6.2 tons/ha), or more, every five years, with a two-year 'rest' maintained or slightly increased it.

The permanent manurial experiment at Serere has been continuously cropped since 1933. Farmyard manure at 0, 10, 20 and 30 tons FYM per acre (0, 25, 50 and 75 tons/ha) has been given with 2 tons per acre (5 tons/ha) of lime every third year. Responses to FYM were of the order of 100 per cent. Lime increased yields significantly but much less than FYM. 10 tons per acre (25 tons/ha) of FYM every three years seemed to maintain soil fertility constant under continuous cropping at Serere.

Results obtained at Serere were confirmed in experiments at Kawanda in Uganda. They showed that without FYM the land must be 'rested' for at least half the time in a rotation, to maintain soil fertility constant, but with 10 tons of FYM per acre (25 tons/ha) every three years, continuous cropping is possible. Invariably, care is necessary to prevent accelerated soil erosion.

Dennison (1961) reviewed the effect of FYM in maintaining fertility in northern Nigeria. Continuous monocropping was possible with some crops, not with others, if FYM was applied Some crops responded better than others to FYM compared with inorganic fertilizers. Guineacorn (*Sorghum vulgare*) responded best and 3 tons/ acre (7.5 tons/ha) FYM was found to be optimum. Dennison (1961) summarized the recommended levels of FYM in northern Nigeria. In west Africa, generally, FYM was best given at planting time rather than during the previous year, or on a grass ley.

In continuous cultivation experiments at Samaru, Nigeria, Jones (1971) found that an organic matter content of 1 per cent in the topsoil could be maintained either by an annual application of 7-8 tons of FYM per ha or by putting the land down to a planted grass fallow for three years out of every six, preferably with the application of mineral fertilizer as well.

Dennison (1961) reported that in southern Guinea, the three-year fallow was unnecessary if 2 tons FYM per acre (5 tons/ha) was given annually, and 3 tons FYM per acre (7.5 tons/ha) gradually built up soil fertility when growing crops such as millet and guineacorn, but not others, e.g., groundnuts.

Farmyard manure has benefited peasants' holdings in Kenya (Webster, 1954) in modified forms of shifting cultivation and in continuous cropping (Djokoto and Stephens, 1961 *a* and *b*). Its persistent residual effects may be caused by its phosphate content or by FYM making soil phosphate more available to plants.

12.3.4 Fertilizers

Much of the fertilizers used in the tropics are for plantation and cash crops but very little for food crops. In India, fertilizers for food crops have increased only since 1952 with the Five-Year Plans (Agarwal, 1965). In Ceylon, plantation crops such as tea, rubber and coconut, were heavily fertilized in the past forty years; the quantities used in 1968 were tea 147,000 tons, rubber 20,000 tons and coconut 63,000 tons. For rice, only 791 tons in 1951 increased to 84,510 tons by 1969 (Kalpagé, 1970). Plant nutrients given to other arable and horticultural crops and pastures almost doubled between 1965 and 1968 but were still only 52,000 tons. This has been so elsewhere in the tropics as well.

Upland arable farms in the tropics use only small amounts of fertilizer partly because peasant farmers are conservative, unfamiliar with fertilizers and not informed in their use, but mainly because to them the returns are uneconomic (Watson, 1964). Agarwal (1965) stated that the Indian farmer generally pays more for fertilizers but gets less return for his crop than the average European farmer. Other constraints against the more intensive use of fertilizers in the tropics are that ready cash and cheap sources of credit are not available, fertilizers cannot be bought at the proper time and irrigation water is scarce.

That fertilizers are profitable and can maintain and increase soil fertility has been shown in the field, in experiment stations (e.g. Djokoto and Stephens, 1961; Watson and Goldsworthy, 1964) and, more recently, in farmers' own plots (Richardson, 1966; Anderson, 1969). In the semi-arid areas of Africa (Russell, in Moss, 1968), phosphorus and nitrogen are considered most likely to limit crop yields, though sulphur may also be needed sometimes. Repeated fertilizing builds up poor soils remarkably but too much fertilizer must not be used because rapid vegetative growth depletes the limited water resources. The crop then suffers from drought if rains are interrupted for long. Residual phosphate is valuable for crops (Boswinkle, 1961 ; Peat and Brown, 1962).

13 Flooded Rice Soils

RICE HAS been grown in Asia from time immemorial in the flood plains of river banks, in alluvial valleys, in the silt and mud of river deltas, wherever rain or irrigation water could be impounded. The technique of permanent wet rice production probably originated in China and spread to other tropical Asian countries. Outside Asia it was introduced into the coastal plains in Surinam and British Guiana (Gasser, 1961), on land reclaimed from mangrove swamps in Sierra Leone, on the flood plains of rivers in northern Nigeria (Higgins, 1964), along the Gambia River, and in inland valleys in Madagascar. Rice is increasingly grown in many tropical American countries. Imle (1967) suggested that parts of the Amazon basin, flooded annually, could become a 'rice bowl' area larger than the Mekong Delta.

Before about 543 B.C., rice was probably grown in Ceylon's uplands. From about 420 B.C., large tanks were constructed to conserve water for irrigating rice fields and for domestic purposes. The ancient tanks dotting the dry zone in the north-east and south-east are considered feats of engineering skill. Foreign invasions and malaria epidemics caused the population to be decimated and the tanks were abandoned. These are now being repaired and restored and the acreage under rice expanded above the present 1.5 million acres.

13.1 Soil Fertility in Flooded Rice Soils

Flooded rice culture is an adaptable system of land use suitable for a wide range of soil types and helps to maintain soil fertility more than any other permanent system of land use in the tropics.

Little soil is lost by erosion from the levelled or gently sloping, bunded fields. Also, soil and nutrients flow in with irrigation water, or in the water washed down from adjacent high land. Little nutrients are lost by leaching. Much of the irrigation water is lost by evapotranspiration. Water percolates slowly through the heavy soil, even more slowly through the structureless cultivated and puddled soil. Organic matter is added regularly by ploughing in crop residues and weeds, and, of course, none is burnt as in shifting cultivation.

Organic matter decomposes anaerobically in flooded 'reducing' rice soils more slowly than by oxidation in dry-land soils. This results in reduced substances such as methane, ammonia and sulphides being formed. Cations such as Ca^{2+} and Mg^{2+} are released by exchange with NH_4^+ and the reducing conditions solubilize and make available increasingly, ions such as Fe^{2+},

Mn^{2+}, phosphate and silicate. Because organic materials decompose slowly, many nutrients are increasingly available to plants and soil fertility is maintained well.

Significant amounts of nitrogen are fixed in waterlogged rice fields. Thirteen species of blue-green algae, common in the rice fields of south-east Asia were shown by Watanabe *et al.* (1951) to fix nitrogen. Azotobacter probably fix nitrogen when these organisms are associated with living rice roots (Uppal *et al.*, 1939), while N-fixing *Beijerinckia*, found in tropical rice soils, are also thought to fix very large amounts of atmospheric nitrogen (Harmsen and Van Schreven, 1955).

13.2 Characteristics of Rice Soils

The rice crop can adapt well to different conditions so long as soil water is adequate for growth. Water, perhaps more important than the nature of the soil, is provided by rain, flooding or irrigation. Medium and heavy soils are best, especially when not too freely drained, but irrigated sandy soils, with much organic matter and fertilizers, frequently give good yields.

Soil analysis does not indicate with certainty whether a soil is suitable for rice because of nutrients present in irrigation waters. Good rice soils are mostly acid, *p*H 5.5 to 6.5, but rice does grow well in some very alkaline soils, *p*H> 8, on the coasts of Bengal, Orissa and Madras. Rice soils in Peru, where water is abundant and paddy is transplanted, are very alkaline. Saline soils, close to the sea and occasionally flooded by sea water, can grow a few salt-resistant varieties of paddy.

Soil structure is of little or no significance in flooded rice soils. The puddled rice soil is a creamy mud, devoid of crumbs. But structure may be important with regard to the behaviour of the soil particles after puddling. Certain soils may be soft directly after puddling but may afterwards become hard and compact, thereby impeding percolation of water and oxygen. Application of organic manures to such soils will improve their physical condition and have a favourable influence on soil productivity. Only on strongly reducing soils should the application of organic manures be withheld. Ahmad (1963) showed that the upward movement of gases and the reduction of iron can affect the structural stability of the drained soil. Aggregation is initiated by gas channels made by the escaping gas. The oxidation of reduced iron, on drying, coats the aggregates and makes them stable.

13.3 Redox Potentials in the Study of Rice Soils

13.3.1 Redox Potential Profile

In the 1930's workers in Japan investigated the gradual decrease of ammoniacal nitrogen added to rice soils as well as a root-rot disease of rice plants caused by sulphide produced in certain submerged soils (Aomine, 1962).

Shiori and Aomine (1937) solved the dentrification problem by measuring the redox potentials of the different layers observed in the profiles of submerged rice soils. These investigations, together with the researches of Pearsall and Mortimer (1939) on soils, muds and natural waters, clarified the E_h profile of a flooded rice soil (Figure 7). The principal layers in a well-drained flooded rice soil are :

(i) the water layer, 15-30 cm deep ;

(ii) an oxidized layer, yellowish-red in colour, 1-10 mm thick, depending on soil type. This surface oxidized layer is characterized by high E_h values, relatively low *p*H values and oxidized ionic forms such as nitrate, ferric, manganic and sulphate. Diffusion of air into the soil from irrigation water, algae growing on the soil surface and air excreted by plant roots are responsible for the oxidized condition of this layer ;

(iii) a reduced layer, greyish in colour and extending right down to the compact plough-sole. Reducing conditions prevail in this layer ; E_h is low and *p*H relatively high. Ferrous, manganous, ammonium and sulphide are the predominant ions. In the presence of sulphide. ferrous ions may react to give black ferrous sulphide ;

(iv) an oxidized subsoil beneath the plough-sole where iron and manganese (Baars, 1950) leached down with the irrigation water from the reduced layer, are usually precipitated as coatings on the soil peds or sometimes form as concretionary layers. This is often black in colour as a result of black coatings on the structural units of the soil ;

(v) a reduced zone above the water table.

In a poorly drained soil (iv) is not present and the entire profile below the surface oxidized layer is in a reduced condition.

Shiori and Aomine (1938-1940) drew attention to the fact that oxidized conditions prevail around living rice roots in submerged soils and that the E_h in such areas was higher. Kumada (1949) and Mitsui (1949) proved that the rhizosphere was oxidized by rice roots.

13.3.2 Changes in a Rice Soil on Flooding

On flooding a rice soil the following changes occur ;

13.3.2.1 Redox Potential : 1. The E_h decreases. The rate of decrease is governed by (*a*) air-drying prior to submergence ; (*b*) the readily decomposable organic matter ; (*c*) the extent to which the soil is agitated by turbulence ; (*d*) soil temperature.

2. The E_h of the plough layer approaches a minimum value within a few weeks after flooding and then gradually rises.
3. On draining a rice field, the E_h rises rapidly. The rate of increase depends on the structure of the soil. The E_h rises rapidly in granular porous soils and in those with a single-grained structure.

4. E_h changes fluctuate daily and with season.

13.3.2.2 *p*H: 1. *p*H values of acid soils increase asymptotically to between 6.5 and 7.2 within four to six weeks of flooding (Ponnamperuma *et al.*, 1966).

2. *p*H values of sodic and calcareous soils decrease to above values.
3. On draining, the *p*H changes are rapidly reversed.
4. The *p*H of soil solutions were less than those of the soil suspensions for alkali and calcareous soils from the start of submergence, and for acid soils after reduction. The apparent inversion of the suspension effect has been attributed to the smaller CO_2 concentration in the soil slurry than in the soil solution (Ponnamperuma *et al.*, 1966).

The increase in *p*H of acid soils was quantitatively related to the potential of the Fe (OH)-Fe^{++} system, while the decrease in *p*H of the alkali and calcareous soils was defined by the partial pressure of CO_2 through the Na_2 CO_3-H_2O-CO_2 and $CaCO_3$-H_2O-CO systems, respectively. (Ponnamperuma (et al., 1966). The *p*H values of reduced acid soils high in iron appeared to be determined by the Fe_3 $(OH)_8$-H_2O-CO_2 system.

13.3.2.3 NUTRIENT AVAILABILITY: Nitrogen is lost gradually in several ways. Nitrate nitrogen, moving downwards through the reduced zone, is denitrified and lost as gaseous nitrous oxide or nitrogen. This fact is of significance in the nitrogen fertilization of flooded rice. If ammoniacal nitrogen is broadcast on the surface, nitrification occurs by the action of autotrophic organisms. Nitrate nitrogen thus produced is denitrified by microbial and chemical processes as it passes through the reduced layer and lost as N_2O or N_2. Ammonium forms of nitrogen are therefore either ploughed into the reduced layer before flooding or placed, in Japan in the form of mud-mixed balls, shortly after transplanting. Ammonium nitrogen placed in the reduced zone to a depth of 5-10 cm is held by the soil colloids for use by the plant. Ammonium fertilizers should be applied shortly before flooding, so that no time is left for nitrification.

The form in which nitrogen is applied to rice is also important. The superiority of ammonium nitrogen is not because the ammonium is preferentially used by the rice plant, which can assimilate both NH_4^+ and NO_3 equally well. Denitrification and loss from the soil as nitrogen gas can be considerably more than that of suitably placed ammonium fertilizer.

Nitrate fertilizers could be effective when applied to the mat of surface roots at primordia initiation as was first recognised by Alberda (1953). These surface roots should be in a position to utilize nitrate applied as a top-dressing to the surface oxidized layer before the nitrate is leached downwards and denitrified.

Submerging a soil results in an increase in the solubility as well as plant availability of phosphorus. The increase in solubility is caused largely by ferric phosphate being reduced to more soluble ferrous phosphate. Aoki

(1941), studying the behaviour of phosphate under flooded and upland conditions, found that soil phosphate was least soluble in the *p*H range 5-7. Under lowland conditions the solubility below *p*H 5 and above *p*H 7 was much more than under upland conditions.

13.3.2.4 TOXIC SUBSTANCES : Very small values of E_h can be harmful to rice. Nishigaki *et al.*, (1960) considered soil E_h values less than 150 mV to be harmful E_h values greater than 300 mV to be good.

At E_h values much less than 200 mV, the concentrations of soluble iron and manganese are likely to be excessive (Ponnamperuma, 1955). Organic acids such as butyric acid can be harmful to rice roots. A particularly serious effect of very low E_h values is the reduction of SO_4^{2-} to S^{2-}.

If sufficient iron is present in the soil, S^{2-} ions are converted into black ferrous sulphide, FeS. In the absence of sufficient iron to immobilize the S^{2-} ions, H_2S is liberated. H_2S is toxic to rice roots (Mitsui, 1954), acting as a respiratory inhibitor and interfering with the uptake of a number of nutrient ions. In Japan the disease is known as 'Akiochi' and was found prevalent in the so-called 'degraded' rice soils, in which Fe^{++} and Mn^{++} had leached out of the reduced zone, as well as on ill-drained soils containing excessive organic matter.

Adding iron-rich soil material from the adjoining uplands is one method of combating 'Akiochi'. The use of fertilizer materials that do not contain the sulphate radicle is also recommended. Attempts have been made (Jeffery, 1960 ; 1961a and 1961b) to work out a range of E_h values considered optimum for rice culture. E_5 values ranging from 300 to 200 mV seem most suitable. At E_h values greater than about 300 mV the solubilization of Fe, Mn and phosphate would be inadequate to supply the requirements of the rice plant for these nutrients. At E_h values much less than 200 mV iron, manganese, sulphide and other toxicities can occur.

Yuan and Ponnamperuma (1966) investigated, in a green house experiment, the chemical retardation of reducing conditions in flooded soils as a remedy for disorders of the rice plant associated with low redox potentials and excess reduced compounds. Manganese dioxide (0.4 per cent by weight of soil), calcium nitrate $Ca(NO_3)_2\,4H_2O$ (0.4 per cent) and calcium chlorate $Ca(ClO_3)_2$ (0.1 per cent) were tried. Manganese dioxide showed promise as a remedy for physiological disorders caused by reducing conditions in soil and as an amendment for acid lowland rice soils. It retarded the decrease in E_h, corrected a physiological disorder of rice and improved growth as well as yield. Calcium nitrate and calcium chlorate were less effective and sometime even toxic.

The reduction by manganese dioxide was retarded by the high E_0 of the system

$$MnO_2 + 2e + 4H^+ = Mn^{++} + 2H_2O$$

$$\text{for which } E_h = 1.23 - 0.0285 \log Mn^{++} - 0.118\,pH.$$

13.3.3 REDOX SYSTEMS IN FLOODED RICE SOILS

Ponnamperuma *et al.*, (1967, 1969) in a quantitative study of redox equilibria in flooded soils, regarded the dominant redox systems in reduced soils, other then the acid sulphate soils, to be the iron hydroxide systems. The bulk of the iron in such soils is ferroso ferric hydroxide, $Fe_3(OH)_8$. The principal redox systems in reduced soils, moderate to poor in sulphide, were shown to be (*a*) $Fe(OH)_3$-Fe^{++}, (*b*) $Fe(OH)_3$-$Fe_3(OH)_8$, and (*c*) the $Fe_3(OH)_8$-Fe^{++} systems. Arguing that the E_h, *p*H and ion activities of the soil solution were the thermodynamically meaningful measures of redox equilibria in flooded soils, Ponnamperuma *et al.*, (1967) measured E_h, *p*H and Fe^{++} activity of the soil solutions from 32 flooded soils sampled weekly for 17 weeks and showed that these values conformed closely to the following equations:

$$E_h = 1.06 - 0.059 \log (Fe^{++}) - 0.177\, pH \qquad (a)$$
$$E_h = 0.429 - 0.059\, pH \qquad (b)$$
$$E_h = 1.373 - 0.0885 \log (Fe^{++}) - 0.236\, pH \qquad (c)$$

Equation (*a*) held for the entire period of submergence and equations (*b*) and (*c*) especially after water-soluble Fe^{++} reached peak concentration.

Studies on the manganese oxide systems (Ponnamperuma *et al.*, 1969) showed that the manganese oxide involved in redox equilibria in flooded soils that undergo seasonal oxidation-reduction are complex non-stoichiometric oxides of variable composition. Manganese is more readily reduced in flooded soils than iron. An E_6 value of 500 mV appears to be transition for Mn^{++} $Mn^{++} \rightarrow Mn^{++++}$ (Aomine, 1962).

13.4 Land Preparation for Rice Culture

Land preparation varies according to local tradition, nature of soil and water supply. In the case of rain-fed rice, the aim is to take full advantage of the rains.

The first field operation is usually that of repairing and clearing irrigation channels and constructing the bunds or levees. The field is divided by bunds according to the contours. On flat land the divisions can be large and the bunds in straight lines whereas on hillsides the divisions are small and the bunds run along the contours. Bunds are constructed of mud and weeds, convenient openings being made for water inlets and outlets from each division. The land within a division is approximately level but may slope very gently from inlet to outlet to allow a slow superficial movement of water.

The soil is ploughed in the wet state to a depth of 5-10 cm with a single-toothed wooden or light iron plough. Deep ploughing is unnecessary and inadvisable. The lumps are broken up and the mud is made soft by using a wooden-ribbed roller. Weeds are turned in by being dragged under the mud. After ploughing, the land is left for three or four weeks, during which time the nursery is planted. Then a second ploughing is given, the land harrowed and the weeds turned in or removed and allowed to die on the bunds or allowed to rot in heaps and then trodden under the mud. The mud is worked into a regular soft consistency, excess water is drained off and the surface smoothened and levelled.

Six-week old seedlings pulled out from the nurseries are transplanted. In Sri Lanka, one seedling is placed at each point, four to six inches apart; in Burma, one to four plants per 'hill' are spaced from four to eight inches apart.

13.5 Irrigation Water

The supply and control of irrigation water is the most important aspect of flooded rice culture under irrigation. Rice is planted in a well-soaked field with little standing water, the depth of the water being increased as the plant grows until the depth is from 15-30 cm. The water layer should never be deeper than necessary for weed control because a thick water layer exerts an unfavourable influence on tillering and reduces yields. During the first three weeks, a relatively low water level is important for the emergence and growth of the seedlings. Too low a water depth may, however, establish a weed growth that is difficult to control afterwards. After this initial period, the water depth is primarily for the purpose of adequate weed control. When the plant flowers, the water should be drained off gradually until the field is dry at harvest time. During growth, the water should not be stagnant but flowing gently. Fields are drained for weeding or for applying top-dressings of fertilizer.

It is advisable to drain the rice fields a few times to aerate the soil and to prevent the formation of toxic reduction products such as H_2S, excess Fe^{++} ions, and low molecular weight organic acids. Under strongly reducing conditions, not only will the soils lack free oxygen but also the free oxygen originating from the special oxygen-conducting system in the rice plant will be made unavailable thereby interfering with the absorption of K and P. Adequate drainage is essential for eliminating these unfavourable soil conditions.

Water consumption depends on land preparation, extent of initial flooding, evapotranspiration and seepage. The water requirements of a 4-month rice variety cultivated in Sri Lanka are given in Table 52.

The quality of irrigation water depends on its origin. Water may contain useful nutrients, toxic substances, and suspended material like silt and clay. A reasonable quantity of coarse silt has a favourable effect on the soil.

13.6 Rotations with Rice

Where the water supply is seasonal and restricted, one crop is grown annually and the land lies fallow, possibly grazed by domestic or draught animals during the rest of the year. Animal dung and the ploughing in of weeds and stubble maintain soil fertility at an adequate level for one crop.

If the supply of water is ample and controlled and drainage is good, two

TABLE 52: WATER REQUIREMENTS OF FLOODED RICE (Murakami, 1968)
(Water expended in two cultivations of a 4-month variety in Sri Lanka)

Period	*Requirement*	*Maha* (November-February)	*Yala* (May-August)
Prior to sowing	For puddling and levelling and maintaining field at saturated moisture condition for 2 days	170 mm	170 mm
From sowing to transplanting	Evapotranspiration (Em/days) (No. of days)	(3.3 mm) (20) = 66 mm	(6.5 mm) (20) = 130 mm
From transplanting to harvest	Transpiration (TR) (DW)	(340) (700 kg/10a) = 238 mm	(450) (700 kg/10a) = 315 mm
	Evaporation (Em/day) (No. of days) (E/Em)	(3.3 mm (110) (0.54) = 196 mm	(6.5 mm) (110) (0.54) = 385 mm
From sowing to harvest	Percolation (Percolation/day) (No. of days)	(7mm) (130) = 910 mm	(7 mm) (130) = 910 mm
From sowing to harvest	Total	1580 mm	1910 mm

: are 100 sq. metres = 140.47 acre

TR : Transpiration ratio

DW: Dry weight

Em : Evaporation from a free water surface
E : Evaporation in the field

crops of rice can be grown each year. This is done in parts of the dry zone of Ceylon where irrigation facilities are available for the *yala* crop (April to September). Short term, non-photoperiodic varieties are necessary for this purpose and a strict regimen of cultural operations are required of the farmer. In the Ton King delta of Vietnam, 50 per cent of the land is double cropped with rice (Gouson, 1953).

Double cropping with rice followed by some other annual requires soils which can be well drained for crops needing aerated soils. The farmers of China have practised this for several centuries. Intensive multiple cropping is practised with crops planted in rows and fertilized with animal manures, night soil, well-rotted composts, crop residues and green manures.

Farmers in the Tangong Karang area of Malaysia have also practised intensive crop rotations with rice. Crops grown include groundnuts, sweet potato, okra, brinjal and cowpea (Allen, 1956). Bradfield (1969), working on multiple cropping in rice fields at the International Rice Research Institute, Los Banos, Philippines, experimented with a number of crops such as sweet potatoes, soya bean, corn, sweet corn and grain sorghum. He suggested cropping sequences for several different rotations.

13.7 Classification of Rice Soils

Rice soils are not a group of natural soils for which the classifications discussed in chapter 4 can be directly applied. Special methods of classification are therefore necessary. Such methods have been developed mostly in Japan where rice soil classification studies began in the 1930's. The first classification of rice soils was proposed by Kamoshita (1940). He considered rice soils to be natural hydromorphic soils and divided them into five categories. When it became necessary to include irrigated rice soils also in a scheme of classification, further criteria were necessary and contributions to rice soil classification came from Uchiyama (1949) and Kanno (1956).

Yamazaki (1960) defined rice soil formation as the process of formation of an eluvial A horizon and an illuvial B horizon from 'parent soil'. He distinguished *groundwater* type soils and *surface water* type soils, with no intermediate type in between. Kanno (1962) proposed a classification system based on the differences in hydromorphism as expressed in differences in gleying and mottling.

Oyama (1962) suggested a classification of rice soils independent of hydromorphism. He classified rice soils according to the most diagnostic horizon and recognised seven Great Soil Groups each with a diagnostic horizon.

In 1963, Japanese pedologists adopted a system of rice soil classification based essentially on the proposals of Kanno (1962), while Matsui (1966) modified this in his classification, based on the characteristics of the original soil, hydrologic conditions, and the resulting hydromorphism.

Tokutome (1970) used a modification of Matsui's system to classify the rice soils of Sri Lanka at six different levels. In Matsui's system, *surface water* rice soils are divided into *series* based on the degree of mottling in the subsoil. In some *surface water* rice soils in Sri Lanka, the difficulty of observing mottles in the subsoil, led Tokutome to classify them into a series according to subsoil colour. Again, Matsui's classification does not provide specific criteria for distinguishing between organic and mineral rice soils or for subdividing organic rice soils into a series. Criteria used by other workers were therefore adopted for this purpose. Rice soils containing more than 20 per cent organic matter in the organic horizon were termed organic rice soils (Thorp, 1935; Buckman and Brady, 1964). Organic rice soils were further characterized as 'thick', 'medium' or 'thin' according as the surface organic matter horizon exceeded 50 cm, was between 20 and 50 cm, or was less than 20 cm respectively (Japanese official method, cited by Kamoshita *et al.*, 1967).

14 Plantation Soils

SOILS ON WHICH crops are grown on a plantation* scale in the tropics are often well-managed and their fertility is maintained at high levels to ensure maximum productivity. Various techniques are often employed on such soils to ensure soil and water conservation, good aeration and drainage, a favourable structure for optimum moisture retention and easy root penetration, a soil reaction which will prevent nutrient deficiencies as well as freedom from harmful organisms. Together with high-yielding planting material, such practices have contributed to a high level of productivity on tropical plantations.

Table 53 indicates the climatic and soil requirements of the more important crops grown in large plantations in the tropics.

14.1 SOIL REQUIREMENTS OF THE PRINCIPAL TROPICAL PLANTATION CROPS

14.1.1 SOIL DEPTH

For the deep-rooting crops an adequate depth of soil is a prime requirement. Tea, for example, has a long tap root and requires a deep well-drained soil for proper root aeration. Deep, pervious soils without any hardpans or other impervious layers are essential for cocoa, coffee, rubber, cotton, coconut, oil palm, sugarcane, banana and citrus. Cocoa has a tap root which bifurcates in the subsoil and reaches down to a depth of 1-2 m. The tap root divides up into fibrous terminals in the subsoil and these serve mainly for the absorption of water. On the surface, a dense mat of fine fibrous roots are found beneath the decaying organic material (Jacob and Uexküll, 1960). *Arabica* coffee has the bulk of its root system in the top 2 m but some roots penetrate to 4 m. Tobacco, on the other hand is a comparatively shallow-rooted crop and will do well on the shallower soils. The existence of a well-developed root system extending into the deeper soil layers will enable the crop to explore the subsoil for moisture and nutrients.

14.1.2 TEXTURE

The water and nutrient requirements of the crop together with the tolerance of the root system to soil air and water conditions will determine the texture of the surface and subsoils for each crop.

For tea, surface layers of loamy sand to sandy loam texture underlain by a sandy or silty clay loam, are most suitable. Soils with finer textures may

*There is no generally accepted definition of a plantation. Usually a plantation is differentiated from a small holding merely on the basis of size but this is not always a valid criterion and varies from country to country.

TABLE 53: Climatic and Soil Requirements of Tropical Plantation Crops

Crop	Climatic requirements				Soil requirements						
	Temperature	*Rainfall*	*Elevation*	*Other factors*	*depth*	*texture*	*structure*	*aeration*	*water*	*humus*	*pH*
Stimulants											
Tea	13-27°C high humidity	moderate to high	moderate	no frost, shade	deep		variable	good	good whc, well-drained		4.0-5.5
Cocoa	25.5°C with a fluctuation not more than 90°C	1250 mm	not more than 600 m	shade	deep	loamy		good	good whc, well-drained		6.5-7.5
Coffee	*arabica*: 16-24°C *robusta*: higher temp.	1700-2000 mm with wet and dry season	higher altitudes, low lands	no frost or cold winds	deep	loamy	friable	good	moist sub-soil	good supply	4.5-7.0
Tobacco	wide range from temperate to tropical					modera-tely heavy		good	well-drained		5.7-7.5
Starch crops											
Potatoes	cooler mountain areas of the tropics	(<17°C)				sandy loams		good	good whc, well-drained		4.8-6.5
Sugarcane	25-28° C cool dry weather for ripening	at least 1500 mm rainfall mostly during hot growing season, or irrigation		sensitive to frost	deep	heavy		good	well-drained, good whc	good supply	6.0-8.0

Agro-industrial crops											
Rubber	26-28°C	1500 mm well distributed	not more than 600 m	susceptible to wind damage	deep			good	good whc		3.5-8.0
Cotton		at low elevations under irrigation		dry weather during flowering, ripening	deep	medium loams	porous	good	good whc		5.0-6.0
Oil crops											
Coconut		on sea coasts of tropical areas		high light intensity.	deep	light or loamy		good	well-drained		6.0-7.5
Oil palm	20°C	1500-1600 mm well-distributed		sensitive to prolonged drought ; flat or gently sloping terrain is preferable	deep	moderately heavy		good	good whc ; well-drained	good supply	5.0-5.5
Fruits											
Pineapple	21-27°C	760 mm well-distributed	100-800 m; sloping land	sensitive to prolonged drought; also to direct sunlight	moderately deep	sandy to loamy		good	well-drained	good supply	5.0-6.5
Banana	25-28°C	1500-2000 mm well-distributed	lowland	damaged by strong winds		moderately heavy		good	well-drained	good supply	6.0-7.5
Citrus	subtropical and tropical; irrigation may be necessary		lowlands		deep, up to 6 ft.	light, sandy to medium loams		good	well-drained	moderate supply	4.0-7.8

whc = water holding capacity

be used if the structure is favourable. Pervious soils of medium to fine texture are best for cocoa, although coarser-textured soils may be satisfactory under irrigation or in areas which do not have long dry spells. A friable soil of medium-loam texture is desirable for coffee ; stiff clays or sandy loams are unsuitable. Tobacco cannot tolerate a high watertable and waterlogging, and quality is affected by soil texture. Sandy loam soils with poor waterholding capacity and soluble mineral content tend to produce large, light-coloured leaves with a weak aroma, in contrast with finer-textured soils. On clay loams and clays, growth may be luxuriant but the quality is poor; the tobacco contains much nitrogen and nicotine but is poor in potassium and carbohydrate.

Rubber, potatoes, bananas and citrus will grow on soils of varied texture. For potatoes, sandy loams are best for early crops while loams, silt loams and organic soils are better for late crops. Commercial plantations of bananas of good quality are restricted to the sandy loams and light clays of alluvial valleys. Deep, well-drained soils are necessary for a satisfactory cultivation of bananas. Simmonds (1966) gave analytical results for a number of soils growing bananas.

For cotton, fine-textured soils were preferred at one time owing to their higher productivity but such soils promote later maturity and greater vegetative growth that is subject to more extensive boll-weevil damage. Sandy loams, loams and well-granulated clay loams are now considered best. Cotton soils should be deep, porous, and well-aerated, so that the roots can extend easily into the subsoil.

Sandy or loamy soils are most suitable for coconut, mainly found on coastal alluvial and colluvial plains. For oil-palm, deep, medium to heavy loam soils, friable, well-aerated and permeable, are preferred. In general, African soils, growing oil-palms, have smaller clay contents and CEC values than Malaysian and Sumatran soils (Hartley in Moss, 1968).

Sugarcane can be grown on a variety of soils but does best on deep soils with a medium-textured surface overlying a permeable, porous subsoil. A somewhat finer-textured subsoil can be useful in preventing leaching of nutrients and in retaining moisture at a suitable depth.

14.1.3 Structure, Consistence and Porosity

Arable crops generally need a well-aerated soil through which water drains freely without stagnation. Granular, blocky or crumb structures would result in the presence of a suitable proportion of larger pores in a soil, while a soft friable consistency will facilitate tillage and cultivation. On both light as well as heavy-textured soils, humus will promote such structures.

The coffee bush needs good soil aeration for maximum root development and root activity and a good porosity is therefore highly desirable. Well-granulated soils are necessary for cocoa, sugarcane, potatoes, citrus, rubber and oil-palm. For tobacco and oil-palm, friable soils with crumb structures are preferred. Tea soils should have fine to medium subangular, blocky structures. Loose sandy or loamy soils without much structure would suffice for coconuts if the annual rainfall is more than 1,500 mm. The oil-palm needs well aerated soils and cannot tolerate stagnant groundwater.

14.1.4 Soil Water

Although good drainage and aeration are essential, soils must retain water sufficient for growth, particularly during periods of drought. For tea, soils with a large water-holding capacity are essential if production is not to be affected during the drier seasons. Coarse-textured soils are therefore not suitable for growing tea. Cocoa, coffee and tobacco require much moisture throughout the growing season and, therefore, soils with a large water-holding capacity. Cotton needs a surface soil with good water-holding capacity and a subsoil that is permeable enough to allow the water to drain away easily. Coconut palms seem to thrive best where there is a plentiful supply of laterally-moving subsoil water, with good drainage in the upper layers.

14.1.5 Humus

Humus maintains good soil structure, a large water-holding capacity and a slowly-released reserve of nutrients, particularly trace elements. It regulates soil temperature and micro-organism activity. The humus content of tropical soils diminishes rapidly on cultivation and management techniques designed to maintain the humus content at a satisfactory level are important.

14.1.6 *p*H

Because tea, coffee, rubber and oil-palm are grown mostly on acid soils, induced micronutrient deficiencies do not show up. Potatoes can be grown over a wide range of soil *p*H from 4.5 to 7.0 but potato scab, a soil-borne disease, often becomes serious above *p*H 5.5. The optimum *p*H for good yield and efficient scab control is between 5.0 and 5.8. Rubber grows well on soils in the *p*H range 3.5 to 7.5 but in Malaysia, growth is poor in the range 6.5 to 7.5. The optimum soil *p*H for tobacco is between 5.5 and 6.5. More acidity decreases the uptake of calcium, magnesium and phosphorus and increases manganese and aluminium uptake excessively, adversely affecting quality. Soil reaction above *p*H 6.5 favours the incidence of black-root disease. Cocoa, sugarcane, cotton, coconut and bananas require soils with *p*H values around neutrality. Irrigated cotton does best at *p*H values in the range 6.5 to 7.0. On very acid soils, bananas are attacked by a soil-borne fungal infection, Panama disease (*Fusarium cubense*), whose incidence is much less on less acid soils. The most favourable soil reaction for the oil-palm is in the slightly acid range, though it does quite well even on neutral to slightly alkaline soils. For citrus, a relatively high base status in the soil is considered desirable to produce fruit of best quality, but excess base uptake tends to produce chlorosis and the optimum *p*H range for citrus lies between 5.5 and 6.5.

14.1.7 Nutrient Content

The nutrient needs of tropical plantation crops vary (Table 60). Thus, the requirements of rubber are less than those of coffee, cocoa and oil palm ; sugarcane takes up very large quantities of nutrients from the soil. However, provided the physical properties of the soil such as depth, aeration, drainage, water-holding capacity, and ability to withstand erosion are satisfactory, inherent plant nutrient deficiencies can be corrected by proper fertilizer application. Fertilizer use in tropical plantations is discussed in Chapter 7.

14.1.8 Special Soils for Particular Crops

Tea is grown mainly on the Red-Yellow Podsolic soils of the hilly regions of central Sri Lanka. The best tea soils in Indonesia are in the areas of volcanic ash and young, slightly weathered Gray-Brown Podsols.

Cocoa plantations are found on a wide range of soils. Charter (1955) showed that the bulk of the cocoa grown in Ghana, the world's largest producer, was found in the Ochrosols—surface soil of *p*H 6 or more, rich in exchangeable Ca and Mg. Little cocoa was grown on the Oxysols—surface *p*H less than 5 and poor in exchangeable Ca and Mg. In South and Central America, volcanic ash river delta soils are found best for growing cocoa. Young soils of volcanic ash origin with lateritic weathering are found suitable for cocoa in Indonesia. Alluvial clays as well as the Montserrat Series, derived from glauconitic sandstone, are some of the most fertile cocoa-growing soils in Trinidad. The best quality cocoa estates in Sri Lanka are found on the Reddish Brown Latosolic soils in the vicinity of Kandy.

The soils most widely used for *coffee* in Brazil are the deep Terra Roxa soils derived from diabase. Volcanic ash-derived soils are used for coffee in Brazil, Colombia and Costa Rica while coffee is grown in Peru on very acid leached lateritic soils. The main coffee soils of west Africa and India are derived from granite and gneiss. In Hawaii, large quantities of fertilizer are necessary on the degraded volcanic material-derived soils. Kenya's main coffee-producing soils are found on red lateritic lava-derived clays. Red Earths are used in Vietnam for *Arabica* coffee.

Subtropical black clays as well as ferruginous lateritic soils are used for growing *tobacco* in South Africa. In the United States, tobacco is grown mostly on the more sandy members of the Red-Yellow Podsolic soils. In Indonesia, young volcanic ash Regosols are used for growing cigar tobacco.

Sugarcane is grown successfully on alluvial soils, as for example in the Indo-Gangetic plain and in the lower Mississippi valley in the USA. In the Philippines, sugar-cane is grown on alluvial and volcanic red and yellow soils. In the West Indies, Red Podsolic soils and even Rendzinas are used in addition to alluvial soils. Sugarcane grown on the 'muck' soils of the Everglades in Florida do not usually require nitrogen but potassium and micronutrients must be supplied.

Reddish Brown Latosolic soils have been found excellent for growing *rubber* in Malaysia. The rubber plantations of Sri Lanka are also on the Reddish Brown Latosolic and Red-Yellow Podsolic soils of the low-country, wet and intermediate zones.

The regur soils (Vertisols) of India are well known for *cotton* but in India cotton is also grown on some alluvial and red soils. The cotton-growing soils of Egypt and Sudan are also mainly Vertisols. In Uganda and Morocco, cotton grows well on Red-Yellow Latosols. Cotton in the West Indies is found on fresh volcanic ash Lithosols, while the main cotton-growing soils of the United States are the Red-Yellow Podsolic and alluvial soils.

The most productive *coconut* plantations in the Philippines are on alluvial and volcanic ash loams of at least 60 cm depth. In Sri Lanka, sandy Regosols, Red-Yellow Latosols, Red-Yellow Podsolic soils of the wet zone lowlands, and Reddish Brown Latosolic soils have all been used successfully for coconut plantations. In many countries, coconuts are grown very close to the sea. Cassidy (1968) found that coconut palms show much tolerance of oceanic salt but can suffer damage at large salt concentrations.

Alluvial soils with good drainage have been found excellent for *oil-palm* in Africa. Oil-palm grown on some sandy Red-Yellow Podsolic soils on the Ivory Coast and other areas of west Africa is very deficient in potassium. African soils that grow oil-palms have less clay and exchangeable cations than the Malaysian and Sumatran soils (Hartley in Moss, 1968). In Nigeria, oil palm is grown mostly on acid, sandy soils, while in Indonesia the liparitic soils of Sumatra's east coast are very suitable. Malaysian soils growing oil-palm were described and classified at great soil group level by Panton (1964) and later by Leamy and Panton (1966). Of these, sedentary and alluvial soils constitute the largest number but lithosols, peats and organic soils are also used.

Commercial *banana* plantations in Central and South America are found on river alluvial sandy loams and light clays; in French Guinea and in the Cameroons on lowland, sometimes peaty, soil. In the West Indies, calcareous shale soils have been proved very satisfactory.

Most of the citrus in Florida is on the sandy Red-Yellow Podsolic soils, while in California alluvial soils, Regosols and others of varying texture are used. In Australia, citrus is grown on Reddish Brown Earths and on some podsolic and solodic soils.

14.2 Management Practices to Promote Soil Fertility

14.2.1 Tillage

Tillage operations on tropical soils are useful for preparing a satisfactory seed bed and to control weeds but the effects on water infiltration are only transitory. Tillage after planting should be no deeper nor more frequent than is necessary to control weeds. Subsoiling may be profitably used to break up a hard pan but if the impervious layer is of clay the effect is short-lived.

Ploughing and cultivation should, wherever possible, be done on the contour to check erosion. In areas of moderate rainfall, tie-ridging will facilitate water infiltration and retention except on very steep land.

A tractor-mounted rotary hoe was both safe and efficient in incorporating grass mulch into a Kenya coffee soil in a fifteen year factorial tillage trial (Jones and Wallis, 1963; Pereira *et al.*, 1964). Grass mulch incorporated with a modified rotary hoe was more successful than hand implements in maintaining soil structure. Without the grass mulch, the rotary hoe did more damage than hand implements.

14.2.2 Weeding

Weeds limit crop production more in the tropics than in temperate climates (e.g. Jones and Wallis, 1963). They grow more vigorously and luxuriantly soon after the rains and are particularly difficult to cope with in the humid tropics that have no marked or prolonged dry season. On the dry savanna soils, weeds are slashed and burned during the dry season.

Cultivation of crops after planting is necessary in the tropics mainly for weed control. But such cultivation should be kept to a minimum and should be as shallow as possible to control weeds. Excessively deep or frequent cultivation can lead to surface sealing and progressive deterioration in soil structure. Early weeding is particularly essential to reduce crop losses and in peasant farming this is seldom done. Mechanization can speed up land preparation and enable tillage for weed control to be carried out sufficiently early. Where farmers cannot afford the expenditure involved and where labour is inexpensive, weeds are periodically slashed.

Weedicides have not yet come into use in tropical agriculture as much as in temperate regions although a number of experiments on weed control are reported in the literature (e.g. Ishag, 1971). Their use is confined to well-managed plantations of perennial crops in areas where labour is costly, and, to a lesser extent, for annual crops on large intensively cultivated farms. Peasant cultivators use little or no weedicides. Weedicides help to weed land speedily at the right time. They are perhaps best used early as post-emergence sprays, soon after germination before the weeds have begun to grow vigorously, or as pre-emergence applications. 2, 4-D or Simazine has been found to be most effective for weed control in maize. Kasasian (1967) reported good weed control in five tropical root crops, with little or no crop injury.

Much basic research is however still necessary in the tropics on the ecology and biochemistry of weeds and on the mode of action of herbicides. Practical research is also required on the susceptibility of various crops, the most suitable chemicals, and how they are affected by crop species and environment. Effective weedicides for the pernicious rhizomatous grasses and hedges are required. The use of chemicals should, however, be regarded as a supplement to good husbandry and proper cultivation methods.

Clean-weeding was widely practised on tropical plantations at one time but even on moderately sloping land adverse effects were soon evident in the form of accelerated decomposition of organic matter, leaching losses and deterioration in soil physical conditions leading to erosion. Rainfall acceptance data showed clean weeding to cause 15 per cent reduction in infiltration from very heavy rain storms compared with minimum weeding treatment (Pereira *et al.*, 1964). Consequently, clean weeding is now restricted to circles around trees or to strips along tree rows and the rest of the soil is protected with a suitable soil cover.

14.2.3 Soil Covers

Considerable loss in soil fertility can occur in tropical plantations if the soil is not adequately protected at all times with a suitable cover. It is generally recognized that leaving a soil bare in the tropics is bad husbandry. In the humid tropics, the effect of battering raindrops during high intensity showers occurs almost continuously and can cause much erosion on even practically level land unless infiltration is increased and surface runoff is reduced. In drier areas, the effect of the first torrential downpours after a dry spell during which the ground cover has been thinned out can be equally disastrous. In such areas, morover, suitable covers can effectively conserve soil moisture and reduce evapo-transpiration. Management practices, in which soil covers are used, are therefore of prime importance in maintaining the fertility of soils in tropical and subtropical regions. Among the methods used to cover the soil and to protect it from the adverse effects of tropical climates are intercropping, the establishment of ground crop covers, mulching, shade trees and windbreaks.

Intercropping—Intercropping the land between the plantation rows can be harmful unless practised with circumspection. Care should be taken to ensure that the intercrops do not unnecessarily shade the plantation crop, or compete for nutrients and water. Excessive cultivation of intercrops may result in a deterioration of soil structure as well as loss of soil moisture.

Intercropping is practised where the return from the intercrops is profitable. Pineapples are grown very successfully in coconut plantations in the low-country wet zone in Sri Lanka. Sufficient fertilizer ensures an adequate nutrient supply to both coconut and pineapples. Elsewhere in the coconut growing areas, pastures are being established for rearing dairy cattle. Experiments are also investigating the feasibility of growing cereals and vegetables as intercrops in coconut plantations. Leguminous crops are most suitable for intercropping. Rainfall should not be limiting, otherwise the moisture supply to the main crop will be affected.

Intercropping is not possible with most other plantation crops. Shade provided by rubber is sometimes used for cocoa grown in rubber estates but tea and coffee are invariably grown in pure stands.

Ground covers—Herbaceous or shrubby species are commonly used as ground covers in tree plantations such as those of rubber (Table 54).

Leguminous creepers such as *Pueraria phaseoloides*, *Centrosema pubescens* and *Stylosanthes gracilis* or shrubs such as *Crotalaria* spp. are usually established as ground covers. Such leguminous covers have been shown to fix considerable amounts of nitrogen in pot trials (Watson, 1957; Watson *et al.*, 1963), but their chief disadvantage is that they compete with the main crop for moisture and nutrients. Where moisture is limiting they can adversely affect yields (Laycock and Wood, 1963), but on poor soils where the rainfall is good they can be beneficial provided the soil is fertilized.

TABLE 54: COVER CROPS USED IN THE TROPICS

Crop species	*Plantation*	*Remarks*
Calopogonium mucunoides	rubber, sisal, coconut, oil palm, cocoa	leguminous creeper; spreads quickly; tolerates poor soil and partial, but not heavy, shade; planted mixed with *Centrosema*
Centrosema pubescens	rubber, coffee cocoa, oil palm, coconut, sisal	leguminous creeper; drought resistant
Desmodium ovalifolium	rubber	shrub; tolerant to shade and drought; popular in Sri Lanka
sandwicense	sisal	
Dolichos argenteus		leguminous creeper; used for drier areas in Kenya
biflorus	sisal	
hosei		
Glycine javanica	sisal, rubber	a perennial, leguminous creeping or twining herb
Indigofera spicata		leguminous creeper
arrecta		shrub
Leucaena glauca	coffee	must be cut back regularly to prevent growth into shade tree
Mimosa invisa		has sharp thorns; is a fire hazard in dry season; competes strongly with cocoa and tree cassava during first 6 months Jordan & Opoku (1966)
Pueraria (phaseoloides =)	rubber, citrus, coconut, sisal, oil palm, cocoa	vigorous leguminous creeper; sheds leaves in the dry season; planted mixed with *Centrosema*; useful for heavier types of soil; shade tolerant
Stylosanthes gracilis		leguminous creeper

Apart from protecting the soil from sun, rain and wind, ground covers have other advantages. They add organic matter to the soil and improve its physical condition. The added organic matter releases plant nutrients as it decomposes and increases the CEC of the soil complex. Because soil structure is improved, root development of the main crop may be promoted and leaching losses minimized. In the case of deep-rooted cover plants, nutrient cycling can provide relatively shallow-rooted plantation crops with nutrients from the subsoil.

Mulching—The use of mulches as soil covers is widely practised in tropical agriculture. Mulching has several advantages and wherever cutting, transport and application costs are not excessive, mulching can maintain and improve fertility considerably. Jacks *et al.*, (1955) have described the various kinds of mulches and discussed their effects on soil properties.

Paper mulches have been successfully used in pineapple culture in Hawaii but they have been employed only to a limited extent elsewhere. In the tropics, mulching materials are generally crop residues such as sugarcane trash and banana leaves or easily grown bulky fodder crops such as elephant grass. In the wet tropics, with a fairly well-distributed rainfall, cover crops rather than mulches are favoured in rubber, oil-palm and coconut plantations. Where there is a pronounced dry season, however, mulches are of the greatest importance, especially in the cultivation of sugarcane and coffee.

Dust mulches are effective in conserving moisture in soils mainly because they inhibit weed growth but they have several disadvantages, especially under tropical conditions. The frequent cultivation that is necessary to keep the mulch intact, exposes the bare soil to erosion hazards, disturbs the root system and accelerates the destruction of soil humus. In some circumstances, the mulching action of naturally occurring stones may more than counteract their drawbacks in cultivation. Organic mulches such as grass, straw and vegetable mulches are the most effective.

Unlike live ground covers, mulches do not compete with the main crop for moisture and nutrients. They improve soil moisture conditions in a number of ways, suppressing weed growth, lowering soil temperature, restricting diurnal temperature fluctuations and protecting the soil from the wind (thus decreasing evapotranspiration losses). Also, by protecting the soil surface from the direct impact of the rain causing puddling, moisture infiltration is increased while surface runoff, and consequently soil erosion, is minimized. Increased earthworm, termite and millipede activity under a mulch provides many channels for the percolation of water. Under a mulch, root development is promoted, soil structure is improved, and the rate of leaching diminished. Experiments in Kenya (Pereira and Jones, 1954) showed that mulching before the rains, by making the soil more receptive to the rainfall, is of greater benefit than mulching after the rains when the mulch only helps to decrease the rate at which the soil dries out during an ensuing dry season.

The favourable effects of a mulch have sometimes been attributed to the nutrients contained in the mulch (Hardy, 1941; Wasowicz, 1952). The most marked effect is an increase in the available potassium content of the soil, suggesting that many common mulching materials have an appreciable content of water-soluble potassium compounds. Jacks *et al.*, (1955) quote P. A. Jones to indicate that studies in Kenya have shown that elephant grass (*Pennisetum purpureum*) has an extremely high potassium content; a standard mulch which adds about 10 tons of dry matter per acre (25 tons/ha), contains 900-1,200 lb (400-540 kg) of potassium. Where the soil does not contain much magnesium a need for magnesium may arise because the potassium in the mulch may set up a state of K/Mg imbalance (Richardson, 1963).

Nitrogen may be expected to be in short supply if the mulch has a large C/N ratio, so that a nitrogen fertilizer is necessary. But Greenland and Nye (1960) found that straws of C/N ratio up to 70 did not immobilize mineral nitrogen.

Jacks *et al.*, (1955) referred to the increased nitrification under paper mulches in Hawaiian pineapple soils and under a mixed organic mulch in Trinidad cocoa soils. In Tanzania, mulching of coffee allowed nitrification to proceed throughout the dry season. Jacob and Uexküll (1960) point out the possible hazard of the mulching material starting a fire after a prolonged dry spell and also the occurrences of rats and mice in mulch layers.

Mulches are widely used in the tropics for crops such as coffee, citrus, bananas, cocoa and pineapples. In the dry East Rift area of Kenya, for example, mulching has a greater effect on coffee yields than any other cultivation or manurial treatment. Pereira and Jones (1954) reported on experiments in Kenya, in which mulching alone increased the yield by 67 per cent. Elephant grass for mulching is often produced on plantations. The standard mulch applied to coffee, an alternate row mulch of 10 tons (dry matter) of elephant grass per acre (25 tons/ha) contains about 900-1200 lb K_2O (400-540 kg) and 500-600 lb P_2O_5 (230-270 kg). Banana trash is commonly used in banana gardens.

The beneficial action of mulches on the 'lunyu' soils of Uganda has been referred to (p. 106). In Hawaii, a high content of active manganese in the pineapple soils of the Wahiawa series was reduced by using a mulch of boards (Sherman and Fujimoto, 1946).

14.2.4 Shade Trees

Cover plants which provide shade for the main crop in addition to protecting the soil are known as shade trees. Shade trees are commonly used in cocoa, coffee and tea plantations as well as in some orchards (Table 55).

Temporary shade is supplied in young plantations until the plants establish themselves. This is provided either by artificial means or by interplanting with quick-growing annuals. Temporary shade is essential for young cocoa and is provided by interplanting with leguminous trees and shrubs, such as *Glyricidia maculata* or with food crops such as cassava, tannia and bananas. It may be beneficial for young coffee if rainfall is good but is not widely used for tea.

There has been much debate about the usefulness of permanent shade (e.g. van Dierendonck, 1959 *b*; Eden, 1961; Joachim, 1961). Shade reduces the light energy available for photosynthesis and influences flower initiation in fruiting crops. Growth and yields are therefore likely to be diminished but the nutrient requirements of shaded crops will be correspondingly lower. This is probably why greater responses are obtained to fertilizer applied without shade rather than with shade (Cunningham and Lamb, 1959). Experiments indicate an interaction between light intensity and crop nu-

TABLE 55 : SHADE TREES USED IN THE TROPICS

Tree species	*Plantation*	*Remarks*
Acacia decurrence	tea	used at elevations less than 1800 m.
Albizzia	tea, coffee	grows fast and easily from seed; sensitive to canker; for loamy soils
chinesis (sau)		
coriana		
falcata		
gummifera		for non-droughty areas
lebbeck		
odoratissima		not suitable for heavy soils; very susceptible to certain pests;
pr cera		erect, long-lived tree; less attacked by canker than *A. stipulata*;
stipulata		commonly used in Assam but is sensitive to canker; for non-droughty areas
zygia		
Derris robusta	tea	
Erythrina	coffee, cocoa, tea	popular for coffee in South America
lithosperma (dadap)		
subumbrans		
indica		
poepigiana		very susceptible to pests and diseases
Glyricidia	tea, coffee, cocoa	common on low-country tea estates in Sri Lanka and in South India
maculata		
sepium		
Grevillea robusta	tea, coffee	dominant shade tree in south India, in Tanzania
Inga	coffee	
inga		
laurina		
Leucaena	coffee, cocoa	used in Indonesia
glauca (leucocephala)		
pulverulenta		
Tephrosia vogelii	coffee, cocoa tea	susceptible to *Poria* root disease of tea

trition. Therefore, cocoa, coffee and tea do not necessarily thrive better under shade and shade *per se* may even depress growth and yields (Laycock and Wood, 1963). But the beneficial effects of shade are also evident.

Surface litter or loppings from shade trees provide a useful mulch that conserves moisture. When leguminous shade trees are used, total soil nitrogen is increased by symbiotic fixation. Deep-rooting shade trees extract nutrients from deeper soil layers and deposit them on the surface in the leaf litter. This recycling of nutrients is increased by pruning. The extent to which recycling of nutrients by shade trees can prove beneficial has been questioned because cocoa, coffee and tea plants can also forage deep into the soil and so bring about nutrient recycling (Nutman, 1933; Kerfoot, 1962). Root formation of shallow-rooting plants is limited if a protective shade cover is missing.

The root systems of shade trees and the increase in soil organic matter as a result of using shade promotes soil structure. Together with the fact that weed growth under shade is likely to be less, evapotranspiration under shade will be less if the shade trees are constantly pruned and transpiration from them is kept down. Damage to plantation crops by winds, hail or sun-scorch is reduced by shade trees. Eden (1961) estimates that tea production can be affected for as much as two months by hail damage to the foliage. The incidence of certain insects, such as thrips in cocoa and coffee, have been found to be less under shade, but shade tends to encourage blister blight in tea.

The principal disadvantages of shade trees are the reduction in light intensity for crop growth and the competition by shade trees for moisture and nutrients. Nor are all the nutrients absorbed by shade trees returned to the soil, because part is locked up in the wood or may sometimes be removed in the foliage. Shade trees must, therefore, be properly managed. They should be planted at a suitable spacing and should be regularly lopped to control the rate of growth and the density of the canopy.

Windbreaks—By reducing wind speed, trees planted as windbreaks diminish mechanical damage to trees, increase air temperature and humidity, and reduce evapotranspiration. They serve to decrease wind erosion of soil, particularly when it is dry and easily blown about.

14.2.5 Mechanical Soil Conservation Measures

The main types of mechanical measures for soil conservation have been discussed in a number of standard works (e.g. Bennett, 1955; Stallings, 1957). They are designed to ensure the interception and safe disposal of runoff from areas under crops as well as to protect such areas from runoff from higher land which may be uncultivated. Such soil conservation measures are common on well-managed tropical plantations. They include simple hillside ditches, often with grass strips planted along the upper side, contour bunds and ridges, crops planted on strips parallel to the contour, terraces of various types constructed on the contour, and storm drains with relatively wide and shallow channels.

14.2.6 Green Manuring

Green manuring in the tropics has been discussed in connection with the intensive arable cropping in the upland areas (see p 152).

14.2.7 Fertilizers

Fertilizer use in tropical plantations is discussed in detail in Chapter 7.

14.2.8 Soil Sterilization

Soil sterilization consists in removing organisms harmful to plant growth, or at least in reducing their numbers so that they are no longer of consequence, without permanently disturbing the balance of those organisms essential for soil fertility. Heat, gamma radiation, and chemicals are used to sterilize soils. In addition to killing harmful organisms, these treatments may also disturb the balance of useful organisms and therefore affect soil properties such as nutrient availability, especially the amount of plant-available nitrogen. In addition, the solubility of other nutrients may be increased and it is likely that heat treatment may improve soil structure, aeration and drainage.

Chemicals such as chloropicrin, formaldehyde, ethylene dibromide and methyl bromide are used in tropical plantations. Methyl bromide is used as a fumigant in controlling the root-lesion nematode (*Pratylenchus loosi*) in tea nurseries in Sri Lanka (Kerr and Vythilingam, 1967) and against the root-knot nematode in tobacco (Daulton, 1965). 1, 2-dibromo-3-chloro-propane (DBCP) is especially effective against root-knot nematodes and has been applied to many established perennial crops including tea, citrus and bananas (Wrigley, 1969). A mixture of 1, 2-dichloropropene and 1, 2-dichloropropane is used on pineapples and on sugarcane in Hawaii.

15 Tropical Grassland Soils

15.1 Natural Tropical Grasslands

MOST OF THE natural grasslands of the tropics occur in regions where the climax natural vegetation is forest or woodland and thus they form relatively unstable sub-climaxes. The total extent of the relatively high-quality grasslands derived from montane evergreen forests is comparatively small, being confined to the Kenya Highlands in east Africa, the higher altitudes in Peru, Equador, Venezuela and Costa Rica. The greater part of the natural grassland in the tropics is found in the drier broad-leaved woodland, thorn woodland and thicket zones. Here the rainfall is poor and unreliable and the vegetation is subjected to annual fires during the severe dry seasons (Table 56). Many of the pastures are dominated by coarse grasses, which are palatable and nutritious only when young, and legumes are scarce. The carrying capacity is low at all times and particularly during the dry season, when growth stops rather abruptly and the grass dries out rapidly and there is a marked lack of dry-season feed. During the rainy season the grass makes rapid growth and at the beginning of the growing season the grasses show a relatively high protein content and a fair nutritive value but these decrease rapidly thereafter. De Geus (1967) has summarized the estimated feeding values derived from the chemical composition of tropical grasses.

15.2 Problems in Establishing Legume-Based Pastures

It was thought at one time (Whyte and Trumble, 1953) that legumes cannot be easily introduced into tropical grassland and that legume-based pastures in the tropics were not as productive as those in temperate areas. Whyte (1962) considered that 'the great potentialities of tropical and subtropical countries are still largely a myth'. Since then, however, grassland scientists in a number of tropical or subtropical regions and particularly in eastern Australia, Hawaii, Nigeria (Okori *et al.*, 1965), the Congo, Uganda (Stobbs, 1969, b, c and d), Tanzania (Andersen and Naveh, 1965, 1968), Kenya (Suttie, 1968) and Malawi (Smith, 1962) have pioneered research on legume-based pastures for tropical regions. The importance of such research can be gauged from the fact that the equatorial, tropical, monsoon, and humid subtropical regions, with a distinct potential for increased cattle production, based on improved pastures cover about 27 per cent of the world's area.

It is now clear that the tropics of Latin America and Africa are rich in indigenous grasses and legumes with potential value for improved pastures (Williams, 1967; Hutton, 1970). Work on indigenous grasses at Kitale in Kenya (Edwards and Bogdan, 1951) has shown the presence of a valuable grass flora, including such grasses as Rhodes (*Chloris gayana*), Star (*Cynodon dactylon*), Molasses (*Melinis minutiflora*), Guinea (*Panicum maximum*), Kikuya (*Pennisetum clandestinum*), and Setaria (*Setaria sphacelata*).

Promising pasture legumes such as *Glycine javanica* and *Desmodium* spp., have been found to be native in South America. In the humid area of Peru, Napier or elephant grass (*Pennisetum purpureum*) was found best for cut feeding or silage; Guinea (*Pannicum maximum*), Pangola (*Digitaria decumbens*) or Napier grasses were best for grazing. Pangola grass was well suited to haying; tropical kudzu (*Pueraria phaseoloides*)—molasses grass (*Melinis minutiflora*) pastures yielded less than well-fertilized grasses but were adapted to steep, inexpensive lands (Vicente-Chandler, 1967). Pangola grass is an important pasture grass in several tropical countries but it is not recommended in Surinam because of the widespread occurrence of stunt virus disease (de Geus, 1967).

Verboom (1965) evaluated the indigenous legumes of Zambia for their potential use in agriculture. The value of certain legume fodder trees was shown by the proximate analysis of their seeds and pods.

The poor productivity of the majority of tropical grasslands, found in the drier regions, was referred to earlier. Research in northern Australia (Davies and Eyles, 1965) and elsewhere (e.g. Stobbs, 1969 *a*, *b*, *c*) showed that legume-based pasture is most economical for developing the cattle industry and that maintaining 40 per cent of a legume responding to phosphate fertilizer in a tropical pasture is the cheapest way to provide nitrogen for the pasture and the grazing animal (Hutton, 1968).

Rapid and effective nodulation of legumes is essential for their establishment, vigorous growth and significant nitrogen and protein contents. Although substantial amounts of nitrogen are fixed by several tropical legumes (Henzell, 1963; Moore, 1962; Jones *et al.*, 1967), they are not quite up to the level of white clover and lucerne in temperate pastures. Russell (in Moss, 1968) considers a rainy season of at least five months necessary for satisfactory nitrogen fixation by legumes in tropical African soils.

That the majority of tropical legumes and their associated rhizobia are adapted to acid soils is of special significance in tropical pasture development. Andrew and Norris (1961) showed that this was partly because they have the ability to extract calcium from acid soils poor in calcium. On such soils, widely prevalent in the tropics, lime applications are unnecessary for vigorous legume growth, and only superphosphate, with or without molybdenum, is usually needed for large yields of legume-based pastures. Norris (1959) found that *Rhizobium* associated with tropical legumes requires more magnesium than calcium, thus resulting in the use of dolomite in mixtures used for pelleting clover seed (Hastings and Drake, 1962). *Leucaena* is exceptional among tropical legumes and is well adapted to the cal-

TABLE 56: NATURAL TROPICAL GRASSLANDS (From information in Webster and Wilson, 1966)

Type of grassland	Climate	Altitude	Characteristic features	Distribution
1. Derived from low-altitude forest	>1400 mm rainfall	low	Coarse grass and scattered shrubs. Dominated by *Imperata Cylindrica*	Malaysia, Sri Lanka, Uganda West Africa.
2. Derived from broad-leaved woodlands	630-1400 mm rain fall; one severe dry season annually	low	Continuous ground cover under widely spaced trees	West and East Africa
			(a) Low tree-high grass *Hyparrhenia* spp. dominant 3-9 m high trees	
			(b) 'Miombo' type*	East Africa, Northern Nigeria Tanzania, Malawi, Zambia, Rhodesia
3. Derived from thorn woodlands and thickets	380-760 mm rainfall; seasonal	low and medium	Perennial grasses less than 1 metre high. Legumes are scarce. Low carrying capacity	Northern Guinea, Sudan, Somalia, Uganda, Kenya, Tanzania, Rhodesia, South Africa, India.
			(a) Scattered tree grasslands Good cover under flat-topped Acacia trees. Dominant grass is *Themeda triandra*	Central and South America

			(b) less productive grasslands associated with *Commiphora Acacia* bush and thicket	East Africa
4. Subdesert scrub and grass	250 mm rainfall	low	Ephemeral grasses and herbs during rains	
5. Derived from montane evergreen forests	1000-5000 mm rainfall	1200-3300 m	Undulating open grassland with short to medium-sized species, eg. Kikuyu grass	Kenya Highlands, Sri Lanka, India, Burma, Peru, Ecuador, Venezuela, Costa Rica
6. Alpine meadows	Near snowline	3000 m	Low-growing herbs and short grasses	Peru, Ecuador, Colombia, Venezuela, Bolivia
7. Swamp grasslands	1000-1800 mm rainfall	lowland river basins	Coarse grasses and unpalatable sedges	South America, Africa

*'Miombo' type consists of woodlands and savanna characterized by trees of the genera *Brachystegia*, *Julbernardia* and *Isoberlenia*. The ground cover is of tall tussocky grasses with *Hyperrhenia* spp. dominant, but the clumps of grass are usually widely spaced.

careous soils and soils of high base status as, for example, in the Caribbean and Hawaiian islands.

Experiments in Hawaii (Younge *et al.*, 1964) demonstrated the importance of adequate nutrients, notably P and Mo, for satisfactory growth and seed production of legumes in the tropics. In northern Australia, soils are very deficient in P and Mo as well as S, all of which are vital to legume growth. Tropical legumes are as much affected as temperate legumes by excess Mn (Andiew and Hegarty, 1969); both manganese and aluminium are often present in excess on the poor acid soils of northern Australia (Andrew, 1963; 1966).

15.3 Fertilizing Tropical Grassland

Most tropical grasses have a capacity for high photosynthetic rates (Ludlow and Wilson, 1968), and dry matter production of as high as 10,000 lb or more per acre (11,200 kg/ha) in response to nitrogen fertilization is usual in the humid tropics and sub-tropics (Oakes and Skov, 1962; Ahmad *et al.*, 1969). de Geus (1967) estimates the yield potential of grasses in the tropics as being much higher, up to 75,000 lb of dry matter per acre (84,000 kg/ha). But, although the dry matter response to nitrogen fertilizer is high, there are problems in uptake and recovery of nitrogen as well as in the maintenance of a sufficiently high nitrogen content. Among the grasses which give high yields of digestible nutrients with nitrogen fertilization are Pangola (*Digitaria decumbens*), Kikuya (*Pennisetum clandestinum*), Signal grass (*B. decumbens*) and Guinea grass (*Pannicum maximum*). Generally, however, except in the young growth stage, nitrogen fertilization of tropical grasses does not increase concentration of digestible nutrients and efficiency of conversion into protein in animal products.

Responses to fertilizer application depend to a large extent on rainfall and on the type of grassland. Fertilizer experiments on grassland in the Kenya Highlands showed that yield increases were directly related to rainfall and that fertilizers were profitable on pastures dominated by the better grasses such as Kikuya and Star grass (Dougall, 1954).

de Geus (1967) has reviewed the results of fertilizer experiments on grassland in tropical countries. The yield and nutritive value of grasses were markedly influenced by nitrogen applications. In most cases, applications of 200 lb N per acre per year (225 kg N/ha/year) were not excessive and it was often profitable to go much higher. Split applications of nitrogen were most effective. In pastures, nitrogen was best applied soon after each cut to obtain maximum yields. The protein content of grasses drops sharply as they mature, even when heavily fertilized with nitrogen but Guinea grass, heavily fertilized with nitrogen and harvested after thirty days, yielded a forage with a protein content exceeding that of most legumes.

The continued use of heavy applications of ammonium sulphate increases acidity and reduces the content of exchangeable bases. For intensive grass management, therefore, either the acidifying effect of ammonium sulphate should be controlled by regular liming (Abruna *et al.*, 1964) or less acidifying nitrogenous fertilizers should be used. Calcium ammonium nitrate (20-23 per cent N) has proved to be excellent for fertilizing pasture. But the sulphur content of ammonium sulphate may prove to be an advantage in areas where sulphur deficiency is liable to occur e.g. in Australia (Walker, 1955) and in Brazil (McClung and Quinn, 1959).

Where large dressings of nitrogen are used to maintain productivity at a high level, both phosphate and potassium fertilizers are required, particularly on soils deficient in these nutrients. Responses to phosphate are widespread. On phosphate-deficient soils, the phosphorus content of the herbage can be increased by the use of phosphate fertilizers to the value of 0.33 per cent in the dry matter, which Whyte *et al.*, (1959) consider as an absolute minimum for maintenance of grazing cattle. Higher levels of phosphorus are required for dairy cows and young growing stock. In calcium carbonate-rich soils, a P_2O_5/CaO ratio of 1.5 : 1 is regarded as being necessary to prevent a calcium-phosphorus imbalance.

Potassium deficiency is most likely on the lighter types of soil. Potassium requirements are also high when grasses are removed for pen-feeding and the farmyard manure is not returned to the field. When fields are grazed, the greater part of the potassium taken up is returned in the form of dung and urine and the dressings of potassium fertilizer can be reduced. Potassium deficiency makes grasses much more sensitive to *Helminthosporium*. An excess of potassium may result in magnesium deficiency.

Sulphur deficiency has been reported in Australia (Walker, 1955) and in Brazil (McClung and Quinn, 1959) where nitrogen applications are high and sulphur is immobilized in the soil organic matter on soils with low levels of soil sulphur and in areas far from industrial centres.

Among the micronutrients, molybdenum is a limiting factor over large areas in Australia and this may be corrected by the application of small dressings, a few ounces per acre, of sodium molybdate. Excess must be avoided as excess molybdenum can be poisonous to livestock. Copper and cobalt deficiencies are fairly common in Australia and South America. Sodium, chlorine and iodine deficiencies, reported in South and Central America, can be remedied by the provision of mineral licks.

In a series of studies conducted in the humid area of Puerto Rico (mean annual temperature 27.7 °C and annual rainfall 1500-2000 mm) on deep, red, acid latosols with kaolinitic clay minerals and high amounts of free iron and aluminium oxides, Vicente-Chandler (1967) reported that intensive management practices, including fertilization and liming, weed control and mowing, cutting and grazing procedures, resulted in yields of over 40,000 lb of good-quality cut forage (135 tons of green forage) per acre yearly (45,000 kg/ha), a carrying capacity of five head per acre on cut grass and of more than two head on pastures, and the production of more than 1,000 lb of beef per acre (1120 kg/ha) yearly on steep mountain pastures.

Cut grasses responded strongly to heavy applications of N, P, K and lime, responding economically to up to 5 tons of 14-4-10 or similar fertilizer per hectare annually. In terms of beef production, pastures responded economically to applications of 2.5 tons of 14-4-10 fertilizer. For maximum yields, the soils had to be limed to about *p*H 5.5, and one ton of limestone applied for each ton of acid residue fertilizer to increase the base saturation of the soil to 50 per cent or more and to prevent the accumulation of more than 2 meq of exchangeable aluminium per 100 g of soil. In the humid coastal region of Puerto Rico, Vicente Chandler, Silva and Figarella (1962) found it best to apply nitrogen immediately after each cutting and to make about four applications a year to Guinea grass pasture. Nitrogen and potassium uptake was more uniform with frequent fertilization.

In the semi-arid regions of the south and north-west of Puerto Rico, applications of nitrogenous fertilizer resulted in marked increases in dry matter and protein production of herbage (Vincente—Chandler and Figarella, 1968), and about 50 per cent of the nitrogen applied was recovered in the forage. For a rotationally grazed Guinea grass pasture, 80 lb N/acre (90 kg N/ha) every four months was considered a good fertilizer practice under the semi-dry conditions prevailing.

The importance of nitrogen as the most limiting factor in pasture production has been shown in experiments conducted elsewhere in tropical America: in Trinidad and Tobago (Wilson, 1960), in Venezuela (Iljin, 1952), in Costa Rica (Blue, 1966), in Mexico (Ferrer, 1963/64), in Colombia (Lotero *et al.*, 1965; Crowder *et al.*, 1964), in Brazil (Quinn *et al.*, 1961), and in the Virgin Islands (Oakes, 1967).

A number of experiments in tropical and subtropical Africa—in South Africa, Rhodesia, Zambia (Brockington, 1964), Tanzania, Kenya and Uganda (Horrell and Bredon, 1963)—have shown large responses to nitrogen in grass production and indicate the great potential for an increased meat and milk production by means of intensive grassland management and fertilization. In South Africa, Rhodes grass (*Chloris gayana*), Napier (*Pennisetum purpureum*) and blue buffalo grass (*Cenchrus ciliaris*) have been used in fertilizer experiments on pastures and substantial increases in dry matter and crude protein yields have been recorded. Phosphate usually showed a yield response only in combination with the higher nitrogen applications. In Rhodesia, Napier grass stands showed consistently large responses to nitrogen applications, while star grass as well as other grasses also showed much promise in their reaction to fertilizers. Experiments at Henderson Research Station (Rodel, 1966) showed that planted pastures could profitably be used by grazing beef steers when heavy dressings of nitrogenous fertilizers, with basal applications of phosphate and potassium, are used; the carrying capacity as well as the return per acre could be increased about five times and the pastures be made to compete successfully with maize. Experience with fertilizer application on grassland in Zambia, Tanzania, Kenya and Uganda has confirmed the Rhodesian experiments. In Uganda, the unit response to nitrogen, however, was low compared with other areas, probably due to deficiencies of other nutrients (Horrell and Bredon, 1963).

In Asia, where about 40 per cent of the land is under arable crops, the emphasis is on food crops. Permanent pasture is only 3 per cent in extent. Fodder crops are grown only to a limited extent and pastures are rare. Trials conducted on irrigated fields in north India (Mehta *et al.*, 1965) showed that green fodder yields of Pusa Giant Napier grass intercropped with berseem (*Trifolium alexandrinum*) were higher than those of the grass alone.

The nitrogen requirements of pastures in Australia are generally lower than elsewhere in the tropics owing to their high content of legumes, but acute nitrogen deficiency restricts the maintenance of productive grasslands in the humid, subtropical coastal region of Queensland. At Samford Pasture Research Station, the yield response was approximately linear between 0 and 400 lb N/acre/year (0-450 kg N/ha/year) (Henzell, 1963 ; Henzell and Strik, 1963). Increasing rates of nitrogen markedly reduced the proportion of legumes.

15.4 Ley-Farming in the Tropics

Ley-farming in temperate regions has been shown to improve soil fertility and to maintain a high level of crop and animal production. Leys containing grasses and legumes together may influence succeeding crops by improving soil structure and by increasing the soil nitrogen reserves. But leys remove soil nutrients. Where they are cut and the herbage removed, or where milk cattle graze on the land for limited periods, there may be a serious loss of nutrients. Properly managed, however, leys can contribute to soil fertility in temperate climates.

In the tropics, the use of leys in rotation is at present confined almost entirely to wet highland areas. Information on the value of the ley at low and medium latitudes is scanty.

Some of the experiments that have been conducted indicate that the effects of pure grass leys (grazed or ungrazed) on soil fertility are no better than those of natural regeneration (Dennison, 1959 ; Clarke, 1962 ; Peat and Brown, 1962). But Kirkham (1947) reported that grazing a grass fallow did not reduce subsequent crop yields. Brockington *et al.*, (1965) presented evidence to show that grazing resting land benefited subsequent arable crops. Organic C and total soil N were built up in the ley and declined with cultivation. Inorganic N and legumes in the ley increased animal production but depressed yields of following arable crops, which was remedied, however, by the addition of superphosphate. Stobbs (1969 *a*) in a long term trial at Ngetta Experimental Station, in northern Uganda, showed that three years of intensive day grazing significantly increased yields of the following test crops compared with yields from ungrazed areas. At Kawanda Research Station in Uganda, the standard rotation is a three-year elephant grass ley followed by three years' cropping. The grass leys, irrespective of management, raised the level of soil nitrogen but, depending on management, a high-yielding ley depleted and hence caused deficiencies of soil phosphate and potassium (Foster, 1971). The soil fertility built up by a ley declined rapidly over two years of cropping, mainly because of a loss of soil nitrogen.

Deep-rooted grasses, such as elephant grass, could considerably improve subsoil structure and water relationships, besides helping to recycle nutrients. But Pereira *et al.*, (1954) demonstrated that the beneficial effects of grass fallows on soil fertility were not caused primarily by their improving soil structure. Grass leys of three years' duration did effect significant improvements in water-stable aggregation, freely-drained pore space, and rates of infiltration of rainfall, but these effects disappeared rapidly on cultivation and were largely lost before the end of the first year. The beneficial effects on crop production, however, continued for some time longer. Compared with continuous arable cropping, a grass ley is also likely to reduce erosion on steep land in high rainfall regions and to decrease weed and pest infestation. The use of leys in a rotation will, in the long run, be satisfactory if crop yields during the arable period and animal production from the ley will together compensate for the loss of cropping during the ley period. For this, fertilizer use and the inclusion of efficient, palatable, nitrogen-fixing legumes in a well-managed mixed sward will be necessary.

The hazards of grass leys in the drier parts of east Africa have been pointed out by Pereira and McCulloch (1962). These workers have determined that if the indigenous vegetation is well supplied with water, it can transpire about 2300 mm in the low-lying areas of the coast falling to about 1150 mm at altitudes around 2450 m. They estimate that a ley can dry the soil profile to wilting point to a depth of 2 m or more during the dry season. Even if the ley is ploughed before the rains begin, the following arable crop will be extremely vulnerable to the characteristic breaks in the rains during the early part of the growing season. In areas with two rainy seasons, this danger can be reduced by breaking the ley before the short rainy season and managing the land in such a way during these rains that the soil profile will be recharged as fully as possible.

16 Fertilizers in Tropical Soil Fertility

16.1 Need for Increased Fertilizer use in Tropical Agriculture

The increase in world population between 1960 and 2000 is likely to be as great as the total growth throughout the entire previous history of mankind (UN, New York 1967). Population growth rates are generally more in the developing areas of the world, including almost the whole tropical belt, than in the developed areas: about 2 per cent per annum in Asia and 3 per cent in Latin America, compared with about 1 per cent in Japan and in western Europe.

The problem of providing adequate food supplies is most acute in tropical regions, in Asia with high population densities and moderate to high rates of population growth and in Latin America and Africa, where population densities are less but population expansion is rapid.

The fuller use of the seas offers some potential for increasing food supplies but agriculture is still the primary source of increased food production. In agriculture, a number of possibilities exist for stepping up production : better systems of farming, shifts in cropping patterns to high yielding crops and varieties, and reduction in crop losses during growth, harvesting, storage and marketing. But the main means of increasing the world food supply are, extending the acreage under production and increasing crop yields on land already under cultivation.

Increasing the area of land under cultivation will require large investments in capital and human effort and the results of such projects will not be realized immediately. The alternative of increasing output through more intensive use of land already under cultivation must therefore be relied on for the urgently needed increased food supply. The urgency of such increased food supply is particularly great in tropical lands where average crop yields are still very small. For example, rice yields in the Philippines and in India in 1960-1962 were 1,180 and 1,520 kg per hectare respectively, compared with 3,830 and 4,870 kg per hectare in the USA and Japan.

More intensive use of cultivated land involves a combination of practices such as the increased use of fertilizers, control of insects and diseases, the use of improved varieties and cultural practices, better water use and land management. Of these, increased fertilizer use has been a major factor in increasing crop yields in the technologically more developed countries ; for example, about one-half of the yield increase in the USA between 1940 and 1955 has been credited to the use of more fertilizers. Equally, upward trends in crop production in a number of tropical countries through increased fertilizer application have demonstrated its potential for increasing productivity and maintaining soil fertility throughout the tropical region. That significant responses are obtained had been apparent from fertilizer trials carried out at experimental stations for a number of years but that the responses are economic as well has been vividly demonstrated by the results obtained on farmers' own plots in the FAO fertilizer programme. On growers' own plots in 19 different countries, average yield increases were 50 per cent from fertilizer use alone (Richardson, 1966).

Richardson (1966) reported on the progress made during the five years of the FAO fertilizer programme under the Freedom from Hunger Campaign. The field programme of fertilizer trials and demonstrations in farmers' fields was conducted in the developing parts of the world, mainly in tropical countries. The FFHC Fertilizer Programme helped to double the normal rate of increase in fertilizer consumption, on the average, in the countries taking part. The total consumption of plant nutrients ($N + P_2O_5 + K_2O$) in 15 of the participating countries increased by 80 per cent in four years from 190,800 tons in 1960/61 to 344,000 tons in 1964—a 20 per cent increase per annum. The average rate of increase in fertilizer consumption during the same period, in 42 other developing countries without a Fertilizer Programme or something similar, was 10 per cent per annum.

Among the general conclusions to be drawn from the field programme in a wide range of countries, climates, soils and crops covered were: (1) nitrogen was effective and profitable almost everywhere except under semi-arid conditions (annual rainfall below about 350 mm); (2) soluble phosphate was also effective and profitable almost everywhere, including low-rainfall conditions, though the average response to phosphate was a little less than that to nitrogen; (3) potassium was frequently effective and profitable, but there were also numerous cases where potassium produced little or no response, or even seemed to depress the yield. Responses to fertilizer potassium were much more variable than those to nitrogen or phosphate; they depended much more on the status of the local soils in available potassium. Potassium responses seemed generally to be larger and more widespread on the calcareous soils of the Near East and northern Africa, and on the leached tropical soils of west Africa, than on the soils, largely derived from volcanic ash deposits, of northern Latin America. Responses to potassium were, however, more widespread in farmers' fields than was expected in some countries on the basis of earlier research carried out mainly at research stations and experimental farms.

16.2 Fertilizer Consumption Patterns and Trends in Tropical Countries

World consumption of fertilizers (Table 57) continues to expand; the total consumption of all commercial fertilizers during 1968/69 was 56.5 million tons of nutrients, an increase of 56 per cent from 1963/64 to 1968/69. Of this, in 1968/69, nitrogen (24,520,000 tons N) was 43 per cent, phosphorus

TABLE 57: WORLD CONSUMPTION OF FERTILIZERS (FAO, 1970a)

Continent	All fertilizers N, P_2O_5, K_2O 1963/64	All fertilizers N, P_2O_5, K_2O 1968/69	N 1963/64	N 1968/69	P_2O_5 1963/64	P_2O_5 1968/69	K_2O 1963/64	K_2O 1968/69
	thousand metric tons							
South America			256	387	269	480	176	299
Africa			444	675	265	512	129	240
Asia			2,006	3,785	979	1,651	817	1,227
North and Central America			4,500	7,291	3,556	4,852	2,688	4,019
Oceania			78	172	894	1,214	128	174
USSR			1,360	3,454	969	1,748	901	2,210
Europe			5,338	8,754	5,319	6,818	5,192	6,544
World total consumption	36,260	56,510	13,980	24,520	12,250	17,280	10,030	14,710

TABLE 58: WORLD PRODUCTION OF FERTILIZERS (FAO, 1970a)

Continent	All fertilizers N, P_2O_5, K_2O 1963/64	All fertilizers N, P_2O_5, K_2O 1968/69	N 1963/64	N 1968/69	P_2O_5 1963/64	P_2O_5 1968/69	K_2O 1963/64	K_2O 1968/69
	thousand metric tons							
South America			284	245	112	161	32	16
Africa			184	271	324	713	—	—
Asia			1,834	3,613	742	1,172	123	344
North and Central America			4,505	7,828	3,791	5,173	3,188	5,476
Oceania			19	74	1,019	1,175	—	—
USSR			1,759	3,750	1,096	1,934	1,400	3,120
Europe			6,330	10,803	5,375	7,138	6,056	6,976
World total production	38,180	59,980	14,920	26,580	12,460	17,470	10,800	15,930

($17,280,000$ tons P_2O_5) and potassium ($14,710,000$ tons K_2O) were 31 and 26 per cent respectively (FAO, 1970 a).

The largest producing and consuming countries are outside the tropical region, in Europe and in North America. Out of a total world production of 26,580,000 tons of nitrogenous fertilizers in 1968/69, 10,803,000 tons were produced in Europe and 7,828,000 tons in North America and Central America (Table 58). Tropical countries in South America, Africa and Asia produced only 4,129,000 tons or 15 per cent of world production. In 1968/69, the USA produced 26 per cent of processed phosphate fertilizers, the USSR (11 per cent) and France (8 per cent) were the next largest producers. Tropical countries produced only 12 per cent of the total. However, more rock phosphate was mined in Morocco and Tunisia, which are outside the tropical zone although in Africa. To a world figure of 15.9 million tons K_2O in 1968/69, tropical countries contributed around 0.36 million tons (2.3 per cent).

Europe used most nitrogen fertilizers (36 per cent), followed by Central America (30 per cent), Asia (15 per cent) and the USSR (14 per cent), leaving only 5 per cent for Africa, South America and Oceania. Europe, North and Central America accounted for more than two-thirds of the total world consumption of phosphate fertilizers and were also the largest consumers of potassium fertilizers in 1968/69.

The rate of fertilizer application was most in Europe, 139 kg of all nutrients (N, P_2O_5 and K_2O) per hectare of arable land in 1967/68 compared with 62 kg in North and Central America, 35 kg in Oceania, 31 kg in the USSR, 19 kg in Asia, 11 kg in South America, and 6 kg in Africa. Taiwan, which is partly within the tropics, had a very high rate of usage (284 kg/ha).

Table 59 shows the fertilizer consumption per hectare of arable land in a number of tropical countries. While a few countries, like Taiwan (284 kg) and the sugar-growing countries like Barbados (273 kg), Mauritius (263 kg), Cuba (165 kg), and Jamaica (105 kg) within the tropical belt, have much higher rates of fertilizer consumption than the world average of around 40 kg/ha and higher even than that of USA (70 kg/ha), the vast majority of tropical countries used very little fertilizer per hectare of arable land.

Fertilizer consumption is however increasing practically everywhere in the tropics though not yet at a rate commensurate with fertilizer needs as indicated by soil surveys and soil fertility studies. Table 59 shows how fertilizer consumption has increased during the period from 1960 to 1967/68.

16.3 Basis of Fertilizer Recommendations

Fertilizer recommendations depend primarily on the nutrient requirements of the growing crop and on the ability of the soil to supply these requirements. Nutrient requirements are determined by soil and climatic factors as well

TABLE 59: FERTILIZER CONSUMPTION IN TROPICAL COUNTRIES (FAO, 1970a)
(kg/ha of arable land)

Country	1960	1965	1967/68
Argentina	1	1	2
Barbados	271	...	274
Brazil	8	...	14
Chile	17	24	26
Colombia	11	...	28
Costa Rica	26	...	52
Cuba	...	...	165
Dominican Republic	10	...	19
Ecuador	6	5	20
El Salvador	32	...	69
Guatemala	10	...	17
Guyana	...	...	49
Jamaica	50	67	105
Mexico	8	...	18
Peru	42	31	31
Trinidad and Tobago	35	...	77
Uruguay	8	...	13
Venezuela	2	...	10
Algeria	8	6	7
Kenya	6	...	22
Madagascar	...	...	3
Mauritius	...	...	263
South Africa	18	...	38
UAR	84	122	102
China	...	...	...
India	2	5	11
Malaysia (west)	...	...	31
Pakistan	3	5	11
Philippines	12	14	14
Sri Lanka	39	14	46
Thailand	2	3	9
Taiwan	201	257	284
Vietnam, Rep. of	8	32	43
Australia	...	29	28
Japan	304	321	384
Netherlands	456	575	626
UK	183	208	253
USA	39	64	77
World			40

as by the physiological needs of the crop concerned. But fertilizer use, based on such recommendations, will depend not only on the needs of the crop for healthy growth and maximum yields but also on the cost of fertilizer and on the margin of profit to the grower. This aspect needs special attention in any programme designed to induce peasant cultivators to use more fertilizers in tropical farming.

PLANT ANALYSIS: Nutrient requirements of crops can be assessed by a number of different methods. Analysis of the crop after harvest indicates the nutrients removed in the harvested portions (Table 60). However, plant tissue analysis at certain critical stages of growth can be used to indicate whether there is a deficiency, a sufficiency, or an excess of nutrient absorption. Visual diagnosis of symptoms indicating abnormalities gives some information of a qualitative nature. These have been described in a number of publications (e.g. Bear and Coleman, 1949; Wallace, 1951).

There has been much interest in plant analysis as a possible method of finding out what elements a soil cannot supply adequately to the plant grown on it. The plant is considered to be the best index of the complex soil-plant-climate system, reflecting all the complex factors that affect its nutrition. Plant analysis gives an idea of amounts and proportions of nutrients actually taken up by the plant.

The leaf is the most suitable part of the plant for analysis because it represents the seat of active growth processes. For some plants, the stem or petiole is preferred. In sugarcane, the leaf sheath is also used. Leaves should be metabolically active when sampled. For comparison, they should be of similar physiological age and should be sampled at the same period during the growing season. Leaves are taken at a definite point so that they are similar in size and age. It is possible to determine critical ranges of values for each nutrient within which supplies are adequate (Table 61).

Leaf analysis is now widely used as a basis for assessing fertilizer requirements for coffee. Foliar diagnosis is used as a guide for fertilizer application on sugarcane and cocoa. In Malaysia and in the Ivory Coast, fertilizer recommendations for oil palm are based on foliar analysis while a combination of soil and leaf analysis is used as a guide for fertilizing mature rubber. Leaf analysis is a valuable tool for nutritional research and fertilizer application on citrus, and gives a useful indication of the nutrient status of the growing potato plant.

SOIL ANALYSIS: Soil tests are designed to extract and determine chemically a quantity of a particular nutrient proportional to that extracted by plant roots. Soil testing methods differ considerably and various chemical solutions are used because soils, crops and climates are different. Chemical soil tests are useful only when the results correlate well with plant responses to fertilizer application in the field, established through field experiments on the same soils.

For chemical tests, only the plough layer (i.e. the top layer from the sur-

TABLE 60: NUTRIENTS REMOVED BY TROPICAL CROPS

Crop	Yield	N	P	K	Ca	Mg	Reference
				kg			
Tea	1121 kg made tea per ha	29	3.2	13.2			Eden (1949)
Cocoa	560 kg dry cocoa per ha	5.4	1.2	3.6			Urquhart (1956)
	1000 kg made cocoa per ha	20	4.2	10.6			Jacob & Uexküll (1960)
Coffee	1.2 tons coffee beans per ha	12.2	1.0	9.1			Jacob & Uexküll (1960)
	1121 kg made coffee per ha	13.6–18.2	1.0–1.2	11.4–19.0			
Tobacco	2500 kg barn-dried tobacco per ha	52	6.1	105	78		Schmid (1951)
Rice	3980 kg grain per ha: entire plant	36.5	8.1	88.4			Anon (1966)
	panicle	19.4	5.3	4.4			
Sugarcane	124 tons per ha	34	10	56			Barnes (1953)
Rubber	2000 kg made rubber per ha	12.6	2.6	8.6			de Geus (1967)
Cotton	785 kg per ha	35	2.5	27	26	7.1	Martin & Leonard (1950)
Sisal	2500 plants per ha	55–64	8.3–10.0	68–75	132–168	18.6–22.5	Jacob & Uexküll (1960)
Coconut	124 palms per ha	74	13.2	114	12.9	19.3	Jacob & Uexküll (1960)
Oil palm	138 palms per ha	41	4	42			Georgi (1931) quoted by Jacob & Uexküll (1960)
Pineapple	55 tons pineapple per ha	43	7.2	109	12.2	6	Martin-Prével (1961)
Banana	30 tons fruit per ha	50–75	6.6–8.8	146–187	7.2–14.3	15.1–18.2	Jacob & Uexküll (1960)
Citrus							
Orange	990 trees per ha: medium crop	170	17.5	120	215		Oppenheim (1932) quoted
Lemon	990 trees per ha: medium crop	183	14.9	116	174		by Jacob & Uexküll (1960)

TABLE 61: Use of Foliar Diagnosis for Assessing Nutrient Requirements of Tropical Crops

Crop	*Leaf analysed*	*Critical levels below which deficiencies are likely to occur*											*Reference*
		N	P	K	Ca	Mg	Fe	Mn	Cu	Zn	Mo	B	
		% dry matter					ppm						
Oil palm	leaf 9 from 2-year old trees	2.70	0.16	1.25	0.5	0.23							Bachy (1964)
	leaf 17 from 3-year old and older trees	2.50	0.15	1.0	0.0	0.24							
	leaf 17 from young & old palms	2.60 -2.70	0.17	1.00 -1.20	0.30 -0.70	0.25 -0.29							Coulter (1958)
	leaf 3 from 1-2 year old trees	2.80 -3.00	0.19 -0.21	1.50 -1.80	0.30 -0.50	0.30 -0.35	80	150	5	18	1	15	
Cocoa	2nd and 3rd leaves from first fully grown flush	1.8 -2.0	0.10	1.0 -1.2			50	11	4	20		8.5 -11	Loué (1962 a & b) Spector (1964)
Coffee	3rd or 4th leaf from the apical bud	2.00	0.09	1.00	0.80	0.10	70	25	10	10		40	Chaveri *et al.* (1957); Loué (1960)
	4th leaf	2.6	0.16	1.8		0.20		80	2-3	15-20		20	Culot *et al.* (1958)
Citrus	4-7 month old leaves	1.90	0.07	0.30	2.0	0.15	50	20	4	15	0.5	15	Cradock and Weir (1964); Weir (1966); Reuther *et al.* (1958)
Sugarcane	3rd leaf: central portion less midrib for ratoon cane crops aged 5 months	1.95	0.21	1.25									Halais (1962)

face to plough depth) is usually sampled. A soil sample from the zone of main root development is also useful. Since different crops have different growing periods, different growing times during the year, and different root systems with different abilities to extract nutrients, a single soil test is not directly applicable to all crops. For a satisfactory interpretation of chemical soil tests, all possible information should be obtained regarding locality, topography, climatic data, parent material, vegetation, ground-water level and the presence of impervious layers in the profile. Nevertheless, they are not easy to interpret and apply to practical situations with the exception, perhaps, of *p*H and conductivity of soil solutions which will show if there is anything radically wrong with the chemical condition of the soil (Bache, 1969).

Rapid chemical tests, or *chemical quick tests,* have been worked out to determine fertilizer requirements (e.g. Weir, 1966 a, b and c). These are simplifications, for rapid practical use, of longer and more tedious laboratory procedures for evaluating available nutrients in soil. They are rapid and relatively inexpensive, they make it possible to quickly analyse several samples from a single field, and they prove useful for limited areas, such as the individual fields in a farm. They must, however, be carried out in a well-equipped laboratory by trained personnel, the procedures laid down must be carefully followed, and their results properly interpreted before dependable recommendations can be made. Soil-test kits are less reliable. The methods used in these are based on standard laboratory procedures, modified so as to permit their use in the field.

FIELD AND POT TRIALS: Fertilizer requirements on any soil for a particular crop can be determined with exactness only by means of properly conducted field trials. This is the oldest method of estimating fertilizer requirements and, from very simple beginnings, the art of field experimentation has been developed with the application of statistical methods. Statistics is applied in the design of field trials as well as in the analysis of results.

In a given location, the effect of the totality of factors responsible for crop growth can be ascertained by measuring the response shown by a crop to the addition of a given nutrient or combination of nutrients in a field experiment. Such experiments are time-consuming, tedious and expensive when compared with plant tests and soil analysis, but they alone can give the most satisfactory answers regarding quantity, time, and method of fertilizer application in a particular field. Results of field trials, properly located and statistically evaluated, can be extended to other areas where soil and climatic factors are similar.

The information gathered in a field experiment can be obtained at less expense and in a more convenient manner by conducting pot experiments using soils from experimental sites (see, for example, Webb, 1954 & 1955). The absence of soil uniformity in the field, eliminated by replicating and randomizing treatments in a field trial, can be overcome to some extent in pot experiments where soil from the field is mixed thoroughly. But, although such uniformity in the soil is secured, it is difficult to make pot experiments to simulate field conditions entirely. Usually, only surface soils are used and the contribution to crop nutrition of the subsoil is ignored. It is difficult to fill several pots to the same degree of compactness, and soil temperature in the field and in the greenhouse are not the same. Care is thus necessary in interpreting the results of pot trials.

A large volume of data is now available on the fertilizer requirements of tropical crops as a result of field experiments conducted over a number of years at research institutes and experimental stations throughout tropical regions. The Tea Research Institute of Sri Lanka the Tocklai Experiment Station in Assam and the Tea Research Institute of east Africa publish the results of field trials in their respective journals. The Rubber Research Institutes of Sri Lanka and Malaysia have contributed much to our knowledge of the fertilizer requirements of rubber. A number of experiment stations in west Africa and in the Congo are working on the nutrient problems of the oil palm, while the oil palm division of the Chemara Research Station in Malaysia is conducting field trials on the nutrient needs of the oil palm under Malaysian conditions. The Coconut Research Institute in Sri Lanka has conducted a number of fertilizer trials with cocount since 1934. Coconut research and experiment stations in the Philippines, in India, and in west Africa have contributed to the knowledge on which fertilizer recommendations for coconut are made in these respective countries. Field trials on cocoa in Ghana, in Trinidad, and in Costa Rica continue to yield valuable data. Information on the fertilization of coffee is based on field trials in several coffee-producing areas such as Brazil, Costa Rica, Colombia and Hawaii. Most countries in Asia have their own rice experiment stations. Much information of fundamental as well as practical importance has emerged from the work of the International Rice Research Institute, Los Baños, Laguna, Philippines. Fertilizer experiments on the fields of rice cultivators have been conducted in some African and Latin American countries under the Freedom from Hunger Campaign of the FAO (Richardson, 1966).

16.4 Manuring of Tropical Crops

Tea (*Camellia sinensis*, L): Tea grows poorly in soils with free calcium carbonate and the *p*H of tea soils should therefore be below 6. In Africa, where ash accumulation after burning has raised soil *p*H, sulphur is mixed with the soil in the planting hole so as to reduce the *p*H (de Geus, 1967). In Africa, bracken (*Pteridium aquilinum*) is considered a good indicator plant for the suitabilty of a soil for growing tea as soils in which bracken grows are usually acid. In Sri Lanka the Comber test has been used for the rapid determination of soil acidity in selecting tea soils.

Nitrogen is the most important fertilizer nutrient for a leaf crop such as tea and, for unshaded tea, the increase in yield is often almost linear up to

120 kg N per ha or more. For shaded tea, fertilizer response is generally more complicated. Responses to nitrogen are cumulative during the pruning cycle and even from cycle to cycle. Phosphorus stimulates new wood and root development and is particularly essential for young plants. Potassium deficiency leads to 'potash scorch' in which the edges of leaves are scorched a light grey or brown, and results in a marked decline in yields.

Magnesium deficiency has developed on the poor liparitic soils of Sumatra and on the leached Red-Yellow Podsolic tea soils of Sri Lanka. It has also been reported from the Congo (Chenery and Schoenmaekers, 1959) and in east Africa (Chenery, 1961). Sulphur deficiency causes 'tea yellows' in Malawi (Storey and Leach, 1933) and in Tanzania, even on the comparatively recently deposited volcanic ash derived soils in the Tukuyu district. Grant and Shaxson (1970) reported that the use of ammonium sulphate once every three to six years was sufficient to prevent the recurrence of sulphur deficiency in established tea in Malawi.

The use of shade trees is a standard cultural practice in the tea gardens of Assam (Wight, 1958) where, without nitrogen, shaded tea was more productive than unshaded tea. Nitrogen increased the yields of both shaded and unshaded tea bushes but the response to nitrogen was greater from the unshaded tea (Dutta, 1961). Under Assam conditions, therefore, maximum yields can be most economically produced by applying nitrogen to shaded tea. This is different to the experience in east Africa (Child, 1960) where yield levels were increased and quality improved by the removal of shade and by increased fertilizer application.

Much of the experience in manuring tea has been obtained in Sri Lanka at the Tea Research Institute where field experimentation began in the late 1920s. Among these is the experiment started by T. Eden in 1931, the oldest fertilizer experiment on tea.

Fertilizer recommendations have changed over the years. At first, the use of balanced NPK mixtures was advocated to supply the nutrients removed by the crop, lost by leaching or fixed in the soil. T 500 was one of the earliest of such mixtures used to provide 8 kg N per 100 kg of made tea (Table 62).

TABLE 62: FERTILIZER MIXTURES FOR TEA IN SRI LANKA (Tea Research Institute, Sri Lanka)

Fertilizer mixture	*Fertilizer level* (lb)			*Nutrient weight ratio*		
	Ammonium sulphate	Saphos phosphate	Muriate of potash	N	: P_2O_5 :	K_2O
T 370	220	110	40	100	71	53
T 500	320	105	75	100	47	57
T 700	500	100	100	100	27	59
T 725	500	100	125	100	27	73
T 750	500	100	150	100	27	88

In 1961, the T 700 series was introduced, with 40 per cent less P_2O_5. Modifications of T 700 contained more K_2O for areas where potash deficiency was likely to occur (Table 62).

Until the 1960s, *ratio manuring*, according to yields of made tea, had been practised but as Gunn and Kanapathipillai (1962) pointed out, actual responses on estates were lower than indicated by the ratios used. Joachim *et. al.* (1963) accordingly recommended a comprehensive system of manuring based on yield category, yield trend and response to manure.

By 1965, experimental evidence was forthcoming that balanced NPK mixtures based on nitrogen requirements may not be economic and Tolhurst (1965) suggested that phosphate applications could be reduced to about 20 lb P_2O_5/acre/year (22.5 kg P_2O_5 /ha/year) or even omitted for one or two years. Potash applications of about 40 lb K_2O (67 kg/ha) for low-yielding and 60-90 lb K_2O (67-100 kg/ha K_2O) for high-yielding tea per acre per annum were deemed sufficient. Bhavanandan (1970) reviewed the results of selected fertilizer experiments on which the latest recommendations for tea were based. These experiments showed that, in the upcountry, for seedling tea responses to nitrogen were best between 90-240 lb N/acre, (100-270 kg N/ha) although occasionally responses up to 270 lb and 360 lb (300 and 400 kg N/ha) N were obtained. For clonal tea, very significant responses were obtained up to the highest level tested namely 300 lb N/acre (335 kg N/ha). Responses to 30 lb P_2O_5 /acre (33.5 kg P_2O_5 /ha) were obtained in some experiments, not in others. Clonal tea responded to potassium only up to 62.5 lb K_2O /acre (70 kg K_2O /ha).

Recommendations depending on the above results were based on yield responses obtained in experiments rather than on a *replacement basis**. No standard mixtures were recommended applicable to all estates under all soil conditions and different types of tea. Estates were advised to formulate their own fertilizer mixtures according to the requirements of each field. For all districts, 20 lb P_2O_5 /acre/year (22.5 kg P_2O_5 /ha) was recommended for mature seedling tea and 30 lb P_2O_5 /acre/year (33.5 kg P_2O_5 /ha) for mature clonal tea, applicable in all years of the pruning cycles. Potassium fertilizer was recommended for each district and for each type of tea, equal amounts for each year of the cycle. Nitrogen was recommended according to the yield pattern over the pruning cycle. In the up-country, where a four-year cycle is practised, more nitrogen was recommended in the second and third years, and the application of fertilizers immediately before and after pruning were to be omitted. In the low-country, on a two-year cycle, more nitrogen was to be given in the second year because the crop yields were more.

Experiments with various forms of nitrogen, begun in 1961, have shown no difference in yield between urea, ammonium sulphate, and calcium ammonium nitrate. Yield responses with respect to ammonium sulphate and urea are summarized in Table 63.

*according to nutrients removed.

TABLE 63: TEA YIELD RESPONSES IN SRI LANKA WITH AMMONIUM SULPHATE AND UREA (Bhavanandan, 1970)

District	*No. of expts.*	*Results for periods of 1-3 years* SA=Urea	**SA>Urea**	Urea> SA	*not analysed*
Up-country	17	14	2	1	—
Low-country	8	7	1	—	—
Uva district	7	6	—	—	1
Mid-country (wet zone)	3	2	—	—	1
Total	35	29	3	1	2

COCOA (*Theobroma cacao*, L.): The amounts of nutrients removed from the soil by harvested cocoa fruits are small (Table 60), much more is removed from the soil by the cocoa tree, and even more by the shade trees under which cocoa is usually grown.

Shade is essential for establishing the cocoa plant properly during the first four to five years (Jordon and Opoku, 1966), but experiments in west Africa (Cunningham and Lamb, 1959 ; Cunningham and Smith 1961 *a* 1961 *b*; Cunningham, Smith and Hurd, 1961) have shown that on the more fertile soils, mature cocoa can be grown without overhead shade provided fertilizers are used. Yield responses were largest when a NPK+Mg fertilizer mixture was used without shade. Hurd and Cunningham (1961) showed that the yield increases were the result of increased vegetative growth and flowering. However, shade can only be safely removed where pests are under effective control and where severe dry-season drought does not lead to defoliation and terminal branch die-back (Hartley, 1968).

When cocoa is cultivated under shade, its nutrient requirements can be met mostly from the decomposing organic matter. The root system of the cocoa tree (Swarbrick, 1964), with a tap root extending deep into the subsoil and a dense surface mat of fine fibrous roots, is capable of absorbing nutrients released by organic matter decomposition, and cocoa under shade can therefore thrive on soils comparatively poor in nutrients. The development of *mycorrhiza* promotes the absorption of nutrients from a well-aerated soil covered by a layer of mulch. Without shade, there is no humus layer, the surface mat of fibrous roots is poorly developed and the roots grow into the lower soil layers. Easily available nutrients are needed under such conditions. The higher light intensity and increased vegetative growth further increases nutrient requirements.

'Sickle-leaf' disease was attributed to zinc deficiency on alkaline soils in west Africa but in Sri Lanka a similar condition was cured by applying dolomitic limestone (Kandiah and Rodrigo 1954). Cunningham (1964) studied cocoa showing symptoms both of iron (chlorosis) and zinc ('sickle-leaf') deficiency, and containing little Mn. Soil *p*H above 7.6 was the main cause of decreased availability of soil Fe, Zn and Mn. Reducing soil *p*H corrected Zn deficiency and increased Mn concentrations in the presence of much organic matter but only foliar spraying with iron corrected iron chlorosis. Magnesium deficiency, induced by excess potassium, caused 'sickle-leaf' disease in Ghana. Boron deficiency in cocoa causes white twisted leaves and death of growing points (Loué, 1952).

Fertilizer experiments on cocoa are difficult because the effect of a treatment shows up only after several years and may even then be affected by other environmental factors more than by fertilizer treatment. The classical fertilizer experiments on cocoa carried out by the Imperial College of Tropical Agriculture in Trinidad showed striking yield responses to potassium, less to nitrogen and phosphorus. Foliar sprays and injections of nutrient solutions into the trunk were effective.

COFFEE (*Coffea* sp.): The amounts of nutrients removed by coffee beans are relatively samll (Table 60) but as the coffee plant has to produce not only mature berries but the fresh wood, leaves and buds of the succeeding crop, the quantities annually needed for sustained growth and production are more than four times those removed by the crop (Van Dierendonck, 1963). Repeated cultivation of coffee even on such fertile soils as the Terra Roxas of Brazil have diminished soil fertility considerably after ten to twenty years.

Organic manuring and shading the soil by mulches or shade trees (Haarer, 1955) forms the basis of manurial treatment as these practices have a beneficial effect on soil structure, in providing a regular supply of humus, in maintaining a uniformly moist soil and in favouring the development of *mycorrhiza*. The use of farmyard manure, compost and grass mulch is usually more effective than a leguminous cover crop and green manuring. The latter practices are only recommended when sufficient water is present. Grass grown separately for mulching is often fertilized with nitrogen and potassium.

Nitrogen and phosphorus are particularly important in the early development of the coffee bush, nitrogen and potassium after the bushes have started bearing. Nitrogen is best applied in two dressings broadcast over the whole area; the first, three to four weeks after flowering, and the second immediately after harvesting. As phosphate is particularly important for the early development of coffee, the soil in the planting holes should be mixed with well rotted compost enriched with phosphate. In established coffee, phosphate fertilizers are mixed with organic manures or compost and worked in a circle around the bushes or in drills between the bushes. Split applications of potassium fertilizers are used, one portion being broadcast with the nitrogen and the other worked into the soil with the phosphate. For uniform ripening of coffee cherries, a balanced relationship between N, P and K is necessary. Coffee grown on some Costa Rican and Colombian coffee soils suffer from calcium deficiency (Haarer, 1962), which has been corrected by liming.

Trace element deficiencies on coffee have been observed in Brazil and in Costa Rica. Magnesium, zinc and boron deficiencies were reported from Brazil (Jacob and Uexküll, 1960) and boron, possibly others, have been observed on the calcareous parent material soils of Costa Rica (Gonzalez, reported in Jacob and Uexküll, 1960).

Foliar analysis is an important method of assessing the nutrient requirements of coffee. The fourth pair of leaves counted from the tip of fruit-bearing twigs is used. Threshold values were given by Machado (quoted by Jacob and Uexküll, 1960). Foliar analysis showed that fertilizers are needed at two critical periods: the beginning of the rainy season and just before, and during, flowering (Ignatieff and Page, 1958).

Fertilizer mixtures used differ in the various coffee growing countries depending on local soil and climatic conditions and also on the age of the plantation. Early investigations were made in Brazil. An exhaustive study of the various fertilizer practices was published by Van Dierendonck (1959 a), who has also reviewed the fertilizer practices in a number of coffee growing countries. Very heavy dressings are used in Hawaii, 10:10:10 mixtures for young bushes and 10:5:20 mixtures for trees in bearing. Mature trees get from 1,700 to 2,000 kg/ha depending on the location. Jacob and Uexküll (1960) gave the following recommendations of complete fertilizers as a rough guide:

1-5 years 225-450 kg/ha of a 10:16:8 fertilizer
5-10 years 450-900 kg/ha of a 12:8:18 fertilizer
over 15 years 900-675 kg/ha of a 12:8:18 fertilizer

TOBACCO (*Nicotiana tabacum* L.): The ash content of tobacco is more than that of most other cultivated plants (Jacob and Uexküll, 1960) and therefore, large amounts of nutrients are removed from the soil (Table 60) varying according to variety and yield. Fertilizer treatment of tobacco is, however, important not only for good yields but increasingly so for better quality characteristics. Jacob and Uexküll (1960) discussed in detail the effects of various nutrients on yield and quality.

Nitrogen affects yield most; deficiency is as unfavourable as an excess. Too little nitrogen causes the leaves to remain small and to acquire a yellowish colour often followed by red spots. Too much nitrogen induces a high nicotine content, delayed ripening, poor irregular colouration and a deterioration of the aroma and taste. The leaf tends to be dark, has a heavy texture, and is more susceptible to various diseases. More nitrogen is usually used for the dark-coloured cigar tobaccos than for the light-coloured cigarette tobaccos. Different inorganic nitrogen fertilizers vary little in their mode of action. However, some nitrogen should be given to the crop in a slowly available form.

Phosphorus promotes the uniform ripening of the leaf, increases carbohydrate formation and, thereby, improves quality. Phosphatic fertilizer may be necessary for tropical soils containing little plant-available phosphate, even though tobacco does not require much phosphate (Table 60).

Potassium affects the yield and quality of tobacco most. The desirable qualities of cigarette, cigar and cut tobaccos are largely influenced by the amount and type of potassium fertilizer used. Chloride should be avoided for cigar tobaccos, but about 2.4 per cent chloride in the fertilizer mixture is considered desirable for cigarette and cut tobaccos to give the leaf greater elasticity. Chloride affects the combustibility and spoils the colour and texture of the leaf. Sulphate is important in regulating the glowing time.

Magnesium, calcium and boron are the other nutrients to be considered in fertilizer application for tobacco (McCants and Woltz, 1967). Magnesium deficiency is most frequently encountered on sandy soils and during heavy rain. The quality, rather than the quantity, of the crop is affected. Tobacco requires much calcium for good quality but, as with magnesium, care is needed because Black Root-rot disease can occur if the *p*H rises much above 6.5. This is also so for boron, and wherever deficiencies of this micronutrient appear, excess boron fertilizer must be avoided.

RICE (*Oryza sativa* L.): Fertilizer application for rice is determined by the physiological needs of the rice plant at different stages of growth as well as by the physico-chemical conditions prevailing in flooded rice soils. The latter aspect has been considered in detail in Chapter 4.

Responses to nitrogen fertilizers have been obtained in every rice-growing country. Rice varieties can be subdivided into low-nitrogen-responsive and high-nitrogen-responsive types. Highly responsive varieties use nitrogen much more efficiently in producing high grain yields, while low-nitrogen-response varieties tiller vigorously, are tall and leafy, have small percentages of effective tillers at harvest, and inefficiently utilize nitrogen to produce grain. To a large extent, high-nitrogen-responsive varieties correspond with the japonicas and low-nitrogen-responsive varieties with the indicas although this is not necessarily so; there are indica varieties, such as Taichung (native) 1 for instance, which are very responsive to nitrogen. The main objectives of current rice breeding programmes have been the production of high-yielding, short and stiff-strawed, lodging and blast-resistant varieties responsive to heavy fertilizer application. The damage caused by blast (*Piricularia oryzae*) is usually aggravated at the higher rates of nitrogen application.

The time and method of fertilizer application is even more important in the case of rice than with upland crops because of the redox potential profile prevalent in flooded rice fields (section 13.3). The need for deep placement of ammonium nitrogen as a basal dressing has already been referred to. The most appropriate times for top-dressing of nitrogenous fertilizer are (*a*) at tillering, to increase the number of tillers and ear heads; (*b*) at the neck-node differentiation stage, to influence the formation of spikelets; and (*c*) at the heading stage, to fill the empty grains and to increase the average weight of a grain.

The actual number of nitrogen applications depends on the variety, climatic conditions, the inherent fertility of the soil and on the practicability of application at a particular growth stage. In Sri Lanka the Department of Agriculture (1967) does not recommend a basal application of nitrogen at transplanting or sowing for local varieties of rice. The nitrogen released during mineralization of organic matter is considered sufficient in most cases to supply the needs of seedlings during the first two weeks after sowing or transplanting (Thenabadu, 1970). Transplanted rice is fertilized at least once in the nursery. For highly-responsive varieties, a basal application of 126 kg/ha ammonium sulphate or 63 kg/ha urea is harrowed into the soil or ploughed under. Two top-dressings are recommended: one at tillering and the other at the flower primordia initiation stage. For local varieties, the first top-dressing is 2 weeks after sowing or transplanting and the second, 6 weeks after sowing for the three-month varieties, 10-11 weeks after sowing (or 8 weeks after transplanting) for the four-month varieties, and 16-17 weeks after sowing (or 14 weeks after transplanting) for the five-month varieties. In the case of a highly-responsive variety like IR-8, a basal application of nitrogen is followed by the first top-dressing 21-24 days after sowing or transplanting and the second top-dressing at 70 days after sowing or 50 days after transplanting.

In the Philippines, long-duration rice varieties give a better response to nitrogen and applications up to 60 kg N per ha can be given only to varieties with a growing period of more than 184 days. Nitrogen dressings of as much 90 kg/ha may be given to lodging-resistant varieties. Split applications are preferred, with half of the nitrogen given at transplanting and half before tillering.

In India, split applications of nitrogen are recommended. Two applications, half before planting and half at tillering, showed less variation in response than any other form of application. Results from Narispur in the Punjab, showed that a split application into three equal doses, one-third before planting, one-third at tillering, and one-third at the pre-flowering stage, gave the best results with both ammonium sulphate and urea. Results from the Central Rice Research Institute, Cuttack (Orissa), showed that split applications of 45 kg N per ha, one-half basal at planting, one-quarter thirty days after planting, and the rest before flowering were superior to a single basal application at planting.

Split applications of nitrogen are adopted in Taiwan (Lee *et al.*, 1960; Sheng, 1962), and in East Pakistan (now Bangladesh) (Vermaat, 1964).

Phosphorus deficiency is widespread in most rice soils of Asia. With the exception of soils in the flood plains of the Walawe Ganga, and Gal-Oya and the western part of the Jaffna peninsula, rice soils in Sri Lanka are generally poorly supplied with plant available phosphorus (Panabokke and Nagarajah, 1964). In Indonesia, phosphate-deficient rice soils belonging mainly to the margalitic, lateritic and bleached dacitic soil groups, cover an area of 800,000-1,000,000 ha (de Geus, 1967). Phosphorus was required nearly everywhere in Bangladesh. In the Rapid Soil Fertility Survey Scheme (Vermaat, 1964), 45 kg P_2O_5 per ha in combination with nitrogen proved superior to 22.5 kg/ha in half out of a total of 46 centres. In Thailand, phosphate alone showed remarkable yield responses and in many cases good responses to nitrogen could be obtained only if a sufficient amount of phosphate was applied.

Flooding increases the availability of soil phosphorus, chiefly because of the reduction of ferric phosphate but also because carbon dioxide and any organic acids liberated on flooding may help in dissolving insoluble phosphates making them more available to plants. In Sri Lanka saphos rock phosphate is recommended for the acid rice soils of the wet zone while concentrated superphosphate is considered better for the more nearly neutral dry zone soils. Bone meal was just as good as rock phosphate for wet zone soils and had the further advantages that it was not easily carried away by wind or water and could be mixed with seed and broadcast uniformly in the field. Phosphate is usually given as a basal dressing.

Rice has a high requirement for potassium but much of this is found in the straw and hulls and, if only the panicles are harvested, the removal of potassium from soils would be low. Potassium increases the resistance of the rice plant to lodging as well as to bacterial and fungal diseases. Heavy-textured rice soils often contain sufficient potassium but alternate flooding and drying may increase fixation. Responses to potassium have been observed mainly on sandy soils but also on calcareous and on lateritic soils. A single basal application of muriate of potash may be sufficient to meet the requirement of rice for potassium but where leaching losses are great and the potassium content of irrigation water is small, a second dressing, applied 5-6 weeks before heading, may prove effective (Mitsui, 1954).

On very acid soils, continuous use of ammonium sulphate may aggravate the acidity and cause injury to rice plants. In such cases, liming has proved beneficial. Wherever severe symptoms of 'bronzing' occur in the low country wet zone of Sri Lanka, ground limestone is recommended to be applied two weeks before planting or sowing at the rate of 1.2 tons per hectare, the application to be continued until no further symptoms of bronzing are observed in the crop. Liming is necessary on the acid sulphate soils of Vietnam, Cambodia and Laos. Soils of the low country wet zone of Sri Lanka, deficient in silica, are treated with repeated seasonal applications of rice husk ash as a basal dressing in amounts up to 502 kg/ha to correct the deficiency.

A number of nutritional disorders of rice caused by micronutrient deficiencies or toxicities have been reported (Anon-IRRI Reporter, 1970). Sulphur deficiency has been reported from Burma (Aiyar, 1945), zinc deficiency from Japan (Tanaka *et al.*, 1969), India and Pakistan (Yoshida and Tanaka, 1969) and the Philippines (Yoshida *et al.*, 1971), aluminium toxicity from British Guiana (Cate and Sukhai, 1964), iron and manganese toxicity from Sri Lanka India, and Malaysia. Soil salinity and alkalinity are problems in arid areas and such soils need irrigation, drainage and the application of suitable soil amendments like gypsum.

SUGARCANE (*Saccharum officinarum* L.): Sugarcane requires large amounts of nutrients (Table 60) and unless adequate fertilizers are given, soil reserves are rapidly depleted. Adequate N, P and K prevent delayed ripening, poor quality of juice and a low sucrose content. Phosphorus is beneficial especially in the early stages of development of the root system and on tillering but less phosphorus is needed than N and K. Sugarcane requires much potassium which increases both cane yields and sugar content (Samuels and Landrau, 1954). By activating enzyme systems, particularly invertase (Samuels, 1955), potassium promotes the formation of sugar, increases the sucrose content and improves the purity of juice. When potassium is deficient, the cane is thin and soft with a tendency to lodge. Sulphur deficiency in sugarcane was reported by Dutt (1962), who corrected the deficiency by foliar application of soluble sulphates.

The level of nitrogen application to sugarcane is determined by water supply, length of the vegetative period, and soil type. Amounts varying between 40 and 450 lb N per acre (45-500 kg/ha) are used in sugarcane growing countries, but in most, 80-150 lb N (90-170 kg/ha/ can be regarded as optimal. Mostly, phosphorus dressings vary between 40-100 lb P_2O_5 per acre (45-110 kg/ha), and occasionally up to 300 lb P_2O_5 per acre (335 kg/ha) is used. Potassium fertilizers range between 100 and 200 lb. K_2O per acre (110-225 kg/ha), although on some Hawaiian sugarcane soils as much as 350 lb. K_2O per acre (400 kg/ha) is not unusual. The proportions of $N:P_2O_5:K_2O$ are usually in the ratio of between 3:2:3 and 3:2:4.

As phosphate is required mostly during the first few months of growth, phosphate fertilizers are best worked into the furrows before planting or placed in drills alongside the plant rows with ratoons. Nitrogen and potassium are given in split applications as the need for these nutrients increases greatly with growth, particularly so during or when side shoots are formed. Only a little of the nitrogen and potassium, together with the whole of the phosphate, is given before planting. A heavy dressing of nitrogen and potassium is given about thirty days after planting, or after cutting, and this is repeated every three months. These top-dressings can be distributed with irrigation water. With unirrigated sugarcane, fertilizer blowers have been employed.

MAIZE (*Zea mays* L.): Maize has a fairly low water requirement per unit of dry matter produced, but it also has a low drought resistance and therefore it is important to maintain an adequate soil moisture regime through water conservation or irrigation. Waterlogging, even over short periods, is most undesirable.

A granular medium-textured soil with a loose, friable consistency and good permeability is best, the optimum *p*H range being from 6.0 to 7.0.

Much nitrogen is needed for maximum growth together with high levels of phosphorus and potassium. High-yielding varieties, and good management practices will enable the best use to be made of fertilizer application. In India application of 80-120 lb N (90-135 kg/ha) and 40-80 lb (45-90 kg/ha) each of P_2O_5 and K_2O per acre was the most economic dose for hybrid maize whereas for open-pollinated maize varieties 40-60 lb N per acre (45-67 kg/ha) were sufficient for optimum yield and net profit (de Geus, 1967).

Under irrigation with an assured water supply, a high plant population as well as large doses of fertilizer give good yields. Split applications of nitrogen are preferred, particularly for larger nitrogen dressings. Nitrogen supply should be particularly plentiful during the tasselling and silking stage. Part of the nitrogen should be given together with all the phosphorus and porassium at planting time so as to provide the early nitrogen needs and to stimulate growth. The rest of the nitrogen may be given when the maize is knee high or later at the pre-tasselling stage. Nitrogen increases the size and number of ears per plant as well as yield and protein content of the grain.

Maize shows a good response to easily available phosphorus in the early stages. Phosphorus stimulates early growth, aids in seed formation and hastens ripening. Potassium deficiency causes weak stalks and depresses yields. The potassium requirement increases as more nitrogen is made available to the crop.

MILLETS: The term 'millets' includes a number of fine-grained cereal crops used as bread grains in the tropics and sub-tropics and comprise the sorghums (jowar, dura etc.) and the millets (*Setaria* sp., *Eleusine* sp., *Pennisetum* sp., etc.). These crops are similar in most respects in their general needs. They can be grown on a variety of soils. When the growing season is moist, the finer-textured soils give the largest yields but with dry seasons the sandy soils are best. A granular structure and porous, friable consistency are desirable. The millets prefer good drainage and good aeration in the root zone. They are drought-resistant crops and can withstand dry periods by interrupting their growth and resuming it after rain. However, in semi-arid areas, where water is a limiting factor, fertilizers are often unprofitable.

The millets can be grown on soils with *p*H values ranging from 5.5 to 7.5 but seem to thrive best on the more alkaline soils. Their tolerance to soluble salts is fairly high but too much calcium has been responsible for an induced iron deficiency.

Some general fertilizer recommendations were discussed by Ignatieff and Page (1958) and by de Geus (1967).

POTATO (*Solanum tuberosum L.*): Successful potato growing in tropical areas is possible only at latitudes with cool temperatures or at the higher elevations. At temperatures above the optimum 17°C, much top growth and excessive respiration result in poor tuberization.

Well drained sandy loams, retentive of moisture and of good granular structure, are best for potatoes. A wide range of soil *p*H, from 4.5 to 7.0, is permissible but potato scab, a soil-borne disease, becomes serious above *p*H 5.5. If potato soils are limed, only light dressings of lime should be applied at any one time and preferably to the preceding crop.

Potatoes respond well to applications of farmyard manure, compost, or green manure, preferably as a basal dressing. In Sri Lanka, on the Red-Yellow

Podsolic soils of the hill country, responses to cattle manure were very high; at least 12 tons of cattle manure per hectare in combination with inorganic fertilizers gave highest profits (Kathirgamathyah and Caesar, 1964). Similar results have been reported from Uttar Pradesh, India (Jakate, 1964) and at the Indian Agricultural Research Institute at New Delhi (Khan, 1956).

In many tropical areas, potatoes have shown striking responses to phosphorus fertilizer application. Amounts varying from 12-224 kg/ha P_2O_5, or even more, are recommended (de Geus, 1967). This is quite high compared with the amount actually taken up by the crop, much of the phosphorus in acid latosolic tropical soils being in unavailable forms. At Rahangala in Sri Lanka, the response to phosphorus was very striking and linear in the range 0-336 kg P_2O_5 per ha.

The potato crop feeds heavily on potassium. The tubers in a high-yielding potato crop may remove as much as 280 kg K_2O per hectare. Potassium is important for carbohydrate synthesis and, as a starch-producing crop, potato has a high potassium requirement. Amounts ranging from 55-220 kg K_2O per ha are used. Potatoes tend to be sensitive to excess chloride, which may lower the quality and depress the starch content of the tubers. Potassium sulphate is therefore recommended, particularly with higher rates of application.

Nitrogen requirements of potatoes vary according to soil type, climate, variety, and level of fertility. Economic dressings may be as high as 135 kg N per ha although lower amounts are usually recommended. Ammonium sulphate is preferred for all soils with a *p*H over 6. Calcium ammonium nitrate has proved to be excellent for potatoes on acid soils in India. Enough nitrogen should be supplied at planting to stimulate early growth and to prevent delayed maturity.

Magnesium deficiency has been reported; the responses obtained from applications of ground dolomite on the strongly acid soils at Rahangala, Sri Lanka (Ponnamperuma, 1958) were probably due to the magnesium present.

CASSAVA (*Manihot utilissima* Pohl.) : Cassava grows even on relatively impoverished soils and is drought resistant. A well-drained, light or sandy loam is most suitable.

Cassava needs much potassium and considerable amounts of nitrogen. Too high a rate of nitrogen application may stimulate and promote vegetative development at the expense of root growth. A satisfactory ratio between nitrogen and potassium is very important. Some experiments in Kerala, India indicated a ratio of N : K=1 : 1.5. Potassium deficiency causes an increase of alkaloids and glucosides and hence an increase in the toxicity due to HCN. On phosphate deficient soils, there can be a response to added phosphate fertilizer. Cassava responds well to farmyard manure (de Geus, 1967).

SWEET POTATO (*Ipomoea batatas*) : Well-drained, coarse-textured soils with well-developed, granular structure and loose, friable consistency are best for sweet potatoes. A *p*H of around 6.0 seems satisfactory. As with other root crops, potassium is most important, nitrogen gives yield increases but should be avoided in excess, and moderate amounts of phosphate are necessary for root development. In Asian countries, organic manures, like farmyard manure and compost, are used as basal applications, whenever available. A general fertilizer recommendation for average soils could be 50-60 kg N, 40-50 kg P_2O_5 and 80-120 kg K_2O per ha.

YAMS (*Dioscorea* spp.): In Africa and in India, the manuring of yams is still mainly based on farmyard manure and compost. Nitrogen is probably a limiting factor in most places. Responses have been observed to applications of both phosphate and potassium in some experiments while on sedimentary soils, derived from tertiary sandstones in western Nigeria, the best results were obtained with a combination of farmyard manure and inorganic fertilizers. De Geus (1967) has summarized the results of fertilizer trials and given general recommendations.

RUBBER (*Hevea brasiliensis* L.): Bellis (1968) reviewed progress in fertilizer use on *Hevea brasiliensis*. Early experiments were difficult and inconclusive because of ill-defined cultural and harvesting methods. Yield increases from fertilizer application were infrequent, and in most cases only prevented the deterioration of yields as the plantation matured. Leaf analysis is now widely employed especially in Malaysia, for evaluating fertilizer needs. Progress has been made (Lynen, 1963) in understanding the biosynthesis of rubber hydrocarbons and the regulation of latex flow.

Carefully selected, high-producing material consisting of selected clones or clonal seedlings have raised rubber yield levels considerably. Proper cultural practices, including a judicious fertilizing programme, can sustain these high yields.

The nutrient requirements of rubber are relatively small (Table 60). Rubber has therefore been usually grown on soils poor in nutrients but of good physical condition. But it is now realized that the nutrients removed in the latex represent only a small fraction of those required for the growth of the tree, that fertilizer application is profitable to advance the tapping stage, and that the vigour and yield of high-producing clonal material are maintained by fertilizer use.

In Malaysia, growth and yield responses to nitrogen are obtained on most soils, to phosphate on inland soils, and to potassium and magnesium on the sandier inland soils (de Geus, 1967).

Leguminous cover crops commonly used on rubber plantations (Table 43) return large quantities of nitrogen to the soil and help to cut down on nitrogen fertilizer applications. Young rubber requires much phosphorus and a basal application of rock phosphate is given in the planting hole, followed by further annual dressings.

Potassium deficiency has been reported from Malaysia, Sri Lanka, Indonesia,

Vietnam and Cambodia (de Geus, 1967) while magnesium deficiency is widespread in Malaysia, Sri Lanka and Nigeria. Manganese deficiency has been observed throughout Malaysia, and boron toxicity on acid coastal alluvial soils associated with poor drainage. Molybdenum, which is important in the symbiotic nitrogen fixation processes of legumes, may be of low availability in acid lateritic rubber soils. The possibility of copper deficiency was mentioned by Beaufils (1957) for rubber in Vietnam and Cambodia. Transient symptoms of zinc deficiency have been reported on young rubber in Sri Lanka (Jeevaratnam, 1959) and in Malaysia (Shorrocks, 1964).

de Geus (1967) discussed fertilizer recommendations in the main rubber growing areas. Recommendations for rubber in Sri Lanka are outlined in Table 64.

TABLE 64. FERTILIZER MIXTURES FOR RUBBER IN SRI LANKA
(Rubber Research Institute, Sri Lanka)

Mixture	*Sulphate of ammonia*	*Rock phosphate* **(pounds)**	*Muriate of potash* (*Kg*)	*Epsom salt*	*Nutrient* *N*	*Content* P_2O_5	K_2O	*MgO*
R.4:6:2+Mg	100(45)	100(45)	20(9)	20(9)	20(9)	30(13.6)	10(4.5)	3(1.4)
R.4:6:3+Mg	100(45)	100(45)	30(13.6)	30(13.6)	20(9)	30(13.6)	15(6.8)	5(2.3)
R.4:6:5+Mg	100(45)	100(45)	50(22.7)	50(22.7)	20(9)	30(13.6)	25(11.4)	8(3.6)

Refers to the Rubber Research Institute, Ceylon

Figures in Kg are within brackets

Field experiments at the Rubber Research Institute, Sri Lanka indicated the widespread deficiency of nitrogen in the rubber-growing districts of Sri Lanka. Phosphorus is essential for growth, particularly during the early stages. Potassium deficiency has been reported in areas poor in potassium-containing minerals and in areas where reserves have been depleted by intensive cultivation. Magnesium deficiency, often aggravated by the use of high potassium mixtures, has been on the increase and severe in clone PB 86. The NPK+Mg mixtures recommended on the basis of these findings are given in Table 64.

Rubber-growing soils in Sri Lanka have been divided into seven series, based on parent material, for fertilizer recommendation (Table 65).

The amount of appropriate fertilizer mixture to be applied depends on the age of the plant (Silva, 1970). The fertilizers are incorporated into the soil by pocketing or forking, at points as far out from the trunk as the furthermost leaves in the main canopy. Where the soils are mulched, the fertilizer mixture is spread on the soil and then covered with mulch. The applications are timed so as to avoid periods of heavy rainfall or times of moisture stress.

TABLE 65: SOILS IN THE RUBBER-GROWING DISTRICTS OF SRI LANKA (Silva, 1970)

Soil series	*Parent material (or other distinguishing feature)*	*Great Soil Group*	*Recommended Mixture*
Matale Parambe	Limestone-derived	Reddish Brown Latosolic and Immature Brown Loam	R. 4:6:5
	Biotite-gneiss-derived (micaceous)	Reddish Brown Latosolic and Immature Brown Loam	R. 4:6:2 + Mg
Ratnapura	Garnetiferous granulite-derived	Red-Yellow Podsolic	R. 4:6:3 + Mg
Homagama	Quartzite-derived	Red-Yellow Podsolic	R. 4:6:5 + Mg
Agalawatta	Granite-derived	Red-Yellow Podsolic	R. 4:6:3 + Mg
Boralu	Lateritic	Red-Yellow Podsolic	R. 4:6:3 + Mg
Deniya	Low-lying, waterlogged	Low-Humic Gley	—

COTTON (*Gossypium hirsutum* L.): The amounts of nutrients removed by cotton are relatively small compared with many other cultivated plants (Table 60) but adequate fertilizers with green manuring and crop rotation are necessary for healthy growth and good yields. Fertilizer given to the preceding crop often shows better results than applying fertilizer directly to the cotton.

Nitrogen is important although the amount required can be reduced if a leguminous green manure crop is ploughed in. Excess nitrogen makes the crop susceptible to disease and big leafy plants develop in which few bolls ripen. Sodium nitrate is preferred to ammonium sulphate but where sulphur deficiency is prevalent, as in parts of Africa, sulphur containing fertilizers may be necessary. In the Sudan, sodium nitrate at the rate of 200-300 kg per hectare has been used; urea as a 5 per cent foliar spray has also been effective. On the Vertisols of India there is a very large yield response to nitrogen, but only a poor response to phosphorus. On alluvial and red soils the combined use of nitrogen and phosphorus gave good results.

Phosphate accelerates ripening and enables the capsules to develop normally. In a number of cotton growing regions, phosphate is the most important fertilizer. In east and west Africa responses to phosphate were observed (King, 1960).

Potassium makes the plant tissues stronger and firmer and enables the plant to utilize soil water more efficiently. It ensures greater uniformity in the ripening of seeds and helps to prevent bursting of capsules, which is associated with fibres of poor quality that are less durable, less curled and absorb dyes irregularly. It is also important for increasing the resistance of cotton plants to frost, disease and insect attacks. Cotton Rust, a physiological disease caused by potassium deficiency, results in loss of yield and deterioration in quality.

Sulphur and magnesium are two other elements which need attention in the nutrition of the cotton plant. In a trial in the Upper Volta, the addition of sulphur to a basic fertilizer treatment increased the yield of rainfed cotton from 2,185 to 3,085 kg of seed cotton per hectare, when the crop was protected from insects (Wrigley, 1969).

The entire fertilizer mixture is best worked into the soil before sowing. In regions with favourable rainfall or good irrigation, a later top-dressing of nitrogen may be given. In India for irrigated cotton, a nitrogen top-dressing at flowering is superior to applications at other times.

A fertilizer mixture that serves as a rough guide to NPK fertilizer recommendations is:

Ammonium sulphate	150-200 kg/ha
Superphosphate	300-450 kg/ha
Potassium chloride (60% K_2O)	50-100 kg/ha

SISAL (*Agava sisalana* Engelm) : A good calcium status of the soil is important for the satisfactory growth of sisal (Table 51). Nitrogen and potassium (in quantities) are also necessary. An optimal Ca : N ratio coupled with a good supply of potassium are necessary for maximum growth. Bole-Rot is associated with a poor soil calcium content while Bandung Disease in Kenya and Tanzania is most prevalent where soil potassium is poor and the N:K ratio of available soil nutrients is less than optimal. Much magnesium is also required; where it is deficient, neither nitrogen or potassium can produce their full effect.

Jacob and Uexküll (1960) recommend 550-800 kg/ha of a 8 : 12 : 12 fertilizer mixture as a rough guide to fertilizer application.

COCONUT (*Cocos nucifera* L.): Field trials in the main coconut growing districts of Sri Lanka have shown that coconut soils are generally deficient in nitrogen, phosphorus and potassium. In the poorer heavily leached lateritic soils of the wet zone with an annual rainfall greater than 2500 mm, the response to manuring was more marked than on the comparatively richer lateritic soils of Bandirippuwa.

Statistically designed experiments at Bandirippuwa have shown potassium to be the dominant requirement. Plants receiving no potash fertilizer soon showed signs of potassium deficiency, a yellowing of foliage followed by the development of 'grey blight', a fungus-induced leaf disease.

The response to phosphorus at Bandirippuwa reached significance only in 1962 in the twenty-seventh year of the experiment at the higher dosage of 1.0 lb P_2O_5 (0.5 kg) per palm (Nethsinghe, 1963). This has been attributed to the residual effect of heavy dressings of bone meal given to the soil before the experiment started (Eden *et al.*, 1963; de Silva *et al.*, 1970). On Ratmalagara Estate, Madampe, response to phosphate has been spectacular. On these soils, phosphate fertilizers have shown a marked effect on leaf development, flowering and yield of nuts.

Nitrogenous fertilizers are beneficial in the early stages of growth of the coconut palm. Throughout the thirty years of the Bandirippuwa experiments there was no significant response to nitrogen, but a significant N-K interaction was observed. With smaller amounts of K, more nitrogen fertilizer decreased yields (Nethsinghe, 1963).

Traditional methods of soil analysis did not account for the different response of the coconut palm to phosphate fertilizer at Bandirippuwa and Ratmalagara. On both soils, phosphorus was little available (Salgado, 1955). Leaf analysis techniques have been useful in studying the nutritional status of coconut palms, but nut-water analysis proved to be a better guide in interpreting the different response patterns to phosphate and potassium fertilizers (Salgado, 1954).

Magnesium deficiency has been observed in all the major coconut growing areas in Sri Lanka particularly in the heavily leached lateritic soils in the southern, western and north-western provinces of the wet zone. Magnesium deficiency is shown by the yellowing of leaves of palms. Yellowing commences at tips of leaflets at the lower end of mature fronds, and spreads gradually to the upper parts of both leaflets and fronds (Nethsinghe, 1961). Magnesium applied to the soil improved leaf colour after three years, magnesium sulphate giving better results than dolomite. Affected palms sprayed fortnightly with 1-2 per cent solutions of magnesium sulphate acted more rapidly but was impracticable on coconut estates. Young eight-year-old palms recovered completely in three months and adult palms in five months.

Magnesium deficiency is accentuated in very acid soils and by ammonium, potassium and calcium ions. Much-leached acid lateritic soils given NPK fertilizer mixtures regularly thus induce conditions for magnesium deficiency. Exchangeable magnesium in soils is not a satisfactory indication of magnesium availability. Nethsinghe (1963) found that in 10^{-2}M $CaCl_2$ extracts of soil, magnesium deficiency was likely to occur if the molar Mg/Ca ratio is less than 20×10^{-3}.

The experiments in Sri Lanka indicate that the type of fertilizer is not important; inorganic fertilizers are at least as efficient as organic manures with equivalent amounts of N, P and K (Salgado, 1957). Experiments using radioactive phosphorus showed that fertilizer placement in the area within a radius of 1.6 m from the tree gave better results than the traditional practice of manuring a metre wide annular ring one metre from the tree. Also, uptake from fertilizer placed in full circles round the palm was 40 per cent more than when put in half circles (Nethsinghe, 1964, 1965). Fertilizers are usually given twice a year and experiments showed that more frequent applications were no better.

In Sri Lanka optimum fertilizer levels have been worked out to give maximum profit though not necessarily maximum yields. These are sufficient to maintain the productivity of the estate without incurring a loss, when prices are low. The optimum fertilizer mixture recommended is 0.50 lb N, 0.50 lb P_2O_5 and 0.75 lb K_2O for Bandirippuwa to give maximum profits (de Silva *et al.*, 1970).

OIL PALM (*Elaeis guineensis* Jacq.): The yield potential of the new hybrids and varieties of oil palm can only be realized under conditions of high soil fertility and good agronomic practices. The use of fertilizers is one of the most effective means of increasing yields. Foliar diagnosis gives valuable indications of deficiencies and has provided the basic information for fertilizer programmes in Africa and Malaysia.

Oil palm has a high requirement for potassium (Table 60). Potassium deficency is common in west Africa and occurs also on the granitic soils of Malaysia. Oil palms absorb and accumulate potassium very rapidly and the rate of absorption increases with the severity of the deficiency and with the rate of fertilizer application. The common symptoms of deficiency are chloritic, rust-brown leaf spots, known as Orange Spot. Potassium is usually applied in the form of KCl but according to Venema (1959), oil formation is promoted by the sulphate ion.

Balanced fertilizer use is of fundamental importance and nutrient imbalance results in poor growth and yields. The need for nitrogen in young plantations has been shown in many areas and there is a special need for nitrogen in the first three or four years. For older palms, nitrogen is a limiting factor on soils derived from sedimentary rocks in Malaysia and on liparitic and Tertiary soils in Sumatra. Organic matter has a beneficial effect and cover crops are usually recommended, in addition to any available organic material. Green manuring is only advisable where rainfall is adequate. In regions with dry seasons, green manuring plants should be cut or worked into the soil before the end of the rainy season. Potassium and phosphate fertilizers are essential for the growth of green manuring plants.

Phosphate deficiency in oil palm is widespread in Malaysia and Sumatra but is much less in west Africa, except for the Aiyinasi area in south-western Ghana.

Magnesium deficiency occurs frequently in west Africa and in the Congo as well as on the alluvial soils in Malaysia. The symptoms of magnesium deficiency in mature plants consist of deep orange tints, known as Orange Frond.

Boron deficiency, causing 'little leaf' symptoms, is a major problem in the southern part of the Congo. In Malaysia, 'hook-leaf' is considered a symptom of boron deficiency. Iron and manganese deficiency symptoms have been observed in the experiments conducted by the West African Institute for Oil Palm Research. Some micronutrient deficiencies can be overcome by returning bunch refuse of ashes in the manuring programme.

Placement of nitrogen and, in most cases, potassium, is unnecessary. On soils where potassium is strongly fixed, placement may be an advantage. A placement of phosphate is usually advised. It is best mixed with compost or peaty soil and then worked into the surface (20-30 cm) layer of the soil, in pockets or in a circle around the trunk.

GROUNDNUT (*Arachis hypogea* L.): Coarser-textured soils, such as sandy and light loams, which will not harden or bake when dry, are best for groundnut, which require well-drained, well-aerated soil conditions. In soils of very fine texture harvesting is difficult.

Better results are often obtained by fertilizing the preceding crop in a rotation than by applying fertilizer directly to groundnut. Tobacco, soya-bean or sweet potato should not be grown immediately before ground-nut as these crops may increase the nematode population and stem rot damage.

Much nitrogen fertilizer is not necessary for a legume like groundnuts but a light nitrogen dressing of 20-30 kg N per ha may be essential for a good start, especially if farmyard manure is not applied as a basal dressing. Application of small amounts of nitrogen will also enable the crop to produce large quantities of nitrogen symbiotically. But the applications should not be too large as this will reduce the amount of nitrogen fixed.

Phosphorus requirements are fairly large. Phosphorus increases the number and density of nodules, nodule growth and the rate of nitrogen fixed per gram of nodule. Phosphorus application may also significantly increase the uptake of other nutrients.

Potassium, calcium and magnesium are also needed in large amounts and the kernels will form and develop well only when plenty of calcium, in particular, is available.

The optimum *p*H range for groundnut is *p*H 5.5-7.0. Soil acidity can reduce nitrogen fixation by *Rhizobia*. But at high *p*H values manganese deficiency may occur. Lime should therefore be applied in split doses and at least three to four months before planting. A top-dressing of gypsum is recommended if the intention is to supply calcium only.

Sulphur is important in groundnut nutrition and sulphur deficiency has been discussed elsewhere. Boron deficiency and toxicity have also been reported.

Fertilizer recommendations for groundnuts in India and in west Africa are discussed by de Geus (1967).

PINEAPPLE (*Ananas comosus* L): The main nutrient requirements of pineapple are nitrogen, even more potassium, and relatively small amounts of phosphorus, calcium and magnesium (Table 60).

Nitrogen deficiency causes delayed growth, small plants with yellowing of the older leaves. Fruit production is affected; the fruits are small and strongly coloured. Phosphate deficient pineapple leaves are dark green with a purple-bluish tint. Excess phosphate hinders nitrogen absorption and affects yield and size of fruits. Potassium requirements of pineapple are high. Potassium counteracts many undesirable effects of nitrogen such as, for instance, a tendency to lodge and a low acid content. Adequate potassium is therefore essential particularly where large applications of nitrogen fertilizer are given. Potassium deficiency is shown by the occurrence of small yellow dots in the green leaf tissue. Soils containing less than 0.5 meq K per 100 g soil should be considered as potash deficient. Magnesium deficiency has been reported from northern Puerto Rico and from Guinea.

Among the trace elements, iron, zinc and copper are important. Iron deficiency is quite common in Hawaii and in Puerto Rico. It can be caused by iron fixation at too high *p*H values and also by an excess of manganese in acid soils. In iron deficient plants, the young, inner leaves of the leaf rosette, become chlorotic and almost white in extreme cases. Zinc deficiency is common in Guinea and a combined deficiency of zinc and copper causes the 'crook-neck' disorder in pineapple growing in peat, sandy peat and sandy soils in Queensland.

Of the nitrogenous fertilizers, ammonium sulphate is most commonly used. On very acid soils ammonium nitrate or calcium ammonium nitrate may be preferable. Urea is about as good as ammonium sulphate, if used as a soil application, but as a foliar spray the biuret can have a toxic effect. Nitrogenous fertilizers having a physiologically alkaline action, such as sodium nitrate, are unsuitable. Ammonium sulphate should be applied at the base of the plant, or possibly in the lower leaf axils, but not in the heart, because this causes burning. Too much nitrogen during the fruiting period can have a detrimental effect on fruit formation. Heavy dressings of nitrogen should be avoided in the two to three months before the appearance of the flower buds.

Moderate quantities of phosphate are required in the early stages of growth and larger amounts during the fruiting period. Potassium increases yields but its main influence is on fruit quality. Potassium sulphate is better than potassium chloride in this respect. In Puerto Rico, higher yields of fruit per acre and a greater mean weight of individual fruit of a more acceptable yellow colour were obtained when K_2SO_4, rather than KCl, was used.

Fertilizer practices in Puerto Rico, Guyana, Guinea, Malaya, Taiwan, Hawaii and Queensland have been reviewed by de Geus (1967).

BANANA (*Musa* sp.): Freiberg (1965) has reviewed the nutrition of the banana, while the principles of banana manuring have been considered in detail by Simmonds (1966).

The banana crop is less demanding in nitrogen than in potassium (Table 60) but nitrogen is usually the first nutrient to which it responds, the need for phosphate and potassium being shown later. The heavy fall of trash and cut psuedostems tends to build up soil organic matter but in intensive cultivation and prolonged heavy cropping the use of nitrogenous fertilizers becomes necessary.

Responses to phosphate fertilizers have varied. Even though soil phosphorus is low in the Uganda banana soils, manurial trials have seldom shown a need for phosphate. In other localities, for example at Melville Hall in Dominica, the West Indies land that had been cleared from secondary forest was so low in phosphate that poor growth and marked deficiency symptoms were shown and unfertilized plantings failed completely. Croucher and Mitchell (1940) estimated 10-20 ppm of P_2O_5 as a level at which banana will probably show response to phosphate.

The potassium requirement of the banana crop is high (Table 60). Soil analysis is not always a reliable guide and potassium deficiency can occur in lime-rich soils containing relatively large amounts of potassium. Pélegrin (1953) found that, for various localities, a series of mixtures approximating in composition to $N:P_2O_5:K_2O$ as 1:1:4 were the most satisfactory. Too high a potassium content can lead to an undesirably high potassium accumulation in the plant and the occurrence of 'Yellow Pulp'—a premature yellowing and softening of the flesh of the banana fruit (Jacob and Uexküll 1960).

Magnesium deficiency causes 'bleu' disease in the dwarf banana, the appearance of bluish stripes on the leaves. Zinc deficiency has been described by Moity (1954).

Banana grown near the sea or on lands recently reclaimed from coastal swamps is adversely affected by toxic amounts of sodium chloride, which can occur sometimes in irrigated soils as well. On some quartzitic soils in Queensland arsenic toxicity has been recorded (Fergus, 1955).

The early stages of growth are critical for later development and an abundant supply of readily available nutrients should be available at the time of planting and at the time of initiation of ratoons. The application of fertilizers should be timed accordingly.

Nitrogenous fertilizers should be applied in small doses at short intervals of time. Croucher and Mitchell (1940) considered that twice-yearly applications of phosphate and potassium were sufficient but these are also often applied more frequently. Farmyard manure is of little importance in banana cultivation in view of its cost and scarcity but if applied it should be worked into furrows or added to planting holes. In flat terrain, fertilizers can be scattered in a ring around each plant. To minimize loss by surface washing on slopes, they may be dug lightly into the soil surface on the uphill side of the plant.

CITRUS (*Citrus* sp.): Citrus is found mainly in subtropical regions although it is also extensively grown in tropical areas. Nitrogen is the most important fertilizing element and most experiments have shown a definite correlation between nitrogen and yield. Young trees with little-developed root systems require easily available phosphorus but an excess should be avoided as the availability of B, Zn and Cu may thereby be reduced, particularly on light soils. Phosphate applications seem, however, to increase the absorption of Mn and Mg. Potassium is the dominant nutrient element in the fruit (Table 60) and is absorbed continuously as the fruit matures. An excess of potassium, however, interferes with Ca and Mg uptake and has an unfavourable influence on fruit quality.

A number of secondary and micronutrient deficiencies of citrus have been reported. *Magnesium* deficiency is corrected by the addition of dolomitic limestone on acid soils and magnesium sulphate (kieserite or Epsom salts) on less acid soils. Magnesium sulphate or magnesium nitrate foliar sprays are quicker acting. In cases of *zinc* deficiency, soil applications of zinc sulphate are often unsatisfactory while sprays are effective. *Copper* deficiency occurs on peaty and acid soils while an excess of copper is a problem in the sandy soils of central Florida. Copper deficiency is corrected more readily by a copper spray than by soil applications of copper sulphate.

Chlorosis resulting from an *iron* deficiency is common in citrus-growing areas and may be caused by a deficiency of iron or, more often, by an unavailability of iron in soils of high *p*H or the antagonistic effect of excess Cu. Soil application of iron chelate of ethylene diamine tetraacetic acid (Fe EDTA) at the rate of 20 g of Fe per tree are recommended for acid soils. At *p*H values above 7, Fe EDTA is easily hydrolysed and the iron precipitated as iron hydroxide. Other iron chelates, such as EDDHA and DTPA, may be used on calcareous soils. Lowering the *p*H of the soil brings more iron into solution and, therefore, mulching with organic materials, the use of acid fertilizers, or the application of sulphur may help to correct the iron deficiency. Iron foliar sprays damage the leaves and are not recommended.

Manganese deficiency occurs on alkaline and over-limed soils and may be corrected with manganese sulphate either as a foliar spray or soil application. Excess manganese on very acid soils may prove toxic. *Boron* deficiency can be corrected by the application of borax to the soil or as a foliar spray. Care must be taken against boron toxicity on sandy soils and where the irrigation water contains more than about 0.5 ppm B (Stolzy *et al.*, 1966). 'Yellow spot' disease caused by molybdenum deficiency is usually associated with acid soils. Maintaining the soil reaction between *p*H 5.5 and 6.5 corrects this deficiency condition while molybdenum foliar sprays are also effective.

Leaf analysis, supplemented by appropriate soil tests and the analysis of irrigation water are used in deciding fertilizer requirements. Chapman (1960) gave fertilizer recommendations under Californian conditions on the basis of leaf and soil analytical data. Some critical nutrient levels are indicated in Table 61. Malavolta (1959) and Malavolta *et al.* (1962) have made fertilizer recommendations for citrus in Brazil, Chang and Wong (1962) for citrus in Taiwan.

Glossary

A Horizon. The surface horizon of a mineral soil having maximum organic matter accumulation, maximum biological activity and/or eluviation of materials such as iron and aluminium oxides and silicate clays.

ABC soil. A soil with a distinctly developed profile, including A, B and C horizons.

AC soil. A soil having a profile, containing only A and C horizons with no clearly developed B horizon.

Accelerated erosion. Erosion more rapid than natural, normal, or geological erosion, resulting from the activities of man or animals.

Acid soil. A soil with a preponderance of hydrogen and aluminium ions in proportion to hydroxyl ions. Specifically, a soil with a *p*H value < 7.0. For most practical purposes a soil with a *p*H value < 6.6. (The term is usually applied to the surface layer or to the root zone unless specified otherwise).

Acidity, active. The activity of hydrogen ions in the aqueous phase of a soil. It is measured and expressed as a *p*H value.

Acidity, potential. The amount of acidity that must be neutralized to bring an acid soil to neutrality or to some predetermined higher *p*H value. It approximates to the sum of the adsorbed hydrogen and aluminium. Usually expressed in milliequivalents per unit mass of soil.

Actinomycetes. A group of organisms intermediate between the bacteria and the true fungi that usually produce a characteristic branched mycelium. Any organism belonging to the order of Actinomycetales.

Adsorption. The attraction of ions or compounds to the surface of a solid. Soil colloids adsorb large amounts of ions and water.

Adsorption complex. The group of substances in soil capable of adsorbing other materials. Colloidal particles account for most of this adsorption.

Aerate. To impregnate with a gas, usually air.

Aeration, soil. The process by which air in the soil is replaced by air from the atmosphere. In a well-aerated soil, the soil air is very similar in composition to the atmosphere above the soil. Poorly aerated soils usually contain a much higher percentage of carbon dioxide and a correspondingly lower percentage of oxygen than the atmosphere above the soil. The rate of aeration depends largely on the volume and continuity of pores within the soil.

Aerobic. (*i*) Having molecular oxygen as a part of the environment. (*ii*) Growing only in the presence of molecular oxygen, as aerobic organisms (*iii*) Occurring only in the presence of molecular oxygen (said of certain chemical or biochemical processes such as aerobic decomposition).

Aggregate (soil). Many soil particles held in a single mass or cluster such as a clod, crumb, block or prism.

Air-dry. (*i*) The state of dryness (of a soil) at equilibrium with the moisture content in the surrounding atmosphere. The actual moisture content will depend upon the relative humidity and the temperature of the surrounding atmosphere.

(*ii*) To allow to reach equilibrium in moisture content with the surrounding atmosphere.

Alkali soil. (*i*) A soil with a high degree of alkalinity (*p*H of 8.5 or higher) or with a high exchangeable sodium content (15 per cent or more of the exchange capacity) or both. (*ii*) A soil that contains sufficient alkali (sodium) to interfere with the growth of most crop plants. See *saline-alkali soil* and *saline soil.*

Alkaline soil. Precisely, any soil that has a *p*H value >7. Practically, a soil with a *p*H >7.3. The term is usually applied to the surface layer or root zone but may be used to characterize a horizon or sample thereof.

Alkalization. The process whereby the exchangeable sodium content of a soil is increased.

Alluvial soil. (*i*) A soil developing from recently deposited alluvium and exhibiting essentially no horizon development or modification of the recently deposited materials. (*ii*) When capitalized, the term refers to a great soil group of the azonal order consisting of soils with little or no modification of the recent sediment in which they are forming (indicated by absence of a B horizon).

Alumino-silicates. Compounds containing aluminium, silicon and oxygen as main constituents. An example is microcline, $KAlSi_3O_8$.

Amendment, soil. Any substance such as lime, sulphur, gypsum, and sawdust used to alter the properties of a soil, generally to make it more productive. Fertilizers are strictly soil amendments, but the term is used most commonly for materials other than fertilizers.

Ammonification. The biochemical process whereby ammoniacal nitrogen is released from nitrogen-containing organic compounds.

Ammonium fixation. The adsorption or absorption of ammonium ions by the mineral or organic fractions of the soil in such a manner that they are relatively insoluble in water and relatively unexchangeable by the usual methods of cation exchange.

Anaerobic. (*i*) The absence of molecular oxygen. (*ii*) Living or functioning in the absence of air or free oxygen.

Anion exchange capacity. The sum total of exchangeable anions that a soil can adsorb. Expressed as milli-equivalents per 100 grams of clay (or another adsorbing material such as clay).

Apatite. A naturally occurring, complex calcium phosphate which is the original source of most of the fertilizers. Formulae such as $3Ca_3(PO_4)_2.CaF_2$ illustrate the complex compounds which make up apatite.

Argillic horizon. A diagnostic illuvial subsurface horizon characterized by an accumulation of silicate clays.

Autotrophic. Capable of utilizing carbon dioxide or carbonate as the sole source of carbon and obtaining energy for life processes from the oxidation of inorganic elements or compounds such as iron, sulphur, hydrogen, ammonium, and nitrites, or from radiant energy. Contrast with *heterotrophic.*

Available nutrient. That portion of any element or compound in the soil that can be readily absorbed or assimilated by growing plants. ('Available' should not be confused with 'exchangeable').

Available water. The portion of water in a soil that can be readily absorbed by plant roots. Considered by most workers to be that water held in the soil against a pressure of up to approximately 15 bars. See *field capacity* and *moisture tension.*

Azonal soils. Soils without distinct genetic horizons. A soil order under the classification of Thorp and Smith (1949).

B horizon. A soil horizon usually beneath the A which is characterized by one or both of the following: (1) an accumulation of silicate clays, iron and aluminium oxides, and humus, alone or in combination; (2) a blocky or primatic structure.

Bar. A unit of pressure equal to one million dynes per square centimetre.

Base saturation percentage. The extent to which the adsorption complex of a soil is saturated with exchangeable cations other than hydrogen and aluminium. It is expressed as a percentage of the total cation exchange capacity.

BC soil. A soil profile with B and C horizons but little or no A horizon. Most BC soils have lost their A horizons by erosion.

Bedrock. The solid rock underlying soils and the regolith in depths ranging from zero (where exposed by erosion) to several hundred feet.

Bog soil. A great soil group of the intrazonal order and hydromorphic sub-order. Includes peat and muck.

Buried soil. Soil covered by an alluvial, loessal, or other deposit, usually to a depth greater than the thickness of the solum.

C horizon. A horizon generally beneath the solum which is relatively little affected by biological activity and pedogenesis and is lacking properties diagnostic of an A or B horizon. It may or may not be like the material from which the A or B have formed.

Calcareous soil. Soil containing sufficient calcium carbonate (often with magnesium carbonate) to effervesce visibly when treated with cold 0.1N hydrochloric acid.

Calcic horizon. A horizon of secondary carbonate accumulation more than 15 cm in thickness, with a $CaCO_3$ equivalence of more than 15 per cent and at least 5 per cent more $CaCO_3$ than the C horizon. (7th approximation).

Calcification. The process or processes of soil formation in which the surface soil is kept sufficiently supplied with calcium to saturate the soil colloid, or the process of accumulation of calcium in some horizon of the profile.

Capillary water. The water held in the 'capillary' or small pores of a soil, usually with a tension$>$60 cm of water.

Carbon cycle. The sequence of transformations whereby carbon dioxide is fixed in living organisms by photosynthesis or by chemosynthesis, liberated by respiration and by the death and decomposition of the fixing organism, used by heterotrophic species, and ultimately returned to its original state.

Carbon-nitrogen ratio. The ratio of the weight of organic carbon (C) to the weight of total nitrogen (N) in a soil or in organic material.

"Cat" clays. Wet clay soils high in reduced forms of sulphur which, upon being drained, become extremely acid due to the oxidation of the sulphur compounds.

Category. Any one of the ranks of the system of soil classification in which soils are grouped on the bases of their characteristics.

Catena. A sequence of soils of about the same age, derived from similar parent material and occurring under similar climatic conditions, but having different characteristics due to variation in relief and in drainage.

Cation exchange. The interchange between a cation in solution and another cation on the surface of any surface-active material such as clay or organic matter.

Cation-exchange capacity. The sum total of exchangeable cations that a soil can adsorb. Sometimes called 'total-exchange capacity', or 'cation-adsorption capacity'. Expressed in milliequivalents per 100 grams of soil (or of other adsorbing material such as clay).

Cemented. Indurated; having a hard, brittle consistency because the particles are held together by cementing substances such as humus, calcium carbonate, or the oxides of silicon, iron and aluminium. The hardness and brittleness persist even when wet.

Chelate. (Gk., claw) A type of chemical compound in which a metallic ion is firmly combined with a molecule by means of multiple chemical bonds.

Chlorosis. A condition in plants relating to the failure of chlorophyll (the green colouring matter) to develop. Chlorotic leaves range from light green through yellow to almost white.

Chroma. The relative purity, strength, or saturation of a colour; directly related to the dominance of the determining wavelength of the light and inversely related to greyness. See *Munsell colour system, hue,* and *value.*

Class, soil. A group of soils having a definite range in a particular property such as acidity, degree of slope, texture, structure, land-use capability, degree of erosion, or drainage. See *soil texture* and *soil structure.*

Classification, soil. The systematic arrangement of soils into groups or categories on the basis of their characteristics. Broad groupings are made on the basis of general characteristics and subdivisions on the basis of more detailed differences in specific properties.

Clay. (*i*) A soil separate consisting of particles $<$ 0.002 mm in equivalent diameter (*ii*) Soil material containing more than 40 per cent clay, less than 45 per cent sand, and less than 40 per cent silt.

Clayey. Containing large amounts of clay or having properties similar to those of clay.

Clay mineral. Naturally occurring inorganic material (usually crystalline) found in soils and other earthy deposits, the particles being of clay size; that is, $<$ 0.002 mm in diameter.

Claypan. A compact, slowly permeable layer in the subsoil having a much higher clay content than the overlying material, from which it is separated by a sharply defined boundary. Claypans are usually hard when dry, and plastic and sticky when wet. Also, they usually impede the movement of water and air, and the growth of plant roots.

Climatic index. A simple, single numerical value that expresses climatic relationships; for example, the numerical value obtained in a precipitation-evaporation ratio.

Climax. A plant community of the most advanced type capable of development under, and in dynamic equilibrium with, the prevailing environment.

Clod. A compact, coherent mass of soil produced artificially, usually by the activity of man by ploughing, digging etc., especially when these operations are performed on soils that are either too wet or too dry for normal tillage operations.

Coarse fragments. Rock or mineral particles $>$ 2.0 mm in diameter.

Coarse sand. One of the soil separates; 2.0 to 0.2 mm in the system of classification recognised by the International Society of Soil Science, and 1.0 to 0.5 mm in the US system.

Coarse texture. The texture exhibited by sands, loamy sands, and sandy loams except very fine sandy loam.

Colloid, soil. (Gk., glue-like) Organic and inorganic matter with very small particle size and a correspondingly large surface area per unit of mass.

Colluvium. A deposit of rock fragment and soil material accumulated at the base of steep slopes as a result of gravitational action.

Compost. Organic residues, or a mixture of organic residues and soil, that have been piled, moistened, and allowed to undergo biological decomposition. Mineral fertilizers are sometimes added. Often called 'artificial manure' or 'synthetic manure' if produced primarily from plant residues.

Concretion. A local concentration of a chemical compound, such as calcium carbonate or iron oxide, in the form of a grain or nodule of varying size, shape, hardness and colour.

Consistence. The combination of properties of soil material that determine its resistance to crushing and its ability to be moulded or changed in shape. Such terms as loose, friable, firm, soft, plastic, and sticky describe soil consistence.

Consumptive use. The water used by plants in transpiration and growth, plus water vapour loss from adjacent soil or from intercepted precipitation in any specified time. Usually expressed as equivalent depth of free water per unit of time.

Contour. An imaginary line connecting points of equal elevation on the surface of the soil. A contour terrace is laid out on a sloping soil at right angles to the direction of the slope and nearly level throughout its course.

Creep. Slow mass movement of soil and soil material down relatively steep slopes primarily under the influence of gravity, but facilitated by saturation with water and by alternate freezing and thawing.

Crotovina. A former animal burrow in one soil horizon that has been filled with organic matter or material from another horizon (also spelled 'krotovina').

Crumb. A soft, porous, more or less rounded ped from 1 to 5 mm in diameter.

Crumb structure. A structural condition in which most of the peds are crumbs.

Crust. A surface layer on soils, ranging in thickness from a few millimetres to perhaps as much as an inch, that is much more compact, hard, and brittle, when dry, than the material immediately beneath it.

Crystal. A homogeneous inorganic substance of definite chemical composition bounded by plane surfaces that form definite angles with each other, thus giving the substance a regular geometrical form.

Crystal lattice. See *lattice structure*.

Crystalline rock. A rock consisting of various minerals that have crystallized in place from magma. See *igneous rock* and *sedimentary rock*.

Decalcification. The removal of calcium carbonate or calcium ions from the soil by leaching.

Deflation. The removal of fine soil particles by wind.

Deflocculate. (*i*) To separate the individual components of compound particles by chemical and/or physical means. (*ii*) To cause the particles of the *disperse phase* of a colloidal system to become suspended in the *dispersion medium*.

Denitrification. The biochemical reduction of nitrate or nitrite to gaseous nitrogen, either as molecular nitrogen or as an oxide of nitrogen.

Deposit. Material left in a new position by a natural transporting agent such as water, wind, ice, or gravity, or by the activity of man.

Desalinization. Removal of salts from saline soil, usually by leaching.

Desert crust. A hard layer, containing calcium carbonate, gypsum or other binding material, exposed at the surface in desert regions.

Desert soil. A zonal great soil group consisting of soils with a very thin, light-coloured surface horizon, which may be vesicular and is ordinarily underlain by calcareous material; formed in arid regions under sparse shrub vegetation.

Desorption. The removal of sorbed material from surfaces.

Diffusion. The transport of matter as a result of the movement of the constituent particles. The intermingling of two gases or liquids in contact with each other takes place by diffusion.

Disintegration. The breakdown of rock and mineral particles into smaller particles by physical forces such as expansion and contraction due to changes in temperature.

Disperse. (*i*) To break up compound particles, such as aggregates, into the individual component particles. (*ii*) To distribute or suspend fine particles such as clay, in or throughout a dispersion medium, such as water.

Drain, to. (*i*) To provide channels, such as open ditches or drain tile, so that excess water can be removed by surface or by internal flow. (*ii*) To lose water (from the soil) by percolation.

Drainage, excessive. Too much or too rapid loss of water from soils, either by percolation or by surface flow. Loss greater than that necessary to prevent the development of an anaerobic condition for any appreciable length of time.

Drift. Material of any sort deposited by geological processes in one place after having been removed from another. Glacial drift includes material moved by the glaciers and by the streams and lakes associated with them.

Dryland farming. The practice of crop production in low-rainfall areas without irrigation.

Duripan (hardpan). An indurated horizon cemented by materials such as aluminium silicate, silica, $CaCO_3$ and iron.

Dust mulch. A loose, finely granular, or powdery condition on the surface of the soil, usually produced by shallow cultivation.

Ecology. The science that deals with the interrelations of organisms and their environment.

Ectodynamomorphic soils. Soils with properties that have been produced, or influenced mainly, by factors other than parent material.

Edaphic. (*i*) Of, or pertaining to, the soil. (*ii*) Resulting from, or influenced by, factors inherent in the soil or other substrate, rather than by climatic factors.

Edaphology. The science that deals with the influence of soils on living things, particularly plants, including man's use of land for plant growth.

Effective precipitation. The portion of the total precipitation which becomes available for plant growth.

Eluvial horizon. A soil horizon that has been formed by the process of eluviation.

Eluviation. The removal of soil material in suspension (or in solution) from a layer or layers of a soil. (Usually, the loss of material in *solution* is described by the term 'leaching'). See *leaching*.

Endodynamomorphic soils. Soils with properties that have been influenced primarily by parent material.

Epipedon. A diagnostic surface horizon which includes the upper part of the soil that is darkened by organic matter, or the upper eluvial horizons, or both.

Erosion. (*i*) The wearing away of the land surface by running water, wind, ice or other geological agents, including such processes as gravitational creep. (*ii*) Detachment and movement of soil or rock by water, wind, ice or gravity. The following terms are used to describe different types of water erosion.

accelerated erosion. Erosion much more rapid than normal, natural geological erosion, primarily as a result of the influence of the activities of man or, in some cases, of animals.

gully erosion. The erosion process whereby water accumulates in narrow channels and, over short periods, removes the soil from this narrow area to considerable depths, ranging from 1 or 2 feet to as much as 75-100 feet.

natural erosion. Wearing away of the earth's surface by water, ice, or other natural agents under environmental conditions of climate, vegetation, etc., undisturbed by man. Synonymous with *geological erosion.*

normal erosion. The gradual erosion of land used by man which does not greatly exceed natural erosion. See *natural erosion.*

rill erosion. An erosion process in which numerous small channels of only several inches in depth are formed; occurs mainly on recently cultivated soils.

sheet erosion. The removal of a fairly uniform layer of soil from the land surface by run-off water.

splash erosion. The spattering of small soil particles caused by the impact of raindrops on very wet soils. The loosened and separated particles may or may not be subsequently removed by surface runoff.

Erosion pavement. A layer of coarse fragments, such as sand or gravel remaining on the surface of the ground after the removal of fine particles by erosion.

Evapotranspiration. The combined loss of water from a given area, and during a specified period of time, by evaporation from the soil surface and by transpiration from plants.

Exchange acidity. The titrable hydrogen and aluminium that can be replaced from the adsorption complex by a neutral salt solution. Usually expressed as milliequivalents per 100 grams of soil.

Exchange capacity. The total ionic charge of the adsorption complex active in the adsorption of ions. See *anion exchange capacity* and *cation exchange capacity.*

Exchangeable-sodium percentage. The extent to which the adsorption complex of a soil is occupied by sodium. It is expressed as follows:

$$ESP = \frac{\text{Exchangeable sodium (meq/100 g soil)}}{\text{Cation exchangeable capacity (meq/100 g soil)}} \times 100$$

Fallow. Cropland left idle in order to restore productivity, mainly through accumulation of water, nutrients, or both.

Family, soil. In soil classification one of the categories intermediate between the great soil group and the soil series.

Ferrisol. A mineral soil found in tropical regions with A-C or A-B-C profile with a faint A_1 horizon, without A_2 and textural B horizon together, having a saturation of the adsorbing complex less than 50 per cent in the B and C horizons.

Ferruginised. Hardened through being impregnated with iron.

Fertilizer. Any organic or inorganic material of natural or synthetic origin which is added to a soil to supply certain elements essential to the growth of plants.

Fertilizer grade. The guaranteed minimum analysis, in per cent, of the major plant nutrient elements contained in a fertilizer material or in a mixed fertilizer. (Usually refers to the percentage of N-P_2O_5-K_2O but proposals are pending to change the designation to the percentage of N-P-K).

Fertilizer requirement. The quantity of certain plant nutrient elements needed, in addition to the amount supplied by the soil, to increase plant growth to a designated optimum.

Field capacity (field moisture capacity). The percentage of water remaining in a soil 2 or 3 days after having been saturated and after free drainage has practically ceased.

Fine texture. Consisting of or containing large quantities of the fine fractions, particularly of silt and clay. (Includes all clay loams and clays; that is, clay loam, sandy clay loam, silty clay loam, sandy clay, silty clay, and clay textural classes).

Fixation. The process or processes in a soil by which certain chemical elements essential for plant growth are converted from a soluble or exchangeable form to a much less soluble or to a nonexchangeable form; for example, phosphate 'fixation'. Contrast with *nitrogen fixation.*

Fixed phosphorus. The phosphorus which has been changed to a less soluble form as a result of reaction with the soil; moderately available phosphorus.

Flood plain. The land bordering a stream, built up of sediments from overflow of the stream and subject to inundation when the stream is at flood stage.

Fluorapatite. A member of the apatite group of minerals rich in fluorine. Most common mineral in rock phosphate.

Foliar diagnosis. An estimation of the extent to which plants are getting certain necessary chemical elements from the soil by examination of the colour and growth habits of the foliage of the plants.

Fragifan. Dense and brittle pan or layer in soils that owe their hardness mainly to extreme density or compactness rather than high clay content or cementation. Removed fragments are friable, but the material in place is so dense that roots cannot penetrate and water moves through it very slowly.

Friable. A soil consistency term pertaining to the ease of crumbling of soils.

Fulvic acid. A term of varied usage but usually referring to the mixture of organic substances remaining in solution upon acidification of a dilute alkali extract from the soil.

Furrow irrigation. See *irrigation methods.*

Genesis, soil. The mode of origin of the soil, with special reference to the processes responsible for the development of the solum, or true soil, from the unconsolidated parent material.

Geological erosion. See *erosion.*

Gilgai. The microrelief of soils produced by expansion and contraction with changes in moisture. Found in soils that contain large amounts of clay which swells and shrinks considerably with wetting and drying. Usually a succession of microbasins and microknolls in nearby level areas or of microvalleys and microridges parallel to the direction of the slope.

Glacial drift. Rock debris that has been transported by glaciers and deposited, either directly from the ice or from the melt-water. The debris may or may not be heterogeneous.

Glacial till. See *till.*

Gley soil. Soil developed under conditions of poor drainage resulting in reduction of iron and other elements and in grey colours and mottles.

Granular structure. Soil structure in which the individual grains are grouped into spherical aggregates with indistinct sides. Highly porous granules are commonly called crumbs. A well-granulated soil has the best structure for most ordinary crop plants.

Gravelly. Containing appreciable or significant amounts of gravel.

Gravitational water. Water which moves into, through, or out of the soil under the influence of gravity.

Great soil group. Any one of several broad groups of soils with fundamental characteristics in common. Examples are Latosols, Reddish-Brown Earths, Red-Yellow Podsolic soils.

Great Group. A category in the new comprehensive classification system adopted in the United States between that of the suborder and the subgroup.

Green Manure. Plant material incorporated with the soil while green, or soon after maturity, for improving the soil.

Groundwater. Water that fills all the unblocked pores of underlying material below the water table, which is the upper limit of saturation.

Groundwater Laterite soil. A great soil group of the intrazonal order and hydromorphic sub-order, consisting of soils characterized by hardpans or concretional horizons rich in iron and aluminium (and sometimes manganese) that have formed immediately above the water table.

Groundwater Podsol soil. A great soil group of the intrazonal order and hydromorphic sub-order, consisting of soils with an organic mat on the surface over a very thin layer of acid humus material underlain by a whitish-grey leached layer, which may be as much as 60 or 90 cm in thickness, and is underlain by a brown, or very dark-brown, cemented hardpan layer; formed under various types of forest vegetation in cool to tropical humid climates under conditions of poor drainage.

Gully erosion. See *erosion.*

Gypsic horizon. A horizon of accumulation of secondary $CaSO_4$, more than 15 cm thick which has at least 5 per cent more gypsum than the C or underlying stratum, and in which the product of thickness in cm and the per cent gypsum is at least 150 per cent-cm.

Half-Bog soil. A great soil group, of the intrazonal order and hydromorphic suborder consisting of soil with dark-brown or black peaty material over greyish and rust-mottled mineral soil formed under conditions of poor drainage under forest, sedge, or grass vegetation in cool to tropical humid climates.

Halomorphic soil. A suborder of the intrazonal soil order, consisting of saline and alkali soils formed under imperfect drainage in arid regions and including the great soil groups Solonchak or Saline soils, Solonetz soils, and Soloth soils.

Halophytic vegetation. Salt-loving or salt tolerant vegetation usually having fleshy leaves or thorns and resembling desert vegetation.

Hardpan. A hardened soil layer, in the lower A or in the B horizon caused by cementation of soil particles with organic matter or with materials such as silica, sesquioxides, or calcium carbonate. The hardness does not change appreciably with changes in moisture content and pieces of the hard layer do not slake in water. See *claypan* and *duripan.*

Heavy soil. (Obsolete in scientific use)—A soil with a high content of the fine separates, particularly clay, or one with a high drawbar pull and hence difficult to cultivate.

Heterotrophic. Capable of deriving energy for life processes only from the decomposition of organic compounds and incapable of using inorganic compounds as sole sources of energy or for organic synthesis. Contrast with *autotrophic.*

Histosols. Soils characterized by their high organic matter content. Bog soils and half-bog soils are included in this soil order. (New comprehensive classification system).

Horizon, soil. A layer of soil, approximately parallel to the soil surface, with distinct characteristics produced by soil-forming processes.

Hue. One of the three variables of colour. It is caused by light of certain wavelengths and changes with the wavelength. See *Munsell colour system, chroma, value* and *colour.*

Humic acid. A mixture of variable or indefinite composition of dark organic substances, precipitated upon acidification of a dilute-alkali extract from soil.

Humic Gley soil. Soil of the intrazonal order and hydromorphic suborder that includes Wisenboden and related soils, such as Half-Bog soils, which have a thin muck or peat O_2 horizon and an A_1 horizon. Developed in wet meadow and in forested swamps.

Humification. The processes involved in the decomposition of organic matter and leading to the formation of humus.

Humus. That more or less stable fraction of the soil organic matter remaining after the major portion of added plant and animal residues have decomposed. Usually it is dark coloured.

Hydromorphic soils. A suborder of intrazonal soils, all formed under conditions of poor drainage in marshes, swamps, seepage areas, or flats.

Hydrous mica. A silicate clay with 2:1 lattice structure, but of indefinite chemical composition as, usually, part of the silicon in the silica tetrahedral layer has been replaced by aluminium and containing a considerable amount of potassium which serves as an additional bonding between the crystal units, resulting in particles larger than normal in montmorillonite and, consequently, in a lower cation-exchange capacity. Sometimes referred to as *illite*.

Hydroxyapatite. A member of the apatite group of minerals rich in hydroxyl groups. A nearly insoluble calcium phosphate.

Hygroscopic coefficient. The amount of moisture in a dry soil when it is in equilibrium with some standard relative humidity near a saturated atmosphere (about 98 per cent), expressed in terms of percentage on the basis of oven-dry soil.

Igneous rock. Rock formed from the cooling and solidification of magma, and has not been changed appreciably since its formation.

Illite. A hydrous mica. See *hydrous mica.*

Illuvial horizon. A soil layer or horizon in which material carried from an overlying layer has been precipitated from solution or deposited from suspension. The layer of accumulation.

Immature soil. A soil with indistinct or only slightly developed horizons because of the relatively short time it has been subjected to the various soil-forming processes. A soil that has not reached equilibrium with its environment.

Immobilization. The conversion of an element from the inorganic to the organic form in microbial or in plant tissues, thus rendering the element not readily available to other organisms or to plants.

Impeded drainage. A condition which hinders the movement of water through soils under the influence of gravity.

Impervious. Resistant to penetration by fluids or by roots.

Inceptisols. Soils with one or more diagnostic horizons that are thought to form rather quickly and that do not represent significant illuviation or eluviation or extreme weathering. Soils classified as Brown Forest, Ando, Sols Bruns Acides, and associated Humic Gley and Low-Humic Gley soils are included in this order. (New comprehensive classification system).

Infiltration. The downward entry of water into the soil.

Infiltration rate. A soil characteristic determining or describing the maximum rate at which water can enter the soil under specified conditions, including the presence of an excess of water

Intergrade. A soil that possesses moderately well-developed distinguishing characteristics of two or more genetically related great soil groups.

Intrazonal soil. A soil with more or less well-developed soil characteristics that reflect the dominating influence of some local factor of relief, parent material, or age, over the normal effect of climate and vegetation. One of the three orders in soil classification.

Ions. Atoms, groups of atoms, or compounds, which are electrically charged as a result of the loss of electrons (cations) or the gain of electrons (anions).

Iron-pan. An indurated soil horizon in which iron oxide is the principal cementing agent.

Irrigation efficiency. The ratio of the water actually consumed by crops on an irrigated area to the amount of water diverted from the source onto the area.

Irrigation methods. The manner in which water is artificially applied to an area. The methods and the manner of applying the water are as follows:

border-strip. The water is applied at the upper end of a strip with earth borders to confine the water to the strip.

check-basin. The water is applied rapidly to relatively level plots surrounded by levees. The basin is a small check.

corrugation. The water is applied to small, closely-spaced furrows, frequently in grain and forage crops, to confine the flow of irrigation water to one direction.

flooding. The water is released from field ditches and allowed to flood over the land.

furrow. The water is applied to row crops in ditches by tillage implements.

sprinkler. The water is sprayed over the soil surface through nozzles from a pressure system.

subirrigation. The water is applied in open ditches or tile lines until the water table is raised sufficiently to wet the soil.

wild-flooding. The water is released at high points in the field and distribution is uncontrolled.

Isomorphous substitution. The replacement of one atom by another of similar size in a crystal lattice without disruption or changing the crystal structure of the mineral.

Kaolin. An aluminosilicate mineral of the 1:1 crystal lattice group; that is, consisting of one silicon tetrahedral layer and one aluminium oxide-hydroxide octahedral layer.

Lacustrine deposit. Material deposited in lake water and later exposed either by lowering of the water level or by the elevation of the land.

Land. Land is a broader term than soil. In addition to soil, its attributes include other physical conditions such as mineral deposits and water supply; location in relation to centres of commerce, populations, and other land; the size of the individual tracts or holdings; and existing plant cover, works of improvement, and the like.

Land classification. The arrangement of land units into various categories based upon the properties of the land or its suitability for some particular purpose.

Land-use planning. The development of plans for the uses of land that, over long periods, will best serve the general welfare, together with the formulation of ways and means for achieving such uses.

Laterite. This is fully explained in the text.

Latosol. A suborder of zonal soils including soils formed under forested, tropical humid conditions and characterized by low silica-sesquioxide ration of the clay fractions, low base-exchange capacity, low activity of the clay, low content of most primary minerals, low content of soluble constituents, a high degree of aggregate stability, and usually having a red colour.

Lattice structure. The orderly arrangement of atoms in a crystalline material.

Leaching. The removal of materials in solution from the soil. See *eluviation.*

Levee. Alluvial deposit on the banks of streams and rivers, usually lighter in texture than that further away from the stream.

Light soil. (Obsolete in scientific use) A coarse-textured soil; a soil with a low draw-bar pull and hence easy to cultivate. See *coarse texture* and *soil texture.*

Lime (agricultural). In strict terms, calcium oxide. In practical terms it is a material containing the carbonates, oxides and/or hydroxides of calcium and/or magnesium used to neutralize soil acidity.

Lime requirement. The mass of agricultural limestone, or the equivalent of other specified liming material, required per hectare to a soil depth of 15 cm (or on 2.2 million kilograms of soil) to raise the *p*H of the soil to a desired value under field conditions.

Limestone. A sedimentary rock composed primarily of calcite ($CaCO_3$). If dolomite ($CaCO_3 . MgCO_3$) is present in appreciable quantities it is called a dolomitic limestone.

Lithosol. A great soil group of azonal soils characterized by an incomplete solum or no clearly expressed soil morphology and consisting of freshly and imperfectly weathered rock or rock fragments.

Loam. The textural class name for soil having a moderate amount of sand, silt, and clay. Loam soils contain 7-27 per cent of clay, 28-50 per cent of silt, and less than 52 per cent of sand.

Loamy. Intermediate in texture and properties between fine-textured and coarse-textured soils. Includes all textural classes with the words 'loam' or 'loamy' as part of the class name, such as clay loam or loamy sand. See *loam* and *soil texture.*

Loess. Material transported and deposited by wind and consisting of predominantly silt-sized particles.

Luxury consumption. The intake by a plant of an essential nutrient in amounts exceeding what it needs. Thus if potassium is abundant in the soil, alfalfa may take in more than is required.

Lysimeter. A device for measuring percolation and leaching losses from a column of soil under controlled conditions.

Macronutrient. A chemical element necessary in large amounts (usually > 1 ppm in the plant) for the growth of plants and usually applied artificially in fertilizer or liming materials ('macro' refers to quantity and not to the essentiality of the element). See *micronutrient.*

Marl. Soft and unconsolidated calcium carbonate, usually mixed with varying amounts of clay or other impurities.

Marsh. Periodically wet or continually flooded areas with surface not deeply submerged. Covered dominantly with sedges, cat-tails, rushes, or other hydrophytic plants. Subclasses include fresh-water and salt-water marshes.

Mature soil. A soil with well-developed soil horizons produced by the natural processes of soil formation and essentially in equilibrium with its present environment.

Maximum water-holding capacity. The average moisture content of a disturbed sample of soil, 1 cm high, which is at equilibrium with a water table at its lower surface.

Mechanical analysis. (Obsolete) See *particle-size analysis* and *particle-size distribution*.

Medium texture. Intermediate between fine-textured and coarse-textured (soils). (It includes the following textural classes: very fine sandy loam, loam, silt loam, and silt).

Mellow soil. A very soft, very friable, porous soil without any tendency towards hardness or harshness. See *consistence*.

Mesophilic bacteria. Bacteria whose optimum temperature for growth falls in an intermediate range of approximately 15° to 45°C.

Metamorphic rock. A rock that has been greatly altered from its previous condition through the combined action of heat and pressure. For example, marble is a metamorphic rock produced from limestone, gneiss is one produced from granite, and slate is produced from shale.

Micas. Primary aluminosilicate minerals in which two silica layers alternate with one alumina layer. They separate readily into thin sheets or flakes.

Microclimate. (*i*) The climatic condition of a small area resulting from the modification of the general climatic conditions by local differences in elevation or exposure. (*ii*) The sequence of atmospheric changes within a very small region.

Microfauna. That part of the animal population which consists of individuals too small to be clearly distinguished without the use of a microscope. Includes protozoa and nematodes.

Microflora. That part of the plant population which consists of individuals too small to be clearly distinguished without the use of a microscope. Includes actinomycetes, algae, bacteria, and fungi.

Micronutrient. A chemical element necessary in only extremely small amounts (<1 ppm in the plant) for the growth of plants. Examples are: B, Mo, Cu, Fe, Mn, Zn and Cl. ('Micro' refers to the amount used rather than to its essentiality). See *macronutrient*.

Microrelief. Small-scale, local differences in topography, including mounds, swales, or pits that are only a few metres in diameter and with elevation differences of up to 2m. See *gilgai*.

Mineralization. The conversion of an element from an organic form to an inorganic state as a result of microbial decomposition.

Mineral soil. A soil consisting predominantly of, and having its properties determined predominantly by, mineral matter. Usually contains <20 per cent organic matter, but may contain an organic surface layer up to 30 cm thick.

Minor element. (Obsolete) See *micronutrient*.

Moderately-coarse texture. Consisting predominantly of coarse particles. (In soil textural classification, it includes all the sandy loams except the very fine sandy loam). See *coarse texture*.

Moderately-fine texture. Consisting predominantly of intermediate-size (soil) particles or with relatively small amounts of fine or coarse particles. (In soil textural classification, it includes clay loam, sandy loam, sandy clay loam, and silty clay loam). See *fine texture*.

Moisture equivalent. The weight percentage of water retained by a previously saturated sample of soil 1 cm in thickness after it has been subjected to a centrifugal force of one thousand times gravity for 30 min.

Moisture tension (*or pressure*). The equivalent negative pressure in the soil water.

It is equal to the equivalent pressure that must be applied to the soil water to bring it to hydraulic equilibrium, through a porous permeable wall or membrane with a pool of water of the same composition.

Mollisols. Soils characterized by a thick, dark mineral surface horizon which is dominantly saturated with bivalent cations and has moderate to strong structure. Includes soils such as Chernozem, Prairie, Chestnut in the earlier classification system.

Montmorillonite. An aluminosilicate clay mineral with a 2:1 expanding crystal lattice; that is, with two silicon tetrahedral layers enclosing an aluminium octahedral layer. Considerable expansion may be caused along the C axis by water moving between silica layers of contiguous units.

Mor. Raw humus; a type of forest humus layer of unincorporated organic material usually matted or compacted or both; distinct from the mineral soil, unless the latter has been blackened by washing in organic matter.

Morphology, soil. The constitution of the soil including the texture, structure, consistence, colour, and other physical, chemical, and biological properties of the various soil horizons that make up the soil profile.

Mottling. Spots or blotches of different colour or shades of colour interspersed with the dominant colour.

Muck. Highly decomposed organic material in which the original plant parts are not recognizable. Contains more mineral matter and is usually darker in colour than peat. See *muck soil, peat.*

Muck soil. An organic soil in which the organic matter is well decomposed.

Mulch. Any material such as straw, sawdust, leaves, plastic film, and loose soil that is spread upon the surface of the soil to protect the soil and plant roots from the effects of raindrops, soil crusting, freezing, evaporation, etc.

Mulch farming. A system of farming in which the organic residues are not ploughed into or otherwise mixed with the soil but are left on the surface as a mulch.

Mull. A humus-rich layer of forested soils consisting of mixed organic and mineral matter. A mull blends into the upper mineral-layers without an abrupt change in soil characteristics.

Munsell colour system. A colour designation system that specifies the relative degree of the three simple variables of colour: hue, value, and chroma. For example: 10YR 6/4 is a colour (of soil) with a hue-10YR, value=6, and chroma=4. These notations can be translated into several different systems of colour names as desired. See *chroma, hue,* and *value, colour.*

Murram. Rounded, hard, iron concretions or rounded, hard fragments of iron-impregnated weathered rock. Soft ferruginous-looking weathered rock is not known as 'murram' in Uganda although it has been referred to by this name in Kenya.

Mycorrhiza. The association, usually symbiotic, of fungi with the roots of seed plants.

Necrosis. Death associated with discoloration and dehydration of all or parts of plant organs, such as leaves.

Nematodes. Very small worms abundant in many soils and important because many of them attack and destroy plant roots.

Neutral soil. A soil in which the surface layer, at least to normal plough depth, is neither acid nor alkaline in reaction. See *acid soil, alkaline soil,* **pH,** and *reaction soil.*

Nitrification. The biochemical oxidation of ammonium to nitrate.

Nitrogen assimilation. The incorporation of nitrogen into organic cell substances by living organisms.

Nitrogen fixation. The conversion of elemental nitrogen (N_2) to organic combinations or to forms readily utilizable in biological processes.

Nodule bacteria. See *rhizobia.*

Nucleic acids. Complex compounds found in the nuclei of plant and animal cells and usually combined with proteins as nucleoproteins.

O horizon. Organic horizon of mineral soils.

Ochrosols. A great soil group used in Charter's scheme for the classification of tropical soils. Ochrosols are associated with moderate to low rainfall areas (250-1800 mm). The surface horizon is slightly acid to neutral while the subsoil is strongly leached. There is a marked concentration of organic matter and bases in the surface horizon. The *p*H decreases with depth.

Order. The highest category in soil classification. The three orders are zonal soils, intrazonal soils, and azonal soils. Ten orders are recognized in the new comprehensive classification system: Entisol, Vertisol, Inceptisol, Aridisol, Mollisol, Spodosol, Alfisol, Ultisol, Oxisol, and Histosol.

Organic soil. A soil which contains a high percentage, (> 20 per cent) of organic matter throughout the solum.

Ortstein. An indurated layer in the B horizon of Podsols in which the cementing material consists of illuviated sesquioxides (mostly iron) and organic matter.

Oven-dry soil. Soil which has been dried at 105°C until it reaches constant weight.

Oxisols. Soils of tropical and subtropical regions characterized by the presence of a horizon (oxic) from which most of the combined silica has been removed by weathering, leaving oxides of iron and aluminium and some quartz. Includes soils referred to as Latosols and some called Groundwater Laterites.

Oxysols. A great soil group used in Charter's scheme for the classification of tropical soils. Oxysols are strongly leached, acid soils found in high rainfall areas (1800-2000 mm)

Pans. Horizons or layers, in soils, that are strongly compacted, indurated or very high in clay content. See *claypan*, *duripan*, *fragipan*, and *hardpan.*

Parent material. The unconsolidated and more or less chemically weathered mineral or organic matter from which the solum of soils is developed by pedogenic processes.

Parent rock. The consolidated rock from which parent material is formed by weathering.

Particle density. The mass per unit volume of the soil particles. In technical work, usually expressed as grams per cubic centimetre.

Particle size. The effective diameter of a particle measured by sedimentation, sieving, or micrometric methods.

Particle-size analysis. The laboratory determination of particle size distribution in an air-dried soil.

Particle-size distribution. The amounts of the various soil separates in a soil sample, usually expressed as weight percentages.

Parts per million (ppm). Weight units of any given substance per one million equivalent weight units of oven-dry soil; or, in the case of soil solution or other solution, the weight units of solute per million weight units of solution.

Peat. Unconsolidated soil material consisting largely of undecomposed, or only slightly decomposed, organic matter accumulated under conditions of excessive moisture.

Peat soil. An organic soil containing more than 50 per cent organic matter. Also used to refer to the stage of decomposition of the organic matter, 'peat' referring to the slightly decomposed or undecomposed deposits and 'muck' to the highly decomposed materials. See *peat, muck* and *muck soil.*

Ped. A unit of soil structure such as an aggregate, crumb, prism, block, or granule, formed by natural processes (in contrast with a clod, which is formed artificially).

Pedalfer. A subdivision of a soil order comprising a large group of soils in which sesquioxides increased relative to silica during soil formation.

Pedocal. A subdivision of a soil order comprising a large group of soils in which calcium accumulated during soil formation.

Peneplain. A once high, rugged area which has been reduced by erosion to a low, gently rolling surface resembling a plain.

Peneplain remnant. Small area of a formerly extensive but now dissected peneplain such as is found on the flat summits of certain hills in the Ghanaian forest zone.

Penetrability. The case with which a probe can be pushed into the soil. (May be expressed in units of distance, speed, force, or work depending on the type of penetrometer used).

Percolation, soil water. The downward movement of water through soil. Especially, the downward flow of water in saturated or nearly saturated soil at hydraulic gradients of the order of 1.0 or less.

Permafrost. (*i*) Permanently frozen material underlying the solum. (*ii*) A perennially frozen soil horizon.

Permeability, soil. The ease with which gases, liquids, or plant roots penetrate or pass through a bulk mass of soil or a layer of soil.

pF. The logarithm of the soil moisture tension expressed in centimetres height of a column of water.

pH, soil. The negative logarithm of the hydrogen-ion activity of a soil. The degree of acidity (or alkalinity) of a soil as determined by means of a glass, quinhydrone, or other suitable electrode or indicator at a specified moisture content or soil-water ratio, and expressed in terms of the *p*H scale.

pH-dependent charge. That portion of the total charge of the soil particles which is affected by, and varies with, changes in *p*H.

Phase, soil. A subdivision of a soil type or other unit of classification having characteristics that affect the use and management of the soil but which do not vary sufficiently to differentiate it as a separate type. A variation in a property or characteristic such as degree of slope, degree of erosion, and content of stones.

Photomap. A mosaic map made from aerial photographs with physical and cultural features shown as on a planimetric map.

Physical properties (of soils). Those characteristics, processes, or reactions of a soil which are caused by physical forces and which can be described by, or expressed in, physical terms or equations. Examples of physical properties are bulk density, water-holding capacity, hydraulic conductivity, porosity, pore-size distribution, etc.

Physical weathering. The breakdown of rock and mineral particles into smaller particles by physical forces such as heating and cooling.

Planosol. A great soil group of the intrazonal order and hydromorphic suborder consisting of soils with eluviated surface horizons underlain by B horizons more strongly eluviated, cemented, or compacted than associated normal soil.

Plastic soil. A soil capable of being moulded or deformed continuously and permanently, by relatively moderate pressure, into various shapes. See *consistence*.

Platy. Consisting of soil aggregates that are developed predominately along the horizontal axes; laminated; flaky.

Plinthite (brick). A highly weathered mixture of sesquioxides of iron and aluminium with quartz and other diluents which occurs as red mottles and which changes irreversibly to hardpan upon alternate wetting and drying.

Podsol. A great soil group of the zonal order consisting of soils formed in cool-temperate to temperate, humid and even tropical humid climates, under coniferous or mixed coniferous and deciduous forest, and characterized particularly by a highly-leached, whitish-gray A_2 horizon.

Podsolization. A process of soil formation resulting in the genesis of Podsols and Podsolic soils.

Pore-size distribution. The volume of the various sizes of pores in a soil. Expressed as percentage of the bulk volume (soil plus pore space).

Porosity. The volume percentage of the total bulk not occupied by solid particles.

Potassium fixation. The process of converting exchangeable or water-soluble potassium to a form not easily exchanged from the adsorption complex with a cation of a neutral salt solution.

Primary mineral. A mineral that has not been altered chemically since deposition and crystallization from molten lava.

Prismatic soil structure. A soil structure type with prismlike aggregates that have a vertical axis much longer than the horizontal axes.

Productivity, soil. The capacity of a soil for producing a specified plant or sequence of plants under a specified system of management. Productivity emphasizes the capacity of a soil to produce crops and should be expressed in terms of yields.

Productive soil. A soil in which the chemical, physical and biological conditions are favourable for the economic production of crops suited to a particular area.

Profile, soil. A vertical section of the soil through all its horizons and extending into the parent material.

Protein. Any of a group of nitrogen-containing compounds that yield amino acids on hydrolysis and have high molecular weights. They are essential parts of living matter and are one of the essential food substances of animals.

Puddled soil. Dense, massive soil artificially compacted when wet and having no regular structure. The condition commonly results from the tillage of a clayey soil when it is wet. Flooded rice soils are usually 'puddled'.

Reaction, soil. The degree of acidity or alkalinity of a soil, usually expressed as a *p*H value.

Extremely acid	Below 4.5
Very strongly acid	4.5-5.0
Strongly acid	5.1-5.5
Medium acid	5.6-6.0
Slghtly acid	6.1-6.5
Neutral	6.6-7.3
Mildly alkaline	7.4-7.8
Moderately alkaline	7.9-8.4
Strongly alkaline	8.5-9.0
Very strongly alkaline	9.1 and higher

Red-Yellow Podsolic soils. A zonal great soil group consisting of soils formed under warm-temperate to tropical, humid climates, under deciduous or coniferous forest vegetation and usually under conditions of good drainage.

Regolith. The unconsolidated mantle of weathered rock and soil material on the earth's surface, loose earth materials above solid rock. (Approximately equivalent to the term 'soil' as used by many engineers).

Regosol. Any soil of the azonal order without definite genetic horizons and developing from or on deep, unconsolidated, soft mineral deposits such as sands, loess, or glacial drift.

Regur. An intrazonal group of dark, calcareous soils high in clay, which is mainly montmorillonitic, and formed mainly from rocks low in quartz; occurring extensively on the Deccan Plateau of India.

Rendzina. A great soil group of the intrazonal order and calcimorphic suborder consisting of soil with brown or black friable surface horizons underlain by light-grey to pale-yellow calcareous material; developed from soft, highly calcareous parent material under grass vegetation or mixed grasses and forest in humid and semi-arid climates.

Residual material. Unconsolidated and partly weathered mineral materials accumulated by disintegration of consolidated rock in place.

Residual soil. A soil formed from, or resting on, consolidated rock of the same kind as that from which it was formed, and in the same location.

Reticulate mottling. A network of streaks of different colour; most commonly found in the deeper profiles of Latosolic soils.

Reversion. The changing of essential plant nutrient elements from soluble to less soluble forms as a result of interaction with, or reactions in, the soil. For example, the conversion of monocalcium phosphate to the less soluble dicalcium phosphate.

Rhizobia. Bacteria capable of living symbiotically with higher plants, usually, legumes, from which they receive their energy, and capable of using atmospheric nitrogen; hence, the term symbiotic nitrogen-fixing bacteria. (Derived from the generic name *Rhizobium*).

Rhizosphere. That portion of the soil in the immediate vicinity of plant roots in which the abundance and composition of the microbial population are influenced by the presence of roots.

Rill. A small, intermittent water course with steep sides; usually only a few inches deep and, hence, no obstacle to tillage operations.

Rill erosion. See *erosion.*

Rock. The material that forms the essential part of the earth's solid crust, including loose incoherent masses such as sand and gravel, as well as solid masses of granite and limestone.

Rough broken land. Land with very steep topography and numerous intermittent drainage channels but usually covered with vegetation.

Runoff. That portion of the precipitation on an area which is discharged from the area through stream channels. That which is lost without entering the soil is called *surface runoff* and that which enters the soil before reaching the stream is called *ground water runoff or seepage flow* from ground water. (In soil science 'runoff' usually refers to the water lost by surface flow; in geology and hydraulics 'runoff' usually includes both surface and subsurface flow).

Saline-alkali soil (or *saline-sodic soil*). A soil containing sufficient exchangeable sodium to interfere with the growth of most crop plants and containing appreciable quantities of soluble salts. The exchangeable-sodium percentage is > 15, the conductivity of the saturation extract > 4 millimhos per centimetre (at 25° C), and the *p*H is usually 8.5 or less in the saturated soil.

Saline soil. A nonsodic soil containing sufficient soluble salts to impair its productivity.

Salinization. The process of accumulation of salts in soil.

Secondary forest. Recent forest which has grown up on a previously cleared site. May include a few trees from the original forest spared during clearing.

Secondary mineral. A mineral resulting from the decomposition of a primary mineral or from the reprecipitation of the products of decomposition of a primary mineral. See *primary mineral.*

Sedentary soil. Soil developed in non-transported parent materials, usually in the products of weathering of the parent rock below.

Sedimentary rock. A rock formed from materials deposited from suspension or precipitated from solution and usually being more or less consolidated. The principal sedimentary rocks are sandstones, shales, limestones, and conglomerates.

Self-mulching soil. A soil in which the surface layer becomes so well aggregated that it does not crust and seal under the impact of rain but instead serves as a surface mulch upon drying.

Separate, soil. One of the individual-size groups of mineral soil particles sand, silt, or clay.

Series, soil. See *soil series.*

Sheet erosion. See *erosion.*

Silica-alumina ratio. The molecules of silicon dioxide (SiO_2) per molecule of aluminium oxide (Al_2O_3) in clay minerals or in soils.

Silica-sesquioxide ratio. The molecules of silicon dioxide (SiO_2) per molecule of aluminium oxide (Al_2O_3) plus ferric oxide (Fe_2O_3) in clay minerals or in soils.

Silt. (*i*) A soil separate consisting of particles between 0.02 and 0.002 mm in equivalent diameter. (*ii*) A soil textural class.

Silting. The deposition of water-borne sediments in stream channels, lakes, reservoirs, or on flood plains, usually resulting from a decrease in the velocity of the water.

Skeletal soil. Shallow, immature soil containing frequent pieces of hard parent rock.

Sodic soil. A soil that contains sufficient sodium to interfere with the growth of most crop plants, and in which the exchangeable-sodium percentage is 15 or more.

Soil. (*i*) A dynamic natural body on the surface of the earth in which plants grow, composed of mineral and organic materials and living forms. (*ii*) The collection of natural bodies occupying parts of the earth's surface that support plants and that have properties due to the integrated effect of climate and living matter acting upon parent materials, as conditioned by relief, over periods of time.

Soil air. The soil atmosphere; the gaseous phase of the soil, being that volume not occupied by solid or liquid.

Soil alkalinity. The degree or intensity of alkalinity of a soil, expressed by a value > 7.0 on the *p*H scale.

Soil association. A group of defined and named taxonomic soil units occurring together in an individual and characteristic pattern over a geographic region, comparable to plant association in many ways.

Soil complex. A mapping unit used in detailed soil surveys where two or more defined taxonomic units are so intimately intermixed geographically that it is undesirable or impractical, because of the scale being used, to separate them. A more intimate mixing of smaller areas of individual taxonomic units than that described under soil association.

Soil conservation. A combination of all management and land use methods which safeguard the soil against depletion or deterioration by natural or by man induced factors.

Soil genesis. The mode of origin of the soil with special reference to the processes or soil-forming factors responsible for the development of the solum, or true soil, from the unconsolidated parent material.

Soil geography. A subspecialization of physical geography concerned with the areal distributions of soil types.

Soil horizon. See *horizon, soil.*

Soil management. The sum total of all tillage operations, cropping practices, fertilizer, lime, and other treatments conducted on or applied to a soil for the production of plants.

Soil map. A map showing the distribution of soil types or other soil mapping units in relation to the prominent physical and cultural features of the earth's surface.

Soil mechanics and engineering. A subspecialization of soil science concerned with the effect of forces on the soil and the application of engineering principles to problems involving the soil.

Soil microbiology. A subspecialization of soil science concerned with soil-inhabiting micro-organisms and with their relation to agriculture, including both plant and animal growth.

Soil monolith. A vertical section of a soil profile removed from the soil and mounted for display or study.

Soil morphology. The physical constitution, particularly the structural properties, of a soil profile as exhibited by the kinds, thickness, and arrangement of the horizons in the profile and by the texture, structure, consistency, and porosity of each horizon.

Soil porosity. See *porosity.*

Soil reaction. See *reaction, soil* and *pH, soil.*

Soil salinity. The amount of soluble salts in a soil, expressed in terms of percentage, parts per million, or other convenient ratios.

Soil science. That science dealing with soils as a natural resource on the surface of the earth including soil formation, classification and mapping, and the physical, chemical, biological, and fertility properties of soils *per se*; and these properties in relation to their management for crop production.

Soil separates. See *separate, soil.*

Soil series. The basic unit of soil classification being a subdivision of a family and consisting of soils which are essentially alike in all major profile characteristics except the texture of the A horizon.

Soil solution. The aqueous liquid phase of the soil and its solutes consisting of ions dissociated from the surface of the soil particles and of other soluble materials.

Soil structure. The combination or arrangement of primary soil particles into secondary particles, units, or peds. These secondary units may be, but usually are not, arranged in the profile in such a manner as to give a distinctive characteristic pattern. The secondary units are characterized and classified on the basis of size, shape, and degree of distinctness into classes, types, and grades, respectively.

Soil survey. The systematic examination, description, classification, and mapping of soils in an area. Soil surveys are classified according to the kind and intensity of field examination.

Soil texture. The relative proportions of the various soil separates in a soil.

Soil type. The lowest unit in the natural system of soil classification; a subdivision of a soil series and consisting of or describing soils that are alike in all characteristics including the texture of the A horizon.

Solodized soil. A soil that has been subjected to the processes responsible for the development of a soloth and having at least some of the characteristics of a soloth.

Solonchak soils. An intrazonal group of soils with high concentrations of soluble salts in relation to those in other soils, usually light-coloured, without characteristic structural form, developed under salt-loving plants, and occurring mostly in a subhumid or semiarid climate.

Solonetz. An intrazonal group of soils having surface horizons of varying degrees of friability underlain by dark hard soil, ordinarily with columnar structure (prismatic structure with rounded tops). This hard layer is usually highly alkaline. Such soils are developed under grass or shrub vegetation, mostly in subhumid or semiarid climates.

Solum (**plural:** *sola*). The upper and most weathered part of the soil profile; the A and B horizons.

Splash erosion. See *erosion.*

Spodosols. Soils characterized by the presence of a spodic horizon, an eluvial horizon in which active organic matter and amorphous oxides of aluminium and iron have precipitated. These soils include most Podsols, Brown Podsolics, and Groundwater Podsols of the old classification system.

Sprinkler irrigation. See *irrigation methods.*

Stratified. Arranged in or composed of strata or layers.

Strip cropping. The practice of growing crops which require different types of tillage, such as row and sod, in alternate strips along contours or across the prevailing direction of wind.

Structure, soil. See *soil structure.*

Stubble mulch. The stubble of crops or crop residues left essentially in place on the land as a surface cover before and during the preparation of the seedbed and at least partly during the growing of a succeeding crop.

Subsoil. That part of the soil below the plough layer.

Subsoiling. Breaking of compact subsoils, without inverting them, with a special knife-like instrument (chisel) which is pulled through the soil at depths usually of 30 to 60 cm and at spacings usually of 0.5 to 1.5m.

Surface runoff. See *runoff.*

Surface soil. The uppermost part of the soil, ordinarily moved in tillage, or its equivalent in uncultivated soils and ranging in depth from 8 to 10 cm to 20 or 25 cm. Frequently designated as the 'plough layer', the "Ap layer" or the "Ap horizon".

Swamp thicket, forest. Thicket or forest consisting mostly of species adapted to poorly drained soils.

Symbiosis. The living together in intimate association of two dissimilar organisms, the cohabitation being mutually beneficial.

Talus. Fragments of rock and other soil material accumulated by gravity at the foot of cliffs or steep slopes.

Tensiometer. A device for measuring the negative pressure (or tension) of water in soil *in situ*; a porous, permeable ceramic cup connected through a tube to a manometer or vacuum gauge.

Tension, soil-moisture. The equivalent negative pressure of suction of water in soil.

Terrace. (*i*) A level, usually narrow, plain bordering a river, lake, or the sea. Rivers sometimes are bordered by terraces at different levels. (*ii*) A raised, more or less level or horizontal strip of earth usually constructed on, or nearly on, a contour and designed to make the land suitable for tillage and to prevent accelerated erosion.

Texture. See *soil texture.*

Thicket. Dense secondary growth of shurb, coppice-shorts, young trees and climbers, usually on sites cleared two to eight years previously and then abandoned.

Thermal analysis (*differential thermal analysis*). A method of analyzing a soil sample for constituents, based on a differential rate of heating of the unknown and standard samples when a uniform source of heat is applied.

Thermophilic organisms. Organisms which grow readily at temperatures above 45°C.

Till. (*i*) Unstratified glacial drift deposited directly by the ice and consisting of clay, sand, gravel, and boulders intermingled in any proportion. (*ii*) To plough and prepare for seeding; to seed or cultivate the soil.

Tilth. The physical condition of soil as related to its ease of tillage, fitness as a seedbed, and its impedance to seedling emergence and root penetration.

Toposequence. A sequence of related soils that differ, one from the other, primarily because of topography as a soil-formation factor.

Topsoil. (*i*) The layer of soil moved in cultivation. See *surface soil.* (*ii*) Presumably fertile soil material used to topdress roadbanks, gardens, and lawns.

Trace element. (Obsolete) See *micronutrient.*

Truncated. Having lost all or part of the upper soil horizon or horizons.

Truff. Volcanic ash usually more or less stratified and in various states of consolidation.

Tundra. A level or undulating treeless plain characteristic of arctic regions.

Type, soil. See *soil type.*

Unsaturated flow. The movement of water in a soil which is not filled to capacity with water.

Value, colour. The relative lightness or intensity of colour and approximately a function of the square root of the total amount of light. One of three variables of colour. See *Munsell colour system, hue,* and *chroma.*

Vertisols. Soils high in swelling clays which crack widely upon drying resulting in shrinking, shearing and soil mass movement. (new comprehensive classification system).

Virgin soil. A soil that has not been significantly disturbed from its natural environment.

Volume weight. (Obsolete) See *bulk density.*

Waterlogged. Saturated with water.

Water-stable aggregate. A soil aggregate stable to the action of water such as falling drops, or agitation as in wet-sieving analysis.

Water table. The upper surface of groundwater or that level below which the soil is saturated with water; locus of points in soil water at which the hydraulic pressure is equal to atmospheric pressure.

Weathering. All physical and chemical changes produced in rocks, at or near the earth's surface by atmospheric agents.

Wilting point (or permanent wilting point). The moisture content of soil, on an oven-dry basis, at which plants (specifically sunflower plants) wilt and fail to recover their turgidity when placed in a dark humid atmosphere.

Windbreak. A planting of trees, shrubs, or other vegetation, usually perpendicular or nearly so to the principal wind direction to protect soil, crops, homesteads, roads etc., against the effects of winds, such as wind erosion, and the drifting of soil and snow.

Xerophytes. Plants that grow in or on extremely dry soils or soil materials.

Yield, sustained. A continual annual, or periodic, yield of plants or plant material from an area; implies management practices which will maintain the productive capacity of the land.

Zonal soil. A soil characteristic of a large area or zone.

References

Abruna, F, Vicente-Chandler, J. and Pearson, R.W. (1964) Effects of liming on yields and composition of heavily fertilized grasses and on soil properties under humid tropical conditions. *Soil Sci. Soc. Amer. Proc.* **28**, 657-661.

Acharya, C.N. and Rajagopalan, K. (1956) *J. Indian Soc. Soil Sci.* **4**, 111.

Acquaye, D.K. and Cunningham, R.K. (1965) Losses of nitrogen by ammonia volatilization from surface-fertilized tropical forest soils. *Trop. Agric. Trin.* 42, 281-292.

Agarwal, R.R. (1965) *Soil Fertility in India.* Asia Publishing House, London.

Agboola, A.A. and Fayemi, A.A. (1971) Preliminary trials on the intercropping of maize with different tropical legumes in western Nigeria. *J. agric. Sci. Camb.* **77**, 219-225.

Agrawal, H.P. and Motiramani, D.P. (1966) Copper status of soils of Madhya Pradesh. *J. Ind. Soc. Soil Sci.* **14**, 161-171.

Ahmad, N. (1963) The effect of evolution of gases and reducing conditions in a submerged soil on its subsequent physical status. *Trop. Agric., Trin.* **40**, 205-209.

Ahmad, N., Jones, R.L. and Beavers, A.H. (1968 *a*) Genesis, morphology, and related properties of West Indian soils. 1. Montserrat Series derived from calcareous Upper Miocene glauconitic sandstone, Central Trinidad. *J. Soil Sci.* **19**, 1-8.

Ahmad, N., Jones, R.L. and Beavers, A.H. (1968*b*) Genesis, morphology, and related properties of West Indian soils. 11. Maracas Series, formed from micaceous schist and phyllite, Northern Range, Trinidad. *J. Soil Sci.* **19**, 9-19.

Ahmad N., Tulloch-Reid, L.I. and Davies, C.E. (1969) Fertilizer studies on Pangola grass (*Digitaria decumbens* Stent.) in Trinidad. I. Description of the experiments and effect of nitrogen. II. Effect of phosphorus, potassium and magnesium. *Trop. Agric., Trin.* **46**, 173-178; 179-186.

Aiyar, S.P. (1945) The significance of sulphur in the manuring of rice. *Indian J. Agric. Sci.* **15**, 283-297.

Alberda, Th. (1953) Growth and root development of lowland rice and its relation to oxygen supply. *Pl. Soil* **5** (1), 1-28.

Alexander, M. (1962) *Introduction to Microbiology.* (2nd ed.) Wiley, New York.

Allen, E.F. (1956) The effect of crop rotation on the growth and yield of padi *Malay. agric. J.* **39**, 133-139.

Alles, W.S. (1967) Soil and water conservation in the dry zone. In *Development of Dry Zone Agriculture* ed. O.S. Peries. Ceylon Assoc. Advmt. Sci. pp. 39-58.

Almeida, L.A.V. and de Miranda, V.H.B. (1954) Les formes de phosphore dans des sols d' Angola. *Trans. 5th int.* **Congr. Soil Sci. 3. 263-267.**

Ancizar-Sordo. (1941) El empobrecimento de los suclos de Colombia y sus repercussiones en la ganaderia. *Revista Facultad Nacional de Agronomia*, Medelin.

Anderson, A.J. (1942) Molybdenum deficiency on a South Australian ironstone soil *J. aust. Inst. agri. Sci.* **8**, 73-75.

Anderson, A.J. (1956) Molybdenum as a fertilizer. *Adv. Agron.* **8**. 163-202.

Anderson, B. (1961) *The Rufigi Basin, Tanganyika. vol VII: Soils of the main irrigation areas.* Tanganyika Dept. of Agriculture. FAO.

Anderson, G.D. (1969) Responses of maize to application of compound fertilizers on farmers' fields in ten districts of Tanzania. *E. afr. agric. For. J.* **34**, 382-397.

Anderson, G.D. and Naveh, Z. (1965) The influence of climate and soil on the productivity of certain promising pastures and legumes in northern Tanzania. *Afr. Soils* **10**, 241.

Anderson, G.D. and Naveh, Z. (1968) Promising pasture plants for northern Tanzania. V. Overall comparisons of promising plants. VI. Practical suggestions for pasture improvement. *E. afr. agric. For. J.* **34**, 84-105; 106-116.

Andrew, C.S (1963) Copper deficiency symptoms of some tropical and temperate pasture legumes. *Aust. J. agric. Res.* **14**, 654-659.

Andrew, C.S. (1966) A kinetic study of phosphate absorption by excised roots of *Stylosanthes humilis*, *Phaseolus lathyroides*, *Desmodium uncinatum*, *Medicago sativa*, and *Hordeum vulgare*. *Aust. J. agric. Res.* **17**, 611-624.

Andrew, C.S. and Hegarty, M.P. (1969) Comparative responses to manganese excess of eight tropical and four temperate pasture legume species. *Aust. J. agric. Res.* **20**, 687-696.

Andrew C.S. and Norris, D.O. (1961), Comparative responses to calcium of five tropical and four temperate pasture legume species. *Aust. J. agric. Res.* **12**, 40-55.

Anghie G.M. (1967) Lift Irrigation. In *Development of Agriculture in the Dry Zone* ed. O.S. Peries. Ceylon Assoc. Advmt. Sci.

Anon. (1964) *Volcanic ash soils in Japan.* Minist. Agric. For., Japanese Govt.

Anon. (1966) *The use of isotopes in soil organic matter studies.* Report of the FAO/IAEA Technical Meeting, 1963. Pergamon Press, London.

Anon. (1966) Rice Information Co-operative Effort 1966. In *Rice Production Manual,* University of the Philippines.

Anon. (1970) Nutritional disorders of rice. In *IRRI Reporter* **6**, (5). Int. Rice Res. Inst. Manila, Philippines.

Aoki, M. (1941) Studies on the behaviour of phosphoric acid under paddy field condition. *J. Sci. Soil Manure*, Tokyo, **15**, 182-202.

Aomine, S. (1962) A review of research on redox potentials of paddy soils. *Soil Sci.* **94**, 6-13.

Arakeri, H.R., Chalam, G.V. and Satyanarayana, P. (1959) *Soil Management in India.* Asia Publishing House, London.

Arnott, G.W. (1957) Soil Survey Reports, No. 6. The Kelantan Deficiency Area. *Malay. agric. J.* **40**, 60-91.

Ashgar, A.G., Dhawan, C.L. and Bhola, K.D. (1949) Preliminary studies on the establishment of colour standards in relation to soil constants. *Indian J. agric Sci.* **19**, 21-30.

Aubert, G. (1964) The classification of soils as used by French pedologists in tropical or arid areas. *Afr. Soils* **9**, 107-115.

Baars, J.K. (1950) Manganese deposits in rice fields. *Trans. 4th int. Congr. Soil Sci* **1**, 192-195.

Bache, B.W. (1969) Chemical soil tests in agricultural practice. *World Crops* **21** (1), 43-45.

Bache, B.W. and Rogers, N.E. (1970) Soil phosphate values in relation to phosphate supply to plants from some Nigerian soils. *J. agric. Sci. Camb.* **74**, 383-390.

Bachy, A. (1963) Controle de la nutrition des plantations. Choix des feuilles pour l' analyse. *Oleagineux* **18**, 471-472.

Bailey, H.H. (1950) Peat formation in the tropics and subtropics. *Proc. Soil Sci. Soc. Am.* **15**, 283-284.

Barnes, A.C. (1953) *Agriculture of the sugarcane.* Leonard Hill, Ltd., London.

Barshad, I. and Rojas-Cruz, L.A. (1950) A pedologic study of a podsol soil profile from the equatorial region of Colombia, South America. *Soil Sci.* **70**, 221-236.

Bates, J.A.R. (1960) Studies on a Nigerian forest soil. I. The distribution of organic matter in the profile and in various soil fractions. *J. Soil Sci.* **11**, 246-256.

Bates, J.A.R. and Baker, T.C.N. (Mrs.) (1960) Studies on a Nigerian forest soil. II. The distribution of phosphorus in the profile and in various soil fractions. *J. Soil Sci.* **11**, 257-265.

Baver, L.D. (1968) The effect of organic matter on soil structure. In *Study Week on Organic Matter in Soil Fertility.* John Wiley and Sons Inc., New York.

Bear, E. (1953) *Soils and Fertilizers* (4th ed.) John Wiley and Sons, Inc., New York.

Bear, F.E. and Coleman, R. (eds.) (1949) *Hunger Signs in Crops—A Symposium. Am. Soc. Agron. and National Fertil. Assoc.*

Beauchamp, R.S.A. (1953) Sulphates in African inland waters. *Nature, Lond* **171**, 769-771.

Beaufils, E.R. (1957) Research for rational exploitation of the hevea using a physiological diagnosis based on the mineral analysis of various parts of the plant. *Fertilite* **3**, 27-38.

Beckett, P.H.T. and Hopkinson, (1961) Some Sarawak soils. I. Soils of the region centred on the Usum Apau Plateaux. *J. Soil Sci.* **12**. 41-51.

Beckinsale, R.P. (1957) The nature of tropical rainfall. *Trop. Agric. Trin.* **34**, 76-97.

Beirnaert, A. (1941) La technique culturale sous l 'Équateur. Influence de la culture sur les reserves en humus et en azote des terres equatoriales. *I.N.E.A.C. Ser. Tech. No.* **26**. Bruxelles, pp. 86.

Bellis, E. (1968) Rubber, an example of progress in fertilizer use in tropical agriculture *Trans. 9th int. Congr. Soil Sci.* **4**, 77-84.

Bennema, J. (1963) The red and yellow soils of the tropical and subtropical uplands. *Soil Sci.* **95**, 250-257.

Bennett, H.H. (1955) *Elements of Soil Conservation.* McGraw-Hill, N.Y. 2nd ed.

Bennison, R.H. and Evans, D.D. (1968) Some effects of crop rotation on the productivity of crops on a red earth in a semiarid tropical climate. *J. agric. Sci.* Camb. **71**, 365-380.

Berger, K.C. and Truog, E. (1945) Boron availability in relation to soil reaction and organic matter content. *Proc. Soil Sci. Soc. Am.* **10**, 113-116.

Berry, L. and Ruxton, B.P. (1959) Notes on weathering zones and soils on granite rocks in two tropical regions. *J. Soil Sci.* **10**, 54-63.

Bhat, K.K.S. and Bouyer, S. (1968) Influence de la matiere organique sur le phosphore isotopiquement diluable dans quelques types de soils tropicaux. *In Isotopes and Radiation in Soil Organic Matter Studies.* IAEA, Vienna.

Bhavanandan, V.P. (1970) Fertilizer application for tea in Ceylon. *Soil Sci. Soc. Ceylon* **1**, 68-88.

Biebl, R. (1952) The effect of different nitrogen fertilizers on the transpiration of some crop plants. *Z.Pf Ernahr. Dung.* **59**, 60-75.

Birch, H.F. (1958) The effect of soil drying on humus decomposition and nitrogen availability. *Pl. Soil* **10**, 9-31.

Birch, H.F. (1960) Soil drying and soil fertility. *Trop. Agric.* Trin, **37**, 3-10.

Birch, H.F. and Friend, M.T. (1956) The organic matter and nitrogen status of East African soils. *J. Soil Sci.* **7**, 156-167.

Bloomfield, C.; Coulter, J.K. and Kanaris-Sotiriou, R. (1968) Oil palms on acid sulphate soils in Malaya. *Trop. Agric.*, Trin. **45**, 289-300.

Blue, W.G. (1966) Fertilizando los pastos tropicales. *Hacienda* **61** (7), 33-40.

Blumenstock, D.I. (1958) Distribution and characteristics of tropical climates. *Proc. 9th Pacific Sci. Congr.* **20**, 3-21.

Blumenstock, D.I. and Thornthwaite, C.W. (1941) Climate and the world pattern. In *Climate and Man.* U.S. Dept. Agric. Yb. Agric.

Bocquier, G. and Maignien, R. (1963) The brown sub-arid tropical soils of West Africa. *Afr. Soils* **8**, 371-381.

Bolle-Jones, E.W. (1964) Incidence of sulphur deficiency in Africa: A review. *Emp. J. exp. Agric.* **32**, 241-248.

Boswinkle, E (1961) Residual effects of phosphorus fertilizers in Kenya. *Emp. exp. Agric.* **29**, 136-142.

Bradfield, R. (1968) The role of organic matter in soil management and the maintenance of soil fertility. In *Study Week on Organic Matter and Soil Fertility.* John Wiley and Sons, Inc., New York.

Bradfield, R. (1969) Training agronomists for increasing food production in the humid tropics. In *International Agronomy Training and Education.* ASA Special Publication No. 15. American Society of Agronomy Inc.

Bramao, D. L. (1963) International soil correlation. *Soil Sci.* **95**, 231-234.

Bramao, D. L. (1968) Discussion reported in *Study Week on Organic Matter and Soil Fertility.* John Wiley and Sons Inc., New York.

Bramao, D. L. and Dudal, R. (1958) Climate, vegetation and rational land utilization in the humid tropics. *Proc.* 9*th Pacific Sci. Congr.*, Bangkok.

Bramao, D. L. and Lemos, P. (1961). **Soil map of South America.** . *Trans. 7th int. Congr. Soil Sci.* **4**, 1-10.

Bramao, D. L. and Riquier, J. (1968) Characteristics of the organic matter in the major soils of the world and its importance to soil fertility. In *Study Week on Organic Matter and Soil Fertility.* John Wiley and Sons Inc., New York.

Brammer, H. (1962) 'Soils'. Chapter 6 in *Agriculture and Land Use in Ghana* ed. J. B. Willis, pp. 86-126. Oxford Univ. Press, London.

Brammer, H. and De Endredy, A. S. (1954) The tropical black earths of the Gold Coast and their associated vlei soils. *Trans 5th int Congr. Soil Sci.* **4**, 70-76.

Bray, R. H. and Kurtz, L. T. (1945) Determination of total, organic, and available forms of phosphorus in soils. *Soil Sci.* **59**, 39-45.

Broadbent, F. E. (1948) Nitrogen release and carbon loss from soil organic matter during decomposition of added plant residues. *Proc. Soil Sci. Soc Am.* **12**, 246-249.

Brockington, N. R. (1964) Fertilizer responses of Giant Rhodes grass (*Chloris gayana,* Kunth.) in northern Rhodesia. *Emp. J. exp. Agric.* **32**, 76-83.

Brockington, N. R., Stobbs, T. H., Newhouse, P.W. and Wadsworth, G. A. (1965) The effects of leys on soil fertility in the annual cropping areas of Uganda. *Afr. Soils.* **10**, 473-481.

Brown, G. (1953) The occurrence of lepidocrocite in some British soils. *J. Soil Sci.* **4**, 220-228.

Broyer, T. C., Carlton, A. B., and Johnston, C. M. (1954) Chlorine—a micronutrient element for higher plants. *Plant Physiol.* **29**, 526-532.

Bryan, W. H. (1939) The red earth residuals and their significance in South Eastern Queensland. *Proc. roy. Soc. Qd.*, **50**, 21-22.

Buchanan, F. (1807) *A Journey from Madras through the countries of Mysore, Canara and Malabar.* Vol. 2, pp. 436-460 (3 volumes in all). East India Company, London.

Buckman, H. C. and Brady, N. C. (1969) *The Nature and Properties of Soils.* 6th ed. pp. 334-356. Macmillan Co., New York.

Bull, R. A. (1954) Studies on the deficiency symptoms of the oil palm. I. Orange Frond disease caused by magnesium deficiency. *J. W. Afr. Inst. Oil Palm Res.* **1**(2), 94-129.

Buringh, P. (1970) *Introduction to the study of soils in tropical and subtropical regions* Centre for Agricultural Publishing and Documentation, Wageningen, 2nd Ed.

Bushnell, T. M. (1942) Some aspects of the soil catena concept. *Proc. Soil Sci. Soc. Am* **7**, 466-476.

Carroll, D. and Jones, N. K. (1947) Laterite developed on acid rocks in southwestern Australia. *Soil Sci.* **64**, 1-15.

Cassidy, N. G. (1968) A note on the salt tolerance of the coconut palm (*Cocos nucifera* L.) *Trop. Agric. Trin.* **45**, 247-250.

Cate, R. B. (Jr.) and Sukhai, A. P. (1964) A study of aluminium in rice soils *Soil Sci.* **98**, 85-93.

Chang, S. C. and Wong, C. M. (1962) Fertilizer use under intensive farming in Taiwan. *Soils Fertil. Taiwan* pp. 12-19.

Chapman, H. D. (1960) Leaf and soil analyses as guides for citrus fertilizer practices in southern California orchards. *Calif. Agric.* **14**, 13-14.

Charter, C. F. (1949) *The detailed reconnaissance soil survey of the cocoa country of the Gold Coast.* Proc. Cocoa Conf., Lond. p 19. The Cocoa, Chocolate, and Confectionery Alliance Ltd., London.

Charter, C. F. (1955) *The nutrient status of Gold Coast forest soils, with special reference to the manuring of cocoa.* Report Cocoa Conf. Lond. pp. 40-48. The Cocoa, Chocolate and Confectionery Alliance, Ltd., London.

Chaverri, G. R., Bornemisza, E. S. and Chaves, F. S. (1957) *Resultados de analisis foliar del cafeto en Costa Rica.* Stica, San Jose, Costa Rica, Information Técnica No. 3.

Chenery, E. M. (1954*a*) **Acid sulphate soils in Central Africa**.. *Trans. 5th int. Congr. Soil Sci.* **15**, 195-198.

Chenery, E. M. (1954*b*) **Minor Elements in Uganda soils.** *Proc. 2nd inter-afr. Soil Conf.* **2**, 1157-1163.

Chenery, E. M. (1958) *A review of the role of minor and trace elements in East African agriculture.* Report of a Conference of Directors and Senior Officers of Overseas Departments of Agriculture and Agricultural Institutions held at Wye College, Kent. Colonial Office Lond. Misc. No. 531, 40-43.

Chenery, E. M. (1960) *An introduction to the soils of the Uganda Protectorate.* Uganda Dept. Agric. Mem. Res. Div. Series I-Soils No. 1.

Chenery, E. M. (1961) *Tea*, **3**, 22.

Chenery, E. M. and Schoenmaekers, J. (1959) Magnesium deficiency in East African tea. *E. afr. agric. J.* **25**, 25-27.

Child, R. (1957) Manuring of tea in East Africa. *Tea Res. Inst. E. Afr. Pamphl.* **14**, 13.

Child, R. (1960) The effects of fertilizer applications on tea soils. *Tea Res. Inst. E. Afr. Pamphl.* **17**, 49-59.

Child, E. C. and Collis-George, N. (1950) The control of soil water. *Adv. Agron.* **2**, 233-272.

Chiu, T. F. (1950) The manganese contents of some Formosan soils and various fertilizers. *Agric. Res. Q. Taiwan* **1**(2), 31-47 (Ch. e).

Cibes, H. and Samuels, G. (1958) *Mineral deficiency symptoms of Red Spanish pineapple plants.* Puerto Rico agr. Expt. Sta. tech. Pap. 25.

Clarke, R. T. (1962) The effect of some resting treatments on a tropical soil. *Emp. J. exp. Agric.* **30**, 57-62.

Cline, M. G. (1949) Basic principles of soil classification. *Soil Sci.* **67**, 81.

Coene, R. de (1956) Banana plantation in rotation in the northern cotton belt. *Bull. Inst. Etud. agron. Congo Belge* **5**, 113-126.

Cook, L. (1939) A contribution to our information on grass burning. *S. afr. J. Sci.* **36**, 270-282.

Cooke, G. W. (1967) *The Control of Soil Fertility.* Crosby Lockwood & Son, Ltd., London.

Corbet, A. A. (1935) *The biological processes of tropical soils.* W. Hoffen & Son, Ltd., Cambridge.

Coulter, J. K. (1950) Organic matter in Malayan soils: a preliminary study of the organic matter content in soils under virgin jungle, forest plantations and abandoned cultivated land. *Malay. Forester* **13**, 189-202.

Coulter, J. K. (1958) Mineral nutrition of the oil palm in Malaya. The use of frond analysis as a guide to manurial requirements. *Malay. agric. J.* **41**, 131-151.

Coulter, J. K. (1964) Soil surveys and their application in tropical agriculture. *Trop. Agric. Trin.* **41**, 185-196.

Cradock, F. W. and Weir, R. G. (1964) Citrus leaf analysis. *Agr. Gaz.* N.S.W. **75**, 1425-1431.

Croucher, H. H. and Mitchell, W. K. (1940) The use of fertilizers on bananas. *Jamaica agric. Soc.* **44**, 138-142.

Crowder, L. V., Michelin, A. and Bastidas, A. (1964) The response of Pangola grass (*Digitaria decumbens* Stent.) to rate and time of nitrogen application in Colombia. *Trop. Agric. Trin.* **41**(1), 21-29.

Culot, J. P., Wambeke, A. Van, and Croegaert, J. (1958) *Contribution a l'etude des deficiences minerales du cafeier d' Arabie au Kivu.* Publ. Inst. Nat. Etud. Agron. Congo Belge, Ser. Sci. **73**, 1-105.

Cunningham, R. K. (1962) Mineral nitrogen in tropical forest soils. *J. Agric Sci. Camb.* **59**, 257-262.

Cunningham, R. K. (1963) The effect of clearing a tropical forest soil. *J. Soil Sci.* **14**, 334-345.

Cunningham, R. K. (1964) Micronutrient deficiency in cacao in Ghana. *Emp. J. exp. Agric.* **32**, 42-50.

Cunningham, R. K. and Arnold, P. W. (1962) The shade and fertilizer requirements of cocoa (*Theobroma cacao*) in Ghana. *J. Sci. Fd. Agric.* **13**, 213-221.

Cunningham, R. K. and Lamb, J. (1959) A cocoa shade and manurial experiment at the W.A.C.R.I., Ghana. I. First year. *J. hort. Sci.* **34**, 14-22.

Cunningham , R. K. and Smith R. W. (1961*a*) Comparison of soil covers during cocoa establishment on clear-felled land. *Trop. Agric. Trin.* **38**, 13-22.

Cunningham, R. K. and Smith, R. W. (1961*b*) *The significance of some recent work in Ghana on the shade and nutrient requirements of Cocoa.* Rep. The Cocoa Conference, Lond. 1961, pp. 119-122. The Cocoa, Chocolate and Confectionery Alliance, Ltd., London.

Cunningham, R. K., Smith, R. W. and Hurd, R. G. (1961) A cocoa shade and manurial experiment at the West Africa Cocoa Research Institute, Ghana. II. Second and third years. *J. hort. Sci.* **36**, 116-125.

Cutting, C. V., Wood, R. A., Brown, P. and Ambrose, H. B. (1959) Assessment of fertility status and the maintenance of productivity of soils in Nyasaland. *Proc. 3rd inter-afr. Soil Conf.* **2**, 817-824.

Da Costa, J.V.B. (1959) Ferallitic, tropical fersiallitic and tropical semiarid soils. *Proc. 3rd inter-afr. Soil Conf.* **1**, 317-319.

Da Costa, J. V. B. Azevedo, A. L., Franco, E. P. C. and Ricardo, R. P. (1959) Non-calcic (quartz-feldspathic) brownish and reddish brown soils of southern Angola. *Proc. 3rd inter-afr. Soil Conf.* **1**, 253-257.

Da Silva **Teixeira** , A. J. (1959) The soils of Portuguese Guinea. *Proc. 3rd inter-afr. Soil Conf.* **1**, 467-477.

Dao, F. (1970) *Climatic influence on the distribution pattern of plant parasitic and soil inhabiting nematodes.* Meded. Landbhogesch. Wageningen **70**, pp. 181.

Datta, N. P. (1964) Review of trace element studies in India. *J. Ind. Soc. Soil Sci.* **12**, 449-466.

Daulton, R. A. C. (1965) Evaluation of chemicals for the control of root-knot nematode. *Rhod. J. agric. Res.* **3**, 59-61.

Davies, E. B. (1956) Factors affecting molybdenum availability in soils. *Soil Sci.* **81**, 209-221.

Davies, J. G. and Eyles, A. G. (1965) Expansion of Australian pastoral production *J. aust. Inst. agric. Sci.* **31**, 77-93.

De, P . K. (1939) *The role of blue-green algae in nitrogen fixation in rice fields.* *Proc. Roy. Soc. B*, 127, 121-139.

De, P. K. and Sulaiman, M. (1950) The influence of algal growth in the rice fields on the yield of crops. *Ind. J. agric. Sci.* **20**, 327-342.

Dean, A. L. (1930) Nitrogen and organic matter in Hawaiian soils. *Soil Sci.* **30**, 439-442.

Dean, L. A. (1938) An attempted fractionation of the soil phosphorus. *J. agric. Sci.* ***Camb.***, **28**, 234-246.

De Coene, R. (1956) Agricultural settlement schemes in the Belgian Congo. *Trop. Agric. Trin.* **33**, 1-12.

De Geus, J. G. (1967) *Fertilizer Guide for Tropical and Subtropical Farming.* Centre d'Etude de l'Azote, Zurich.

Dennison, E. B. (1959) The maintenance of soil fertility in the southern Guinea Zone of northern Nigeria. *Trop. Agric. Trin.* **36**, 171-176.

Dennison, E. B. (1961) The value of farmyard manure in maintaining fertility in northern Nigeria. *Emp. J. exp. Agric.* **29**, 330-336.

De Silva, M. A. T. Balakrishnamurthi, T. S. and Abeywardena, V. (1970) Fertilizer requirements of the coconut palm. *J. Soil Sci. Soc. Ceylon* **1**, 97-106.

Dhar, N. R. (1957) The nitrogen problem and improvement of land fertility by organic matter and phosphates. *Afr. Soils* **4**(2), 6-26.

D'Hoore, J. L. (1960) The soils map of Africa south of the Sahara. (1:5,000,000). *Trans. 7th int Congr. Soil Sci.* **4**, 11-19.

D'Hoore, J. L. (1964) *Soil map of Africa Scale* **1** *to* **5,000,000**. Explanatory monograph Publication No. 93 Joint Project No. 11. Commission for Tech. Co-operation in Africa, Lagos.

D'Hoore, J. L. (1968) The classification of tropical soils. In *The Soil Resources of Tropical Africa* (ed. R. P. Moss). Cambridge, University Press.

Dias, J. S., Povoas, J. C. and Macedo, J. B. (1959) A note on the origin of the colour of tropical black clay soils (gravinigra). *Proc. 2nd inter-afr. Soils Conf.* **1**, 259-263.

Djokoto, R. K. and Stephens, D. (1961*a*). Thirty long term fertilizer trials under continuous cropping in Ghana. I. Crop yields and responses to fertilizers and manures. *Emp. J. exp. Agric.* **29**, 181-195.

Djokoto, R. K. and Stephens, D. **(1961 b)** Thirty long term fertilizer trials under continuous cropping in Ghana. II. Soil studies in relation to the effects of fertilizers and manures on crop yields. *Emp. J. exp. Agric.* **29**, 245-258.

Dobby, E. H. G. (1960) *Southeast Asia*. Univ. Lond. Press.

Dobritskaya, Y. I. (1961) Vestn **Sel' skokhoz** . Nauki Vses Akad. Sel' skokhoz, Nauk' **6**, 109-116.

Dougall, H. W. (1954) Fertilizer experiments of grassland in the Kenya Highlands. *E. afr. agric. For. J.* **19**, 171 ; 212.

Doyne, H. E. and Watson, W. A. (1933) Soil formation in southern Nigeria. The Ilepa profile. *J. Agric. Sci.* ***Camb.*** **23**, 208-215.

Drosdoff, M. *et al.* (1972) *Soils of the Humid Tropics*. National Academy of Sciences, Washington D. C.

Duchaufour, Ph. (1963) Soil classification. A comparison of the American and the French systems. *J. Soil Sci.* **14** , 149-155.

Duchaufour, Ph. and Dommergues, Y. (1963) A study of the humic compounds of some tropical and subtropical soils. *Afr. Soils* **8**, 25-39.

Dudal, R. (1963) Dark clay soils of tropical and subtropical regions. *Soil Sci.* **95**, 264-270.

Dudal, R. and Bramao, D. L. (1965) *Dark clay soils of tropical and subtropical regions.* FAO, Rome.

Dutt, A. D. (1962) Sulphur deficiency in sugarcane. *Emp. J. exp. Agric.* **30**, 257-262.

Dutta, S. K. (1961) Efficient use of shade trees. *Newsletter Tocklai Exp. Sta.* **8**(4), 16-24.

Dyer, B. (1894) On the analytical determination of probably available 'mineral' plant food in soil. *J. chem. Soc.* **65**, 115-167.

Edelman, C. H. and Van der Voorde, P. K. L. (1963) Important characteristics of alluvial soils in the **tropics.** *Soil Sci.* **95**, 258-263.

Eden, T. (1949) *The work of the Agricultural Chemistry Department of the Institute,* 1927-48. Monogr. on Tea Prod., Tea Res. Inst. Ceylon, No. 1.

Eden, T. (1961) The conservation of soil fertility in tropical plantations. *Outl. Agric.* 3, 62-68.

Eden. T. (1965) 2nd ed. *Tea,* Longmans Green and Co. Ltd., London.

Eden, T., Gower, J. C. and Salgado, M. L. M. (1963) A factorial fertilizer experiment on coconuts. *Emp. J. exp. Agric.* **31**, 283-295.

Edmunds, J. E. (1970) Effect of fallowing on banana nematodes and on crop yields. *Trop. Agric. Trin.* **47**, 315-319.

Edwards, D.C. and Bogdan, A. V. (1951) *Important Grassland Plants of Kenya.* Pitman, London.

Ellis, B. S. (1958) Soil genesis and classification. *Soils Fertil.* **21**, 145-147.

Ellison, W. D. (1952) Raindrop energy and soil erosion. *Emp. J. exp. Agric.* **20**, 81.

Ensminger, L. E. and Pearson, R. W. (1951) Soil nitrogen. *Adv. Agron.* **3**, 81-111.

Eriksson, E. (1952) Composition of atmospheric precipitation. I. Nitrogen compounds. II. Sulphur, chloride, iodine compounds. *Tellus* **4**, 215-232; 280-303.

FAO. (1967) *Report of the Meeting on Savanna Development, Khartoum, Sudan.* 25 Oct.-6 Nov. 1966. FAO, Rome.

FAO. (1970a) *Fertilizers. An annual review of world production, consumption, trade and prices,* 1969. FAO, Rome.

FAO. (1970b) *Provisional Indicative World Plan for Agricultural Development Volume* 1. FAO, Rome.

FAO. (1970c) *The State of Food and Agriculture.*

FAO (1970d) *Soil Map of the World FAO/UNESCO Project. Key to Soil Units for the Soil Map of the World.* September 1970. FAO, Rome.

Farrow, W. M. (1954) Tropical soil fungi. *Mycologia* **46**, 632-646.

Fergus, I. F. (1955) A note on arsenic toxicity in some Queensland soils. *Queensland. J. agric. Sci.* **12**, 95-100.

Ferrer, F. M. (1963/64) Fertilization del zacate pangola. *Agricultura tec.* Mex. **2**(3), 116-118.

Finck, A. (1961) Classification of a Gezira clay soil. *Soil Sci.* **92**, 263-267.

Florenzano, G., Balloni, W. and Materassi, R. (1968) Nitrogen-fixing bacteria of the genus Beijerinckia in Venezuelan soils, *Trans. 9th int. Congr. Soil Sci.* **2**, 125-128.

Foster, H. L. (1970) Liming continuously cultivated soils in Uganda. *E. afr. agric. for. J.* **36**, 58-69.

Foster, H. L. (1971) Crop yields after different elephant grass ley treatments at Kawanda Research Station, Uganda. *E. afr. agric. For. J.* **37**, 63-72.

Fox, R. L., Plucknett, D. L. and Whitney, A. S. (1968) Phosphate requirements of Hawaiian latosols and residual effects of fertilizer phosphorus. *Trans. 9th int.* **Congr.** *Soil Sci.* **2**, 301-310.

Fox, R. L., Silva, J. A., Plucknett, D. L. and Teranishi, D. Y. (1969) Soluble and total silicon in sugarcane. *Pl. Soil* **30**, 81-92.

Freiberg, S. R. (1965) Banana nutrition. In *Fruit Nutrition* Chap. 3. ed. Childers, New York.

Fujimoto, G. and Sherman, G. D. (1950) Cobalt content of typical soils and plants of the Hawaiian Islands. *Agron. J.* **42**, 577-581.

Fujimoto, G. and Sherman, G. D. (1951) Molybdenum content of typical soils and plants of the Hawaiian Islands. *Agron. J.* **43**, 424-429.

Fuller, W. H., Cameron, R. E. and **Raica, N. (Jnr.)**(1960) Fixation of nitrogen in desert soils by algae. *Trans. 7th int, Congr. Soil Sci.* **2**, 617-624.

Garcia, J. A. S. and Cardoso, J. C. (1960) The soils of S. Tomé and Principe Islands. *Trans. 7th int. Congr. Soil Sci* **4**, 88-96.

Gasser, J. K. R. (1961) Investigations of rice-growing in British Guiana. I. Characteristics of some rice solis of the coastland. *J. Soil Sci.* **12**, 234-241.

Gerakis, P. A. and Tsangarakis, C. Z. (1969) Underground mulches in central Sudan. *E. afr. agric. For. J.* **35**, 254-256.

Ghani, M. O. **(1943)** Fractionation of soil phosphorus. I. Method of extraction. *Ind. J. Agric. Sci.* **13**, 29-45.

Ghani, M. O. and Aleem, S. A. (1943) Chemical nature of phosphorus fraction. *Ind. J. agric. Sci.* **13**, 142-147.

Gilbert, S. M. (1945) The mulching of *Coffea arabica*. *E. afr. agric. J.* **11**, 75-79.

Glinka, K. D. (1927) Translated by C. F. Marbut. *The Great Soil Groups of the World and their Development*. Murby, London.

Goldschmidt, V. M. (1954) In *Geochemistry* (ed. A. Muir) Oxford Univ. Press (Clarendon), London:

Goldsworthy, P. R. and Heathcote, R. (1963) Fertilizer trials with groundnuts in northern Nigeria. *Emp. J. exp. Agr.* **31**, 124; 351-366.

Goos, R.D. 1960 *Soil fungi from Costa Rica and Panama*. Mycologia **52**, 877-883.

Gourou, P. (1953) (4th ed. 1966) *The Tropical World* translated by S. H. Beaver and E. D. Laborde, Longmans, Green, London.

Grange, L. L. (1949) Soils of some South Pacific Islands. *Commonw. Bur. Soil Sci Techn. Commun.* **46**, 45-48.

Grant, P. M. and Shaxson, T. F. (1970) The effect of ammonium sulphate fertilizer on the sulphur content of tea garden soils in Malawi. ***Trop. Agric. Trin.***, **47**, 31-36.

Greenland, D. J. (1958) Nitrate fluctuations in tropical soils. ***J. agric. Sci. Camb.***, **50**, 82-92.

Greenland, D. J. (1959) Nitrogen gains and losses in tropical soils. *Proc. 3rd inter-afr. Soils Conf.* **1**, 531-535.

Greenland, D. J. (1962) Denitrification in some tropical soils. *J. agric. Sci. Camb.* **58**, 227-233.

Greenland, D. J. and Nye, P. H. (1959) Increases in the carbon and nitrogen contents of tropical soils under natural fallows. *J. Soil. Sci.* **10**, 284-299.

Greenland, D. J. and Nye, P. H. (1960) Does straw induce nitrogen deficiency in tropical soils? *Trans. 7th int. Congr. Soil Sci.* **2**, 478-485.

Greenwood, M. (1951) Fertilizer trials with groundnuts in northern Nigeria. *Emp. J. exp. Agric.* **19**, 225-241.

Greenwood, M. (1954) Sulphur deficiency in groundnuts in northern Nigeria. *Trans. 5th int. Congr. Soil Sci.* **3**, 245-250.

Griffith, G. (1949) Fertility problems in Uganda. *Commonw. Bur. Soil Sci. Tech. Commun.* No. **46**, 160-164.

Gunn, D. L. and Kanapathipillai, P. (1962) A new look at fertilizer (nitrogen) dosage. *Tea Q.* **33**, 122-130.

Haarer, A. G. (1955) Shade or no shade for Arabian coffee. *World Crops* **7**, 271-273.

Haarer, A. E. (1962) 2nd rev. ed. *Modern Coffee Production*. Leonard Hill. Ltd., London.

Halais, P. (1962) The detection of NPK deficiency trends in sugarcane crops by means of foliar diagnosis run from year to year on a follow-up basis. *Proc. 11th Congr. Int Soc. Sugarcane Techn. Mauritius.*

Handawela, J., Kalpagé, F. S. C. P. and Thenabadu, M. W. (1970) The nature of organic matter in the high organic matter content soils of the hill country wet zone of Ceylon. *Trop. Agricst.* **126**, 127-135.

Hardon, H. J. (1936) Factoren die het organische stoffen het stikstofgehalte van tropische gronden beheerschen. *Meded. Alg. Proefste,* Landbouw. Buitenzorg.

Hardon, H. J. (1937) Padang soil, an example of podsol in the tropical lowlands. *Verh. Akad. Wet.* Amst. **40**, 530-538.

Hardy, F. (1941) *Trop. Agric. Trin.* **18**, 75-84.

Hardy, F. (1949) Soil classification in the Caribbean region (a review). *Commonw. Bur. Soil Sci Tech. Commun* **46**, 64-75.

Harmsen, G.W. and Van Schreven, D.A. (1955) Mineralization of organic nitrogen in soil. *Adv. Agron.* **7**, 299-398.

Harris, N.V. (1966) *The Tropical Pacific.* Univ. Lond. Press.

Harris, W.V. (1954) Termites and tropical agriculture. *Trop. Agric. Trin.* **31**, 11-18.

Harris, W.V. (1961) *Termites. Their Recognition and Control.* Longmans, Green and Co. Ltd., London.

Hart, M.G.R., Carpenter, A.J. and Jeffery, J.W.O. (1965) Problems in reclaiming saline mangrove soils in Sierra Leone. *Afr. Soils* **10**, 71-75.

Hartley, C.W.S. (1968) The soil relations and fertilizer requirements of some permanent crops in West and Central Africa. Ch. 8 In *The Soil Resources of Tropical Africa.* ed. R.P. Moss. Camb. Univ. Press.

Hayward, H.E. (1956) Plant growth under saline condition. In *Utilization of Saline Water.* UNESCO, Paris.

Healy, W.B. (1952) Note on zinc deficiency of citrus at Aitutaki, Cork Islands. *N.Z. Jl Sci. Tech.* **34A**, 228-229.

Heintze, S.G. (1957) Studies on soil manganese. *J. Soil Sci.* **8**, 287-300.

Henzell, E.F. (1963) Nitrogen fertilizer responses of pasture grasses in south-eastern Queensland. *Aust. J. exp. Agric. Anim. Husb.* **3**, 290-299.

Henzell, E.F. and Stirk, G.B. (1963) Effects of nitrogen deficiency and soil moisture stress on growth of pasture grasses at Samford, South-East Queensland. I. Results of field experiments. *Aust. J. exp. Agric. Anim Husb.* **3** (11), 300-306.

Higgins, C.M. (1964) Soils of some rice growing areas in northern Nigeria. *Afr. Soils* **9**, 247-265.

Hodgson, J.F. (1963) Chemistry of the micronutrient elements in soils. *Adv. Agron.* **15**, 119-159.

Holm, L.G. (1966) The role of weed control in agricultural development. ***Proc. 8th*** *British Weed Control Conf.* 689-701.

Horrell, C.R. and Bredon, R.M. (1963) Management studies with *Panicum maximum* in Uganda. I. Effect of cutting interval and nitrogen fertilizer on yield. *Emp. J. exp. Agr.* **31**, 334-342.

Hurd, R.G. and Cunningham, R. K. (1961) A cocoa shade and manurial experiment at the W.A.C.R.I. III. Physiological results. *J. hort. Sci.* **36**, 126-137.

Hutton, E.M. (1968) Australia's pasture legumes *J. aust. Inst. agric. Sci.* **34**, 203-218.

Hutton, E.M. (1970) Tropical pastures. *Adv. Agron.* **22**, 1-73.

Ignatieff, V. and Lemos, P. (1963) Some management aspects of more important tropical soils. *Soil Sci.* **95**, 243-249.

Ignatieff, V. (1949) *Efficient use of fertilizers.* FAO Agric. Studies 9, pp. 182.

Ignatieff, V. and Page, H. J. (ed.) (1958) *Efficient use of fertilizers.* FAO Agric. Studies, No. 43, Rome.

Iljin, W. S. (1952) Influencia de abonos fosfatados y nitrogenados sobre la composición quimica y el rendimiento del pasto Guinea (*Panicum maximum*). *Agronomia trop.* Maracay **2**, 145-184.

Imle, E. P. (1967) The region and its problems. In *Rural Development in Tropical Latin America.* ed. K. L. Turk and L. V. Crowder. New York State College of Agriculture.

Ishag, H. M. (1971) Weed control in irrigated groundnuts (*Arachis hypogaea* L.) in the Sudan Gezira. *J. Agric. Sci. Camb.* **77**, 237-242.

Ivanova, E. N. and Rozov, N. N. (1960) Classification of soils and the soil map of the USSR. *Trans. 7th int. Congr. Soil Sci.* **4**, 77-87.

Jackman, R. H. (1955) Organic soils in New Zealand. Soils under pasture II. Relation between organic phosphorus content and some soil characteristics.

Jacks, G. V., Brind., W. D. and Smith, G. A. (1955) *Mulching.* Commonw. Bur. Soil Sci. Tech. Commun. No. 49.

Jackson, J. E. and Burnham, H. O. (1968) Rotation responses of cotton in the Sudan Gezira. II. The effect of fertilizer nitrogen on the response to rotation. *J. agric Sci. Camb.* **70**, 257-263.

Jacob, A. and Uexküll, H. von (1960) *Fertilizer Use. Nutrition and Manuring of Tropical Crops.* Verlagsgesselschaft fur Ackerbau mbH, Hannover.

Jaiyebo, E. O. and Moore, A. W. (1964) Soil fertility and nutrient storage in different soil-vegetation systems in a tropical rainforest environment. *Trop. Agric. Trin.* **41**, 129-139.

Jakate, P. N. (1964) Potash improves yield and quality of potato. *Fertilizer News* **9**, 9-15.

Jameson, J. D. (ed.) (1970) *Agriculture in Uganda.* Oxford University Press.

Jeevaratnam, A. J. (1959) *Rep. Rubb. Res. Inst. Ceylon Agron Div.*

Jeffery, J. W. O. (1960) Iron and the E_h of waterlogged soils with particular reference to paddy. *J. Soil Sci.* **11**, 140-148.

Jeffery, J. W. O. (1961*a*) Defining the state of reduction of a paddy soil. *J. Soil Sci.* **12**, 172-179.

Jeffery, J. W. O. (1961*b*) Measuring the state of reduction of a waterlogged soil. *J. Soil Sci.* **12**, 317-325.

Jenny, H. (1930) A study on the influence of climate upon the nitrogen and organic matter content of the soil. *Missouri agr. exp. Sta. res. Bull.* 152.

Jenny, H. (1948) Great soil groups in the equatorial regions of Colombia, South America. *Soil Sci.* **66**, 5-28.

Jenny, H. (1950) Causes of high nitrogen and organic matter content of certain tropical forest soils. *Soil Sci.* **69**, 63-69.

Jenny, H., Bingham, F. and Padilla-Saravia, B. (1948) Nitrogen and organic matter contents of equatorial soils of Colombia, South America. *Soil Sci.* **66**, 173-186.

Jenny, H., Gessel, S. P. and Bingham, F. T. (1949) Comparative study of decomposition rates of organic matter in temperate and tropical regions. *Soil Sci.* **68**, 419-432.

Jenny, H. and Raychaudhuri, S. P. (1960) *Effect of Climate and Cultivation on Nitrogen and Organic Matter Reserves in Indian* Soils. Ind. Counc. agric. Res., New Delhi.

Jensen, H. L. (1940) Contributions to the nitrogen economy of Australian wheat soils with particular reference to New South Wales. *Proc. Linn. Soc. N.S.W.* **65**, 1.

Jewitt, T. N. (1942) Loss of ammonia from ammonium sulphate applied to alkaline soils. *Soil Sci.* **54**, 401-409.

Jewitt, T.N. (1952) The distribution of organic matter in depth in some tropical seasonally flooded soils. *J. Soil Sci.* **3**, 63-67.

Joachim, A. W. R. (1931) *A Manual of Green Manuring.* Dept. of Agric., Ceylon.

Joachim, A. W. R. (1961) The shade tree question and green manures. *Tea. Q.* **32**(2), 63-68.

Joachim, A. W. R *et al.* (1963) Suggested schemes for levels of manuring mature tea. *Tea Q.* **34**, 106-110.

Joffe, J. S. (1955) Green Manuring viewed by a Pedologist. *Adv. Agron.* **7**, 142-188.

Johnson, W. M., Cady, J. G. and James, M. S. (1962) Characteristics of some brown grumusols of Arizona. *Proc. Soil Soi. Soc.* ***Am.*** **26**,. 389-393.

Jones, L. H. P. and Handreck, K. A. (1967) Silica in soils, plants and animals. *Adv. Agron.* **19**, 107-149.

Jones, M. J. (1971) The maintenance of soil organic matter under continuous cultivation at Samaru, Nigeria. *J. agric. Sci. Camb.* **77**, 473-482.

Jones, P. A. and Wallis, J. A. N. (1963) A tillage study in Kenya coffee. III. The long-term effects of tillage practices upon yield and growth of coffee. *Emp. J. exp. Agric.* **31**, 242-254.

Jones, R. J., Davies, J. G. and Waite, R. B. (1967) *Aust. J. exp. Agric. Anim. Husb.* **7**, 57-65.

Jones, T.A. (1957) Nitrogen studies on the irrigated soils of the Sudan Gezira. *J. Soil Sci.* **8**, 211-224.

Jordon, D. and Opoku, A.A. (1966) The effect of selected covers on the establishment of cocoa. *Trop. Agric. Trin.* **43**, 155-166.

Kalpagé, F.S.C.P. (1967) Manganese in soils and tea under *Grevillea* shade. *Trop. Agric., Trin.* **44**, 209-214.

Kalpagé, F.S.C.P. (1970) Fertilizer use in Ceylon-patterns and trends. *J. Soil Sci. Soc.* ***Ceylon*** **1**, 15-30.

Kalpagé, F.S.C.P. and Thenabadu, M.W. (1969) Some characteristics of the high organic matter content montane soils of the wet zone. *Trop. Agricst.* **125**, 1-13.

Kamoshita, Y. (1940) The soil types of the Tsugaru plain, Aomori Prefecture. Japan. *Jour. agric. expt. Sta.* ***Tokyo*** **3**, 401-420.

Kamoshita, Y. Kosaka, J., Suzuki K. and Okamoto, H. (1967) *Dojo no shuri to schi giijutsu. Soil type and technique of fertilizer use.* 4th ed. (in Japanese). Nogyo-gijutsu Kyokai, Tokyo, pp. 268.

Kamprath, E.J. (1967) Residual effect of large applications of phosphours on high **phosphorus** fixing soils. *Agron. J.* **59**, 25-27.

Kandiah, S. (1943) Soil fertility studies—the fixation and availability of phosphates in some Ceylon soils *Trop. Agricst.* **104**, 17.

Kandiah, S. and Rodrigo, P.M. (1954) Evidence of magnesium deficiency in Ceylon soils. *Trop. Agricst.* **110**, 21-24.

Kanno, I. (1956) A scheme for soil classification of paddy fields in Japan with special reference to mineral paddy soils. *Bull. Kyushu Agric. Expt. Sta.* **4**, 261-272.

Kanno, I. (1962) A new classification system of rice soils in Japan. *Pedologist Tokyo,* **6**, 2-10.

Kasasian, L. (1967) Chemical weed control in tropical root crops. ***Trop. Agric. Trin.***, **44**, 143-150.

Kathirgamathyah, S. and Caesar, K. (1964) Fertilizer Response Studies on Potato. *Trop. Agrist.* **120**, 87-118.

Kelley, W.P. (1951) *Alkali soils, their formation, properties and reclamation.* Reinhold Publishing Corporation. New York.

Kellogg, C.E. (1941) *The Soils that Support Us.* The Macmillan Co., New York.

Kellogg, C.E. (1948) Lands of Shifting Cultivation, Chapter 7 in *Soil Conservation: An International Study.* Washington D.C. FAO Agricultural Studies No. **4** pp. 110-129.

Kellogg, C.E. (1949) Preliminary suggestions for the classification and nomenclature of great soil groups in tropical and equatorial regions. *Commonw. Bur. Soil Sci. Tech. Commun.* No. **46**, 76-85.

Kellogg, C.E. (1950) Tropical soils. *Trans. 4th int. Congr. Soil Sci.* **1**, 266-276.

Kellogg, C.E. (1959) Soil classification and correlation. In *Soil Survey.* USDA Soil Conservation Service.

Kellogg, C.E. (1963) Shifting cultivation. *Soil Sci* **95**, 221-230.

Kellogg, C.E. and Orvedal, A.C. (1969) Potentially arable soils of the world and critical measures for their use, *Adv. Agron.* **21**, 109-170.

Kendrew, W.G. (1961) 5th ed. *The Climates of the Continents.* Oxford Univ. Press.

Kerfoot, O. (1962) Tea root systems. *World Crops,* **14**, 140-144.

Kerr, A. and Vythilingam, M.K. (1967) The occurrence and control of the root-lesion eelworm (***Pratylenchus Loosi***) in nurseries. *Tea Q.* **38**, 22-28.

Khan, A.R. (1956) Potatoes fare better with fertilizers. *Indian Farming* **6**, 31-32.

Khan, A.M. and Siddiqi, M.A. (1962) Fungal population in the soil and on the roots of sugar-cane. *Ind. J. Sug. cane Res. Dev.* **6**, 176-178.

King, H.E. (1960) *Survey of Progress Report for the season* 1958-59. Emp. Cott. Growing Crop., Lond. 10 pp.

King, L.C. (1962) *Morpholgy of the Earth.* Oliver and Boyd, Edinburgh.

Klinge, H, (1965) Podsol soils in the Amazon basin. *J. Soil. Sci.* **16**, 95-103.

Klintworth, H. (1954) Organic matter and soil structure. ***Proc. 2nd inter-Afr. soils*** *Conf.* **1**. 357-364.

Kobayashi, M., Takahashi, E. and Kawaguchi, K. (1967) Distribution on nitrogen-fixing microorganisms in paddy soils of southeast Asia. *Soil Sci.* **104**, 113-118.

Koenigs, F.F.R. (1950) A "Sawah" profile near Bogor (Java.) *Contr. gen. agric Res. Sta. Bogor,* No. **105**, 7.

Kumada, K. (1949) Investigation on the rhizosphere of rice seedlings. *J. Sci. Soil Manure,* Tokyo **19**, 119-124.

Kung, P. (1969) Multiple cropping in Taiwan. *World Crops* **21**, 128-130.

Lam, H. J. (1945) Fragmenta Papuana. Observations of a naturalist in Netherlands New Guinea. Translated by L. M. Perry. *Sargentia,* **5**.

Lawes, D. A. (1961) Rainfall conservation and the yield of cotton in northern Nigeria. *Emp. J. Expt. Agric.* **29**, 307-318.

Laycock, D. H. and Wood, R. A. (1963) Some observations on soil moisture use under tea in Nyasaland. III. The effect of a cover crop; IV Water use in tea nurseries. *Trop. Agric. Trin.* **40**, 121-128.

Leamy, M. L. (1966) *Malayan Soil Survey Report No. 3/1966* Soil Sci. Div. Dept. Agric., West Malaysia.

Leamy, M. L. and Panton, W. P. (1966) Soil Survey Manual for Malayan Conditions. *Bull. Minist. Agric. Fed. Malaya.* **119**, 190-202.

Lee. K. E. (1967) Microrelief features in a humid tropical lowland area, New Guinea, and their relation to earth-worm activity. *Aust. J. Soil. Res.* **5**, 263-274.

Lee, L. T., Lin, C. T., Lin, K. C. and Hsu, S. C. (1960) An experiment on the method of application of urea on rice. *Soils Fertil. Taiwan.*

Le Mare, P. H. (1954) Tie-ridging as a means of soils and water conservation and of yield improvement. *Proc. 2nd inter-Afr. Soils Conf.* **1**, 595-606.

Little, E. C. S. (1967) Progress report on transpiration of some tropical water weeds. *P.A.N.S.* (*C*), **13**(2), 127-132.

Lotero, C. J. *et al.* (1965) Respuesta de una pradera natural a la aplicacion des fertilizantes. *Agricultura trop.* **21**(4). 229-232.

Loué, A. (1952) Étude biochimique de la croissance du caféier. *Bull. Centre Rech. Agron. Bingerville*, **6**, 8-33.

Loué, A. (1960) Nouvelles observations sur les oligo-elements dans la nutrition du caféier (*Coffea canephora* Peirre). *Cafe, Cacao, The*, **4**, 133-149.

Loué, A. (1962*a*) Mineral deficiencies and fertilising cacao. *World Crops*, **14**, 421-426.

Loué, A. (1962*b*) Estudio de las carencias minerales del cacao obtenidas mediante experimentos en medios artificiales. *Fertilite* **14**, 5-41.

Ludow, M. M. and Wilson, G L. (1968) Studies on the productivity of tropical pasture plants. 1. Growth analysis, photosynthesis, and respiration of Hamil Grass and Seratro in a controlled environment. *Aust. J. agr. Res.* **19**, 35-45.

Lutz, J.F. (Chairman) (1965) Committee Report: Glossary of Soil Science terms. *Proc. Soil Sci. Am.* **29**, 330-351.

Lynen, F. (1963) *Revue gen.* Caoutch. **40**, 83.

Malavolta, E. (1959) *Manual de Qiumica Agricola.* Editora Agronomica "Ceres" Ltda., Sao Paulo.

Malavolta, E., Haag, H. P., Mello, F. A. F. and Brasil Sobr., M.O.C. (1962) *On the mineral nutrition of some tropical crops*, Intern. Potash Inst. Berne.

Mann, P. J. G., and Quastel, J. H. (1946) Manganese metabolism in soils. *Nature* **158**, 154.

Marbut, C. F. and Manifold, C. B. (1926) The soils of the Amazon basin in relation to agricultural possibilities. *Geogr. Rev.* **16**, 414-442.

Marbut, C. F. (1928) A scheme for soil classification. *Proc. 1st Int. Congr. Soil Sci.* **4**, 1-31.

Marbut, C. F. (1932) Morphology of laterites. *Proc. 2nd int. Congr. Soil Sci.* **5**, 72-80.

Martin, G. (1969) Some boron-deficiency symptoms in oil palms. *Oleagineux* **24**, 613-614.

Martin, J. H. and Leonard, W. H. (1950) *Principles of Field Crop Production.* Macmillan Co., New York pp. 335.

Martin-Prével, P. (1961) *Bilan mineral de l' ananas au stade recolte*, R. A. IFAC. Doc. 91

Mason, B. (1960) *Principles of Geochemistry*. Wiley, New York.

Massey, R. E. (1915) The agricultural environment. *Sudan Govt. Educ. Dept. Bull. No.* 1 (Central Res. Farm, Khartoum, North), pp. 43-85.

Matsu, T. (1966) A proposal on a new classification system of paddy soil in Japan. *Pedologist Tokyo*, **10**, 68-87.

Matsusuka, Y. and Sherman, G. D. (1950) Titration curves and buffering capacities of Hawaiian soils. *Tech. Bull. Hawaii agric. Exp. Sta. No.* **11**, p. 36.

McCaleb, S. B. (1959) The genesis of the Red-Yellow Podsolic soils. *Proc. Soil Sci. Soc. Am.* **23**, 164-168.

McCants, C. B. and Woltz, W. G. (1967) Growth and mineral nutrition of tobacco. *Adv. Agron.* **19**, 212-265.

McClung, A. C. and Quinn, L. R. (1959) Sulphur and phosphorus responses of Batatais grass (*Paspalum notatum*). *Bull. IBEC Res. Inst.* **18**.

Mehta, R. K. *et al.* (1965) Intercrop Pusa Giant berseem with Napier. *Indian Fmg. Delhi* **15**, 4-5.

Meiklejohn, J. (1953) The nitrifying bacteria: a review. *J. Soil Sci.* **4**, 59-68.

Meiklejohn, J. (1954) Notes on nitrogen-fixing bacteria from East African soils. *Trans. 5th int. Congr. Soil Sci.* **3**, 123-125.

Meiklejohn, J. (1955) The effect of bush burning on the microflora of a Kenya upland soil. *J. Soil Sci* **6**, 111-118.

Meiklejohn, J (1962) Microbiology of the nitrogen cycle in some Ghana soils. *Emp. J. exp. Agric.* **30**, 115-126.

Meiklejohn J. (1968) New nitrogen fixers from Rhodesian soils. *Trans. 9th int. Congr. Soil Sci.* **2**, 141-149.

Mestanza, S. S. and Lainez, J. C. (1970) The correction of boron deficiency in cacoa in Ecuador. *Trop. Agric. Trin.* **47**, 57-61.

Mikami, D. T. and Kanehiro, Y. (1968) Native fixed ammonium in Hawaiian soils. *Proc. Soil Sci. Soc. Am.* **32**, 481-485.

Middelburg, H. A. (1950) Tentative scheme for classification of tropical and subtropical soils. *Trans. 4th int. Congr. Soil Sci.* **4**, 139-142.

Miller, M. H. and Ohlrogge, A. J. (1958) Water-soluble chelating agents in organic materials I. Characterization of chelating agents and their reactions with trace metals in soils; II. Influence of chelate-containing materials on the availability of trace metals to plants. *Proc. Soil. Sci. Soc. Am.* **22**, 225-228; 228-231.

Mills, J. T. and Vlitos, A. J. (1965) The rhizosphere of sugarcane. Preliminary isolations of fungi from an uncultivated heavy clay soil and from similar soil in which sugarcane is being cultivated. *Proc. Int. Soc. Sug. Cane Tech.* **12**, 125-136.

Mills, J. T. and Vlitos, A. J. (1967) Studies on the rhizosphere of sugarcane in Trinidad. *Trop. Agric. Trin.* **44**, (2), 151-157.

Mills, W. R. (1953) Nitrate accumulation in Uganda soils. *E. afr. Agric. J.* **19**, 53.

Mills, W. R. (1954) A review of recent trials with fertilizers in Uganda. *Proc. 2nd inter-Afr. Soils Conf.* **2**, 1133-1142.

Milne, G. (1935*a*) Some suggested units of classification and mapping, particularly for East African soils. *Soil Res.* **4**, 183-198.

Milne, G., (1935*b*) Composite units for the mapping of complex soil associations. *Trans. 3rd int. Congr. Soil Sci.* **1**, 345-347.

Milne, G., Beckley, V. A., Gethin Jones, G. H., Martin, W. S. Griffith, G. and Raymond, L. W. (1936) *A provisional soil map of East Africa.* Amani Memoirs.

Miracle, M. P. (1967) *Agriculture in the Congo Basin.* Univ. Wisconsin Press.

Mishra, M. N., Prasad, R. and Bhan, S. (1968) Arid zone agriculture. *World Crops.* 20(6), 66-68.

Mitchell, R. L. (1964) Trace elements in soils. Chapter 8 in *Chemistry of the Soil.* F. E. Bear. Reinhold Publ. Co.

Mitsui, S. (1949) Nature of akiochi (autumn-decline of rice plants) and its amelioration. *Agriculture and Horticulture* (Nogyo oyobi Engei) **24**, 173-176.

Mitsui, S. (1954) *Inorganic Nutrition, Fertilization and Soil Amelioration for Lowland Rice.* Yokendo Ltd., Tokyo, Japan.

Mitsui, S. and Takatoh, H. (1963) Nutritional study of silicon in graminaceous crops. Part I and Part 2 *Soil Sci. Pl. Nutr. Tokyo* **9**, 49-53; 54-58.

Mohr, E. C. J. (1922) *De grond van Java en Sumatra.* J. H. de Bussy, Amsterdam.

Mohr, E. C. J. and van Baren, F. W. (1954) *Tropical soils.* Uitgeverij W. van Hoeve, The Hague pp. 498.

Moity, M. (1954) La Curence en zinc sur le Bananier. *Fruits* **9**, 354.

Moore, A. W. (1962) The influence of a legume on soil fertility under a grazed tropical pasture. *Emp. J. exp. Agr.* **30**, 239-248.

Moore, A. W. and Jaiyebo, E. O. (1963) The influence of cover on nitrate and nitrifiable nitrogen content of the soil in a tropical rain forest environment. *Emp. J. Expt. Agric.* **31**, 189-198.

Moorman, F. R. (1963) Acid sulphate soils (cat clays) of the tropics. *Soil Sci.* **95**, 271-275.

Moorman, F. R. and Panabokke, C. R. (1961) Soils of Ceylon: a new approach to the identification and classification of the most important soil groups of Ceylon. *Trop. Agricst.* **117**, 5-65.

Morgan, M. F. (1937) The universal soil testing system. *Conn. Agr. Exp. Sta. Bull.* 392.

Morison, C. G. T. (1949) The catena concept and the classification of tropical soils. *Commonw. Bur. Soil Sci. Tech. Commun.* No. **46**, 124-128.

Moss, R. P. (ed.) (1968) *The Soil Resources of Tropical Africa.* Cambridge. Univ. Press.

Murakami, T. (1968) *Rice Culture in South-East Asia.* Edited by Crop Sci. Soc., Japan p. 157, pp. 195.

Nagelschmidt, G., Desai, A. D. and Muir, A. (1940) The minerals in the clay fraction of a Black Cotton Soil and a Red Earth from Hyderabad (Deccan) State, India. *J. agric. Sci.* **30**, 639.

Nethsinghe, D. A. (1961) Magnesium deficiency in coconut palms. *Ceylon Cocon. Plant. Rev.* II, 1-2, 3-6.

Nethsinghe, D. A. (1963) Report of the Soil Chemist Section B. *Ceylon Cocon. Q.* **14**, 16-18.

Nethsinghe, D. A. (1964) Report of the Soil Chemist for 1963. *Ceylon Cocon Q.* **15**, 26-40.

Nikitin, A. A. (1954) Technological aspects of trace element usage. *Adv. Agron.* **6**, 183-197.

Nishigaki, S., Shibuya, M. and Hanaoka, I. (1960) *Examination Methods for crops* (Continued), pp. 497-540. Edited by Y. Togari *et al.*, Tokyo.

Norris, D. O. (1959) The role of calcium and magnesium in the nutrition of *Rhizobium*. *Aust. J. agr. Res.* **10**, 651-698.

Nutman, F. J. (1933) The root system of *Coffea arabica*. *Emp. J. exp. Agric.* **1**, 271-284, 285-296 and **2**, 293-302.

Nye, P. H. (1952) Studies on the fertility of Gold Coast soils III. *Emp. J. exp. Agric.* **20**, 47-55.

Nye, P. H. (1954) Some soil-forming processes in the humid tropics I. A field study of a catena in the West African forest. *J. Soil Sci.* **5**, 7-21.

Nye, P. H. (1955*a*) Some soil-forming processes in the humid tropics II. The development of the upper-slope member of the catena. *J. Soil Sci.* **6**, 51-62.

Nye, P. H. (1955*b*) Some soil-forming processes in the humid tropics IV. The action of the soil fauna. *J. Soil Sci.* **6**, 73-83.

Nye, P. H. (1959) The level of humus under the system of shifting cultivation. *Proc. 3rd. inter-Afr. Soils Conf.* **1**, 525-529.

Nye, P. H. and Bertheux, M. H. (1957) The distribution of phosphorus in forest and savannah soils of the Gold coast and its agricultural significance. *J. Agric. Sci. Camb.* **49**, 141-159.

Nye, P. H. and Greenland, D. J. (1960) *The Soil under Shifting Cultivation*. Commonw. Bur. Soils, Tech. Commun. No. 51.

Nye, P. H. and Stephens, D. (1962) Soil Fertility. In *Agriculture and Land Use in Ghana*. ed. J. B. Wills. Oxford University Press.

Oakes, A. J. (1967) Effect of nitrogen fertilization and plant spacing on yield and composition of Napier Grass in the dry Tropics. *Trop. Agric. Trin.*, **44**, 77-82.

Oakes, A. J. and Skov, O. (1962) Response of Four Pasture Grasses to Nitrogen in the Dry Tropics, *Agron. J.* **54**, 176-178.

Oakes, H. and Thorp, J. (1951) Dark-clay soils of warm regions, variously called rendzina, black cotton soils, regurs and tirs. *Proc. Soil Sci. Soc. Am.*, **15, 347-354.**

Obeng, H. B. and Quagraine, (1960) Characteristics of some latosols and associated soils from the north western savannah zone of Ghana, West Africa. *Trans. 7th int. Congr. Soil Sci.* **4**, 251-256.

Odu, C. T. I. and Vine, H. (1968) Transformation of **^{14}N-tagged ammonium** and nitrate in some savanna soil samples, in *Isotopes and Radiation in Soil-Organic Matter Studies. Proceedings of a Symposium 1968*. IAEA, Vienna.

Oertel, A. C. (1961) Relation between trace-element concentrations in soil and parent material. *J. Soil Sci.* **12**, 119-128.

Ofomata, G. E. K. (1964) Soil erosion in the Enugu region of Nigeria. *Afr. Soils* **9**, 289-319.

Okorie, I. I., Hill, D. D. and McIlroy, R. J. (1965) The productivity and nutritive value of tropical grass/legume pastures rotationally grazed by N'Dama cattle at Ibadan, Nigeria. *J. agric. Sci. Camb.* **64**, 235-245.

Ollagnier, M. and Ochs, R. (1971) Chlorine a new essential element in oil-palm nutrition. *Oleagineaux* **26**, 1-15.

Olsen, S. R., Cole, C. V., Watanabe, F. S. and Dean, L. A. (1954) Estimation of available phosphorus in soils by extraction with sodium bicarbonate. *U.S. Dep. Agric. Circ.* No. 939.

Owen, G. (1953) Studies on the phosphate problem in Malayan soils. *J. Rubb. Res. Inst. Malaya* **14**, *Commun.* 283, 121-132.

Oyama, M. (1962) A classification system of paddy-rice field soils based on their diagnostic horizons. (in Japanese with English summary) *Bull natn. Inst agric. Sci., Tokyo Ser.* B 12, 303-372.

Oyot, P. O. (1970) Evaluation of phosphorus soil test methods by greenhouse studies and laboratory tests. *E. afr. agric. For. J.* **35**, 336-339.

Panabokke, C. R. and Nagarajah, S. (1964) The fertility characteristics of the rice-growing soils of Ceylon. *Trop. Agricst.* **120**, 3-30.

Panton, W. P. (1964) The 1962 soil map of Malaya. *J. Trop. Geogr.* **18**, 118-124.

Papadakis, J. (1961) *Climatic Tables for the World.* Cordoba, Buenos Aires.

Parish, D. H. and Feillafe, S. M. (1960) A comparison of urea with ammonium sulphate as a nitrogen source for sugarcane. *Trop. Agric. Trin.* **37**, 223-225.

Parker, C. A. (1954) Non-symbiotic nitrogen-fixing bacteria in soil. I. Studies on *Clostridium butyricum. Austr. J. agric. Res.* **5**, 90-97.

Paul, H. and Shariff, M. A. A. (1954) The rates of nitrification of peat soils in British Guiana. *J. agric. Sci.* **44**, 377-382.

Peachey, J. E. ed. (1969) *Nematodes of Tropical Crops* Tech. Commun. commonw. Bur. Helminth. No. 40.

Pearsall, W. H. and Mortimer, C. H. (1939) Oxidation and reduction potentials in waterlogged soils, natural waters and muds. *J. Ecol.* **27**, 483-501.

Peat, J. E. and Brown, K. J. (1962) The yield responses of rain-grown cotton at Ukiriguru in the Lake Province of Tanganyika. I. The use of organic manure, inorganic fertilizers, and cotton seed ash. *Emp. J. exp. Agric.* **30**, 215.

Peat, J. E. and Brown, K. J. (1962) The yield responses of rain-grown cotton at Ukiriguru in the Lake Province of Tanganyika. II. Land-resting and other rotational treatments contrasted with the use of organic manure and inorganic fertilizers. *Emp. J. exp. Agric.* **30**, 305-314.

Pelegrin, P. (1953) L'utilisation des engrais en culture bananiere. *Fruits* **8**, 453-458.

Pendleton, R. L. (1943) Land use in north-eastern Thailand. *Geogrl. Rev.* **33**, 14-41.

Pendleton, R. L. (1949) The Classification and Mapping of Tropical soils. *Commonw. Bur. Soil. Sci. Tech. Commun.* **46**, 93-97.

Pereira, H. C. (1954) Moisture measurements in East African soils. *Proc. 2nd inter-Afr. Conf. Soils.* **1**, 425-430.

Pereira, H. C. (1955) The assessment of structure in tropical soils. *J. agric. Sci.* **45**, 401-410.

Pereira, H. C. (1957) Field measurements of water use for irrgation control in Kenya coffee. *J. agric. Sci.* **49**, 459-466.

Pereira, H. C., Chenery, E. and Mills. W. R. (1954) The transient effect of grasses on the structure of tropical soils. *Emp. J. exp. Agric.* **22**, 148-160.

Pereira, H. C., Dagg, M. A., and Hosegood, P. H. (1964) A tillage study in Kenya coffee. IV The physical effects of contrasting tillage treatments over thirty consecutive cultivation seasons. *Emp. J. exp. Agric.* **32**, 31-34.

Pereira, H. C., Hosegood, P. H. and Thomas, D. B. (1961) The productivity of tropical semi-arid thorn-scrub country under intensive management. *Emp. J. exp. Agric.* **29**, 269-286.

Pereira, H. C. and Jones, P. A. (1954) Tillage studies in Kenya coffee. *Emp. J. exp. Agric.* **22**, 231-240; 323-331.

Peries, O. S. ed. (1967) *The Development of Agriculture in the Dry Zone.* Proc. Symposium, Ceylon Assoc. Advmt. Sci., Colombo.

Pinkerton, A. (1967) Copper deficiency of wheat in the Rift Valley, Kenya. *J. Soil Sci.* **18**, 18-26.

Ponnamperuma, F. N. (1955) *The Chemistry of Submerged Soil in Relation to the Growth and Yield of Rice.* Ph. D. Thesis Cornell University.

Ponnamperuma, F. N. (1958) Response of potato to fertilizers, manure, lime and trace elements. *Trop. Agrist.* **114**, 99-113.

Ponnamperuma, F. N., Loy, T. A. and Tianco, E. M. (1969) Redox equilibria in flooded soils. II. The manganese oxide systems. *Soil Sci.* **108**, 42-57.

Ponnamperuma, F. N., Martinez, E. and Loy, T. (1966) Influence of redox and partial pressure of carbon dioxide on pH values and the suspension effect of flooded soils. *Soil Sci.* **101**, 421-431.

Ponnamperuma, F. N., Tianco, E. M. and Loy, T. (1967) Redox equilibria in flooded soils. I. The iron hydroxide systems. *Soil Sci.* **103**, 374-382.

Prescott, J. A. and Pendleton, R. L. (1966) *Laterite and Lateritic Soils.* Commonw. Agric. Bur. England.

Prevot, P. and Ollagnier, M. (1964) Sulphur deficiency of peanuts in Africa. *Agrochimica* **8**, 211-221.

Quinn, L. R., Mott, G. O. and Bisschoff, W. V. A. (1961) Fertilization of Colonial Guinea grass pastures and beef production with Zebu steers. *Bull. IBEC Res. Inst.* 24.

Ramdas, L. A. and Dravid, R. K. (1936) Soil temperatures in relation to other factors controlling the disposal of solar radiation at the earth's surface. *Proc. natn. Inst. Sci. India* **2**, 131-143.

Ramdas, L. A. and Katti, M. S. (1934) Agricultural meteorology. Preliminary studies on soil moisture in relation to moisture in the surface layers of the atmosphere during the clear season at Poona. *Indian J. agric. Sci.* **4**, 923-937.

Ramsay, J. M. and Innes, R. R. (1963) Some quantitative observations on the effects of fire on the Guinea Savannah vegetation of Northern Ghana over a period of eleven years. *Afr. Soils* **8**, 41-86.

Raychaudhuri, S. P. and Biswas, N. R. D. (1964) Trace element status of Indian soils. *J. Indian Soc. Soil Sci.* **12**, 207-214.

Raychaudhuri, S. P. and Murthy, R. S. (1960) Genetic classification of some saline-alkali soils of the Indo-Gangetic alluvium. *Trans. 7th int. Congr. Soil Sci.* **4**, 278-282.

Reuszer, H. W. (1962) Axenic techniques in the determination of root functions and the interrelationships of microorganisms and plant roots. *Soil Sci.* **93**, 56-61.

Reuther, W., Embleton, T. W. and Jones, W. W. (1958) Mineral nutrition of tree crops. *Ann. Rev. Plant Physiol.* **9**, 175-206.

Richards, P. W. (1941) Lowland tropical podsols and their vegetation. *Nature, Lond.* **148**, 129-131.

Richards, P. W. (1964) (2nd repr.) *The Tropical Rain Forest.* Camb. Univ. Press.

Richardson, H. L. (1963) The fertility potentialities and needs of tropical soils. *Trop. Sci.* **5**, 166-177.

Richardson, H. L. (1966) The Freedom From Hunger Campaign—Five years of the FAO fertilizer programme. *Outl. Agric.* **5**(1), 3-16.

Ripado, M. F. B. (1954) Peat soils of southern Mozambique. *Proc. 2nd inter-Afr. Soils Conf.* **2**, 927-932.

Roberts, R. C. (1942) Soil survey of Puerto Rico. *U.S.D.A. Bur. Plant Ind. Ser.* 1936 No. 8, pp. 495.

Robison, B. M. (1970) Micro-fungi of sugarcane roots and soil in Jamaica. *Trop. Agric.* Trin. **47**, 23-29.

Robinson, G. H. (1971) Exchangeable sodium and yields of cotton on certain clay soils of Sudan. *J. Soil. Sci.* **22**, 328-335.

Robinson, J. B. D. (1958) Some chemical characteristics of 'termite soils' in Kenya coffee fields. *J. Soil Sci.* **9**, 58-65.

Robinson J. B. D. and Gacoka, P. (1962) Evidence of the upward movement of nitrate during the dry season in the Kikuya red loam coffee soil. *J. Soil. Sci.* **13**, 133-139.

Robinson, J. B. D. and Semb, G. (1968) Advisory soil or plant analysis and fertilizer use. I. Comparison of soil analysis methods with maize yield data. *E. afr. agric. For.* J. **34**, 117-127.

Robinson, J. B. D,(1968) Advisory soil or plant analysis and fertilizer use. II. Evaluation of soil analysis methods with maize yield data. *E. afr. agric. For. J.* **34**, 140-152.

Robinson. J. B. D. (1969) Advisory soil or plant analysis and fertilizer use. III. Relationships between soil pH and maize yield responses to nitrogen and phosphorus fertilizers. *E. afr. agric. For. J.* **34**, 436-440.

Robinson, J. B. D. (1969) Advisory soil or plant analysis and fertilizer use. IV. Evaluation of plant analysis with maize yield data. *E. afr. agric. For. J* **34**, 449-456.

Robinson, W. O. and Alexander, L. T. (1953) Molybdenum content of soils. *Soil Sci.* **75**, 287-291.

Rodel, M. G. W. (1966) Intensive use of summer pastures. *Rhodesian Farmer* Oct. 7.

Rodrigues, G. (1954) Fixed ammonia in tropical soils. *J. Soil. Sci.* **5**, 264-274.

Russell, E. W. (1952) Soils and fertilizers, *J. Roy. agric. Soc.* **113**, 145-156.

Russell, E. W. (1958) The fertility of tropical soils. In *Report of a Conference of Directors and Senior Officers of Overseas Departments of Agriculture.* Colonial Office, London, Misc. No. 531, 33-40 H. M. Stationery Office.

Russell, E. W. (1960) The level of humic carbon and nitrogen in tropical soils. *Afr. Soils* **5**, 101-105.

Russell, E. W. (1961) *Soil Conditions and Plant Growth.* Longmans Green & Co. Ltd., London.

Russell, E. W. ed. (1962) *The Natural Resources of East Africa.* D. A. Hawkins Ltd., Nairobi.

Russell, E. W. (1968) Some agricultural problems of semi-arid areas. In *Soil Resources of Tropical Africa.* Moss, R. P. ed. Camb. Univ. Press.

Ruthenberg, H. (1970) *Definition of Types of Farming in the tropics.* Economics Institute for Scientific Cooperation, Tubingen **1**.

Salgado, M. L. M. (1954) The nutrient content of nut water in relation to available soil nutrients as a guide to the manuring of the coconut palm (*Cocos nucifera*). A new diagnostic method VIII. In *Plant Analysis and Fertilizer Problems,* 8th Intern. Congr. Botany, Paris pp. 217-237 (IRHO. Paris)

Salgado, M. L. M. (1957) *Ann. Rep. Coc. Res. Inst. Ceylon for* 1955 *Sessional Paper X.*, 1957 p. 21.

Salgado, M. L. M. (1965) Report of the Soil Chemist. *Ceylon Cocon. Q.* **16**, 9-24.

Samuels, G. (1955) Influence of fertilizers on the components of available sucrose-per cent-cane. *J. Agric. Univ. Puerto Rico.* **39** (2), 84.

Samuels, G., and Diaz, H. G. (1960) Effects of potassium chloride and sulphate on pineapple yields and quality. *J. Agr. Univ.* ***Puerto Rico* 44, 16-20.**

Samuels, G. and Landrau, P. (1954) The influence of potassium on the yields and sucrose content of sugarcane. *J. Agric. Univ.* ***Puerto Rico* 38, 171.**

Schmid, K. (1951) Grundsatzliches zur Dungung im Qualitatstabakbau. ***Der Deutsche*** *Tabakbauer* **4**, 27-29.

Schofield, R. K. (1955) Can a precise meaning be given to 'available' soil phosphorus? *Soils Fertil.* **18**, 373-375.

Schutte, K. (1955) A survey of plant minor element deficiencies in Africa *Afr. Soils* **3**, 4-17.

Schutte, K. H. (1957) Micro-nutrient deficiency and wilting in the radish. *Plant & Soil* **9**, 186-188.

Schutte, K. H. (1959) Trace element problems in nature. *Trace Elements in Nature.* A Symposium. Univ. Cape Town, 1-15.

Schutte, K. H. and Amdurer, S. (1960) Some causes and effects of trace elements deficiencies in South African soils. *Trans. 7th int. Congr. Soil Sci.* **3**, 20-26.

Scotto La Massése (1969) The principal plant nematodes of crops in the French West Indies. In *Nematodes of Tropical Crops*, Peachey, J.E. ed. Tech. *Commun commonw. Bur. Helminth No.* 40, 164-183.

Semb, G. and Robinson, J. B. D. (1969) The natural nitrogen flush in different arable soils and climates in East Africa. *E. Afr. agric. For. J.* **34**, 350-370.

Sharma, R. B. and Motiramani, D. P. (1969) Zinc status of the soils of Madhya Pradesh. *J. Indian Soc. Soil. Sci.* **17**, 19-26.

Sheng, C. Y. (1962) Crop response to nitrogen in Taiwan. *Soils Fertil. Taiwon*, 20-31.

Sherman, G. D. 1952, *The titanium context of Hawaiian soils and its significance.* Soil Sci. Soc. Amer. Proc. **16**, 15-18.

Sherman, G. D. and Alexander, L. T. (1959) Characteristics and genesis of low humic latosols. *Proc. Soil Sci. Soc. Am.* **23**, 168-170.

Sherman, G. D. and Fujimoto, C. K. (1946) The effect of the use of lime, soil fumigants and mulch on the solubility of manganese in Hawaiian soils. *Proc. Soil Sci. Soc. Am.* **11**, 206-210.

Sherman, G. D. and Ikawa, H, 1968. **Soil sequences in the Hawaiian Islands**, Pacific Sci **22**, 458-464.

Shiori, M. and Aomine, S. (1937) Change of ammoniacal nitrogen in flooded soils. *J. Sci. Soil Manure, Tokyo* **11**, 389-392.

Shiori, M. and Aomine, S. (1938-1940) Nitrogen transformation in paddy soils. *J. Sci. Soil Manure, Tokyo* **12**, 406-410; **13**, 284-287; ,**14** 369-370.

Shorrocks, V. M. (1964) Mineral deficiencies in *Hevea* and associated cover plants. *J. Rubb. Res. Inst. Malaya.*

Simmonds, N.W. (1966) (2nd ed.) *Bananas.* Longmans, London.

Simonson, R. W. (1954*a*) The regur soils of India and their utilization. *Proc. Soil Sci. Soc. Am.* **18**, 199-203.

Simonson, R. W. (1954*b*) Morphology and classification of the regur soils of India *J. Soil Sci.* **5**, 275-288.

Simpson, J. R. (1960) The mechanism of surface nitrate accumulation on a bare fallow soil in Uganda. *J. Soil Sci.* **11**, 45-60.

Singh, A. (1963) A critical evaluation of green manuring experiments on sugarcane in north India. *Emp. J. expt. Agric.* **31**, 205-212.

Singh, S. (1954) A study of the black cotton soils with special reference to their colouration. *J. Soil Sci.* **5**, 289-299.

Singh, S. (1956) The formation of dark-coloured clay-organic complexes in black soils. *J. Soil Sci.* **7**, 43-58.

Singh, S. B. and Agarwal, R. R. (1953) *Agric. Anim. Husb. U P. India* **4**, 70-80.

Sivarajasingham, S., Alexander, L. T., Cady J. G. and Cline, M. G. (1962) Laterite. *Adv. Agron.* **14**, 1-60.

Smith, C. A. (1962) Tropical grass/legume pastures in Northern Rhodesia. *J. Agric. Sci. Camb.* **59**, 111-118.

Smith, G. D. (1965) *Soil Classification.* Cour. Franqui, University of Ghent, Belgium.

Smith, R. M., Samuels, S. G. and Cernuda, C. F. (1951) Organic matter and nitrogen build-ups in some Puerto Rican soil profiles. *Soil Sci.* **72**, 409-427.

Sombroeck, W. G (1966) *Amazon Soils: A reconnaissance of the soils of the Brazilian Amazon region.* Centre for Agric. Publications and Documentation, Wageningen.

Southern, P. J (1967) *Sulphur deficiency in coconuts. Coconut Research in the South Pacific.* Papers prepared for a Technical Meeting on Coconut Production. Rangiora, August, 1967. South Pacific Commission.

Spector, J. (1964) Trace elements. *Ann. Rep. Cacao Res. Trinidad* 47-49.

Spurr, A. M. M. (1954) **A basis of classifications of the soils of areas of composite topography in central Africa**. *Proc. 2nd inter-Afr. soils Conf.*, **175-192.**

Stallings, J. H. (1957) *Soil Conservation.* Prentice-Hall Inc.

Stamp, L. D. (1953) *Our Undeveloped World.* Faber and Faber, London.

Stamp, L. D (1962) *Asia: A Regional and Economic Geography.* Methuen, London.

Stanford, G and Jordan, H.V. (1966) Sulphur requirements of sugar, fibre and oil crops. *Soil Sci.* **101**, 258-266.

Stenberg, M., Ekman, P., Lundblad. K., and Svanberg, O. (1948) Nord. Jordbrugsforskn., Proc. 7th Congr. III, pp. 689-700, 709.

Stephens, C. G. (1950) Comparative morphology and genetic relationships of certain Australian, North American and European soils. *J. Soil Sci.* **1**, 123-149.

Stephens, C. G. (1953) *A manual of Australian soils.* C.S.I.R.O. Melbourne.

Stephens, C. G. (1961) Laterite at the type locality, Angadipuram, Kerala, India. *J. Soil Sci.* 12, 214-217.

Stephens, D. (1960) Fertilizer trials on peasant farms in Ghana. *Emp. J. exp. Agric.* **28**, 1-15.

Stewart, I. and Leonard, C. D. (1952) Molybdenum deficiency in Florida citrus. *Nature* **170**, 714-715.

Stobbs, T. H. (1969 *a*) The effect of grazing resting land upon subsequent arable crop yields. *E. afr. agric. For. J.* **35**, 28-32.

Stobbs, T. H. (1969 *b*) The influence of inorganic fertilizers upon the adaptation, persistency and production of grass/legume swards in eastern Uganda. *E. afr agric. For. J.* **35**, 112-117.

Stobbs, T. H. (1969 *c*) Animal production from *Hyparrhenia* grassland oversown with *Stylosanthes gracilis*. *E. afr. agric. For. J.* **35**, 128-134.

Stobbs, T. H. (1969 *d*) The value of *Centrosema pubescens* (Benth.) for increasing animal production and improving soil fertility in northern Uganda. *E. afr. agric. For. J.* **35**, 197-202,

Stolzy, L. H. Harding, R. B. and Branson, R. L. (1966) Foliar absorption of boron in sprinkler-irrigated citrus. *Calif. Agric.* **20**, 6-7.

Storey, H. H. and Leach, R. (1933) A sulphur deficiency of the tea bush. *Ann. appl. Biol.* **20**.

Strahler, R. (1960) *Physical Geography*. Wiley, New York.

Su, N. R. and Li, C. Y. (1962) Comparison of two potassium salts as regards their effects on yield and quality of pineapple fruits. *Soils Fertil. Taiwan* pp. 82-83.

Suttie, J. M. (1968) Pasture legume research in Kenya. *E. afr. agric. For. J*, **33**, 281-285.

Swaine, D. J. (1955) *The Trace Element Content of Soils*. Commonw. Bur. Soil Sci. Tech. Commun. No. 48.

Swarbrick, J. T. (1964) The growth and root distribution of some temporary shade plants for cocoa. ***Trop. Agric. Trin.***, **41**, 311-323.

Sys, Ch. (1955) The importance of termites in the formation of latosols in the region of Elisabethville. *Afr. Soils* **3**, 393-395.

Sys, Ch. (1960) Principles of soil classification in the Belgian Congo. *Trans. 7th int. Congr. Soil. Sci.* **4**, 112-118.

Sys, Ch. (1967) The concept of ferrallitic and fersiallitic soils in Central Africa. Their classification and their correlation with the 7th approximation. ***Pedologist***, **17**, 284-325.

Tamhane, R. V. (1950) "Regur" soils of India. *Trans. 4th int. Congr. Soil Sci.* **3** 131-134.

Tamhane, R. V., Motiramani, D. P., Bali, Y. P. and Donahue, R. L. (1964) *Soils: Their Chemistry and Fertility in Tropical Asia*. Prentice Hall of India (Private) Ltd., New Delhi.

Tan, K. H. (1965) The andosols in Indonesia. *Soil Sci.* **99**, 375-378.

Tan, K. H., Perkins, H. F. and McCreery, R. A. (1970) The characteristics, classification and genesis of some tropical spodosols. *Proc. Soil Sci. Soc. Am.* **34**, 775-778.

Tanaka, A., Shimono, K., and Ishizuka Y. (1969). Zinc deficiency as the cause of 'Akkagare' in the rice plant, *J. Sci. Soil Manure, Japan* **40**, 415-419.

Thenabadu, M. (1970) Fertilizers for rice. *J. Soil Sci. Soc.* ***Ceylon*** **1**, 128-145.

Theron, J. and Van Niekerk, P. Le R. (1934) The nature of origin of black turf soils. *S. Afr. J. Sci.* **31**, 320-346.

Thompson, L. M., Black, C. A. and Zoellner, J. A. (1954) Occurrence and mineralization of organic phosphorus in soils, with particular reference to associations with nitrogen, carbon and pH. *Soil Sci.* **77**, 185-196.

Thorp, J. (1935) *Instructions to Chinese soil survey field parties*. Nat. Geol. Surv. China, pp. 122.

Thorp, J. and Baldwin, M. (1938) Nomenclature of higher categories of soil classification as used in the Department of Agriculture. *Proc. Soil Sci. Soc. Am.* **3**, 260-268.

Thorp, J. and Baldwin, M. (1940) Laterite in relation to soils in the Tropics. *Ann. Assoc. Amer. Geog.* **30**(3), 163-183.

Thorp, J. and Smith, G. D. (1949) Higher categories of soil classification: Order, Suborder and Great Soil Group. *Soil Sci.* **67**, 117-126.

Timonin, M. I. (1946) Microflora of the rhizosphere in relation to the manganese-deficiency disease of oats. *Proc. Soil Sci. Soc. Am.* **11**, 284-292.

Tinker, P. B. H. and Smilde, K. W. (1963) Dry matter production and nutrient content of plantation oil palms. II. Nutrient content. *Pl. Soil* **19**, 350-363.

Tinker, P. B. H. and Ziboh, C. O. (1959) Soil analysis and fertilizer response. *J. W. Afr. Inst. Oil Palm Res.* **3**, 52-73.

Tiurin, I. V. (1961) Man stages of the development of the soil classification problem in the USSR. *In Symposium on Soil Classification, Ghent.*

Tokutome, S. (1970) *Report of the Survey and Classification of Paddy Soils of Ceylon.* Mimeographed by the Central Agric. Res. Inst. Dept. Agric., Peradeniya, Ceylon.

Tolhurst, J. A. H. (1965) Economising on phosphate and potash for mature tea. *Tea Q.* **36**, 45-47.

Touré, E. L. H. (1964) Maintenance of the fertility of savannah zone soils. Effects of annual or biennial mineral fertilizer applications under conditions of continuous cropping in the Sudan zone of West Africa. *Afr. Soils* **9**, 221-246.

Truog, E. (1930) Determination of readily available phosphorus of soil. *J. Am. Soc. Agron.* **22**, 874-882.

Turk, K. L. and Crowder, L. V. (ed.) (1967) *Rural Development in Tropical Latin America.* New York State College of Agriculture, New York.

Uchiyama, N. (1949) Morphology of paddy soils. (in Japanese) *Chikyu-Shuppansha, Tokyo.*

UN. (1967) *Fertilizer Manual.* United Nations Industrial Development Organization UN, New York.

Uppal, B. N., Patel, M. K. and Daji, J. A. (1939) Nitrogen fixation in rice soils. *Ind. J. Agric. Sci.* **9**, 689-702.

Urquhart, D. H. (1956) *Cocoa 2nd. ed.* Longmans, Green and Co., London.

USDA. (1938) *Soils and Men: Yearbook of Agriculture.* USDA, Washington.

USDA. (1951) *Soil Survey manual.* USDA, Washington.

USDA Soil Conservation Service, (1960) *Soil Classification: a Comprehensive System, 7th Approximation* (with complements, 1964) USDA, Washington.

Vageler, P. (1910) Die Mkatta-Ebene. *Beih. z. Tropenpfl.* **9**, 251-395.

Van Baren, J. (1927) Profiles of limestone soils from the Tropics. *Proc. 1st int. Congr. Soil Sci.* **4**, 173-193.

Van Baren, J. (1928) *Microscopical, physical and chemical studies of limestones and limestone soils from the East Indian Archipelago.* Commun. Geol. Inst. Agric. Univ. Wageningen, Nr. 14, pp. 195.

Vander Merwe, C. R. (1954) Kalahari and Sahara sandy Soils. *Trans. 5th int. Conf. Soil Sci.* **4**, 117-122.

Van Dierendonck, F. J. E. (1959*a*) *The Manuring of Coffee, Cocoa, Tea and Tobacco.* Centre d'Etude de l'Azote, Geneva.

Van Dierendonck, F. J. E. (1959*b*) Response of tropical soils to manuring. *Afr. Soils* **5**, 149-156.

Van Dierendonck, F. J. E. (1963) Coffee production and fertilizer treatment. *Outl. Agric.* **4**, 13-21.

Venema, K. C. N. (1959) Deficiency symptoms of some elements as manifested by sugarcane. *Potash and Trop. Agric.* **2** (4).

Verboom, W. C. (1965) Legumes in Zambia. *Trop. Agric. Trin.* **42**, 229-242.

Vermaat, J. G. (1964) *Report to the Government of Pakistan on soil fertility investigations.* FAO, Rome.

Vermaat, J. G. and Bie, G. J. Van Der (1950) On the occurrence of copper in tropical soils. *Pl. Soil* **2**, 257-282.

Vicente-Chandler, J. (1967) Intensive pasture production. In *Rural Development in Tropical Latin America* ed. K. L. Turk & L. V. Crowder. pp. 272-295. New York State College of Agriculture.

Vicente-Chandler, J. and Figarella, J. (1958) Growth characteristics of Guinea grass, on the semiarid South Coast of Puerto Rico, and the effect of nitrogen fertilization on forage yields and protein content. *J. Agric. Univ. Puerto Rico* **42**, 151-160.

Vicente-Chandler, J., Silva, S. and Figarella, J. (1962) Effect of frequency of application on response of Guinea grass to nitrogen fertilization. *J. Agr. Univ. Puerto Rico.* **46**, **342-349.**

Vine, H. (1941) A soil catena in the Nigerian cocoa belt. *Farm and Forest* **2**, 139-141.

Vine, H. (1949) Nigerian soils in relation to parent materials. *Commonw. Bur. Soil Sci. Tech. Commun.* **46**, 22-29.

Vine, H. (1953) Experiments on the maintenance of soil fertility at Ibadan, Nigeria. *Emp. J. exp. Agric.* **21**, 65-85.

Vine, H. (1954) Is the lack of fertility of tropical African soils exaggerated? *Proc. 2nd inter-Afr. Soils Conf.* **1**, 389-412.

Vinogradov, A. P. (1959) *The Geochemistry of Rare and Dispersed Chemical Elements in Soils.* Consultants Bur., New York (translated from the Russian).

Wadia, A. (1960) *Geology of India.* Macmillan and Co., London.

Walker, T. W. (1955) Sulphur responses on pasture in Australia and New Zealand. *Soils Fertil.* **18**, 185-187.

Wallace, T. (1951) *The Diagnosis of Mineral Deficiencies in Plants by Visual Symptoms.* H. M. Stationery Office, London, pp. 107.

Wasowicz, T. (1952) The effect of mulching on the status of major nutrients and the growth of cacao seedlings. ***Trop. Agric. Trin.,*** **29**, 163-169.

Watanabe, A., Nishigaki, S. and Konishi, C. (1951) Effects of nitrogen-fixing blue-green algae on the growth of rice plants. *Nature, Lond.* **168**, 748-749.

Watson, G. A. (1957 Nitrogen fixation by *Centrosema pubescens* ***J. Rubb. Res. Inst. Malaya,*** **15**, 168-174.

Watson, G. A. (1964) Maintenance of soil fertility in the permanent cultivation of *Hevea Brasiliensis* in ***Malaya.*** *Outl. Agric.* **4**, 103-109.

Watson, G. A. Chin, T. T. and Wong, W. W. (1962) Loss of ammonia by surface dressings of urea in *Hevea* cultivation. *J. Rubb. Res. Inst.* ***Malaya*** . **17**, 77-90.

Watson, G. A. Wong, P. W. and Narayan, R. (1963) Effect of cover plants on growth of *Hevea* IV. Leguminous cover crops compared with grasses. *J. Rubb. Res. Inst.* ***Malaya*** . **18**(1), 28-37.

Watson, J. P. (1962) The soil below a termite mound. *J. Soil Sci.* **13**, 46-51.

Watson, J. P. (1964) A soil catena on granite in Southern Rhodesia. *J. Soil Sci.* **15**, 238-257.

Watson, K. A. (1964) **Fertilizers in northern Nigeria, current utilization and recommendations for their use.** *Afr. Soils* **9**, 5-20.

Watson, K. A. and Goldsworthy, P. R. (1964) Soil-fertility investigations in the middle belt of Nigeria. *Emp. J. expt. Agric.* **32**, 290-302.

Webb, R A. (1954) The mineral deficiencies of some Gambian soils. *Proc. 2nd inter-afr. Soils Conf.* **1**, 419-424.

Webb, R. A. (1955) A new approach to the study of the fertility of tropical soils. *Afr. Soils* **3**, 379-391.

Webster, C. C. (1954) Land utilisation and soil fertility maintenance in Kenya. *Proc. 2nd inter-Afr. Soils Conf.* **2**, 1425-1433.

Webster, C. C. and Wilson, P. N. (1966) *Agriculture in the Tropics.* Longmans Green & Co. Ltd., London.

Webster, R. (1960) Soil genesis and classification in Central Africa. *Soils Fertil.* **23**, 77-79.

Webster, R. (1965) A catena of soils on the northern Rhodesia plateau. *J. Soil Sci.* **16**, 31-43.

Webster, R. (1968) Fundamental objections to the seventh approximation. *J. Soil Sci.* **19**, 354-366.

Wehrmann, J. and Johannes, L. W. (1965) Effect of "Guie" on soil conditions and plant nutrition. *Afr. Soils* **10**, 129-136.

Weir, C. C. (1962) Evaluation of chemical soil tests for measuring available phosphorus on some Jamaican soils. *Trop. Agric. Trin.* **39**, 67-72.

Weir, C. C. (1966 *a*) Rapid chemical determination of iron, manganese, zinc and copper in citrus leaves. *Trop. Agric. Trin.* **43**, 149-152.

Weir, C. C. (1966 *b*) Leaf sampling for foliar analysis of citrus in Trinidad. *Trop. Agric. Trin.* **43**, 153-154.

Weir, C. C. (1966 *c*) The phosphorus and potassium status of some Trinidad soils. *Trop. Agric. Trin.* **43**, 315-321.

Weir, C. C. (1971) Correction of magnesium deficiency of citrus trees in the Caribbean area. *Trop. Agric. Trin.* **48**, 351-356.

Whyte, R. O. (1962) The myth of tropical grasslands. *Trop. Agric. Trin.* **39**, 1-11.

Whyte, R. O., Moir, T. R. G. and Cooper, J. P. (1959) *Grasses in Agriculture.* FAO, Rome.

Whyte, R. O. and Trumble, H. C. (1953) *Legumes in Agriculture.* FAO Agricultural Studies No. 21, Rome.

Wight, W. (1958) The shade-tree tradition in the tea gardens of Northern India, *Tocklai Expt. Sia. Report.* p. 75.

Willatt, S. T. (1970) A comparative study of the development of young tea under irrigation. I. Establishment in the field. *Trop. Agric. Trin.* **47**, 243-249.

Williams, C. H. (1950) Studies on soil phosphorus. I. A method for the partial fractionation of soil phosphorus. *J. Agric. Sci. Camb.* **40**, 233-242.

Williams, M. A. J. (1968) Soil salinity in the West Central Gezira, Republic of the Sudan, *Soil Sci.* **105**, 451-464.

Williams, W. A. (1967) The role of leguminosae in pasture and soil improvement in the Neotropics. *Trop. Agric. Trin.*, **44**, 103-115.

Willimott, S. G. and Anthony, K. R. M. (1958) Agricultural research and development in the southwest Sudan. Part II. The problem of soil fertility. *Trop. Agric. Trin.* **35**, 16-33.

Willis, J. B. ed., (1962) *Agriculture and Land Use in Ghana.* Oxford University Press, London. 503 p.

Wilson, P. N. (1960) Grass, the world's most important crop. *J. Agric. Soc. Trinidad Tobago* **60** (4), 483-521.

Winchester, J. A. (1968) Plant parasitic nematodes of three widely separated areas in the American tropical rain forest. Abstr. in *Progress in Soil Biodynamics and Soil Productivity*, (A. Primavesi, ed.) Pallotti, Santa Maria, Brazil.

Wood, T. W. W. and Beckett, P. H. T. (1961) Some Sarawak soils. II. Soils of the Bintulu coastal area. *J. Soil Sci.* **12**, 218-233.

Wrigley, G (1969) *Tropical Agriculture.* Faber and Faber, London.

Yamazaki, K. (1960) Studies on the pedogenetic classification of paddy soils in Japan (in Japanese). *Bull. Toyoma Pref. Agric. Expt. Sta., Special Publication No.* 1, pp. 105.

Yoshida, S., Forno, D. A. Cock J. H. and Gomez, K. A. (1971) Zinc deficiency of the rice plant on clacareous and neutral soils in the Philippines. *Soil Sci. Plant Nutr. Tokyo* **17**, 83-87.

Younge, O. R., Plucknett, D. L. and Rotar, P. P. (1964) *Hawaii Agr. Exp. Sta. Tech. Bull.* 59.

Yuan, W. L. and Ponnamperuma, F. N. (1966) Chemical retardation of flooded soils and the growth of rice. *Pl. Soil* **25**, 347-360.

Author Index

Subject Index